Nanoionics

Scrivener Publishing
100 Cummings Center, Suite 541J
Beverly, MA 01915-6106

Publishers at Scrivener
Martin Scrivener (martin@scrivenerpublishing.com)
Phillip Carmical (pcarmical@scrivenerpublishing.com)

Nanoionics

Fundamentals and Applications

Edited by

Inamuddin
Dept. of Applied Chemistry, Zakir Husain College of Engineering and Technology, Faculty of Engineering and Technology, Aligarh Muslim University, Aligarh, India

Tariq Altalhi
Dept. of Chemistry, College of Science, Taif University, Taif, Saudi Arabia

Mohammad Luqman
Dept. of Chemical Engineering, College of Engineering, Taibah University, Yanbu, Saudi Arabia

and

Jorddy Neves Cruz
Dept. of Pharmaceutical Sciences, Federal University of Pará, Pará, Brazil

Scrivener Publishing

WILEY

This edition first published 2025 by John Wiley & Sons, Inc., 111 River Street, Hoboken, NJ 07030, USA and Scrivener Publishing LLC, 100 Cummings Center, Suite 541J, Beverly, MA 01915, USA

For more information about Scrivener publications please visit www.scrivenerpublishing.com.

Wiley Global Headquarters
111 River Street, Hoboken, NJ 07030, USA

For details of our global editorial offices, customer services, and more information about Wiley products visit us at www.wiley.com.

The manufacturer's authorized representative according to the EU General Product Safety Regulation is Wiley-VCH GmbH, Boschstr. 12, 69469 Weinheim, Germany, e-mail: Product_Safety@wiley.com

Library of Congress Cataloging-in-Publication Data

ISBN 978-1-394-31391-4

Cover image: Generated with AI using Adobe Firefly
Cover design by Russell Richardson

Set in size of 11pt and Minion Pro by Manila Typesetting Company, Makati, Philippines

Printed in the USA

10 9 8 7 6 5 4 3 2 1

Contents

Preface

Nanoionics, as a field of study, has emerged at the intersection of various disciplines and industries, revolutionizing our understanding of ion transport and its applications at the nanoscale. It combines principles from materials science, chemistry, physics, electrical engineering, and nanotechnology to explore the behavior and manipulation of ions in nanoscale systems. This multidisciplinary approach has led to significant advancements and opened new possibilities for technological innovation across several sectors.

In terms of disciplinary development, nanoionics has provided valuable insights into the fundamental processes governing ion transport and electrochemical phenomena at the nanoscale. It has allowed researchers to delve deeper into ion migration, diffusion, and interactions with materials, leading to the development of novel materials and structures with enhanced properties. The field has contributed to the advancement of energy storage systems such as lithium-ion batteries, fuel cells, and supercapacitors by enabling improved ion transport and electrode performance.

Moreover, nanoionics has paved the way for innovative electronic devices with enhanced functionalities. It has enabled the development of ion-based transistors, memory devices, and sensors that exploit the unique properties of ions for information storage, signal processing, and sensing applications. These advancements have the potential to transform electronics by offering alternative pathways for computing and data storage beyond conventional semiconductor-based technologies.

In terms of industrial development, nanoionics has found applications in various sectors. In the energy industry, it has contributed to high-performance batteries and supercapacitors, supporting efficient energy storage for portable electronics, electric vehicles, and renewable energy systems. In the biomedical field, nanoionics has facilitated the development of implantable devices, biosensors, and drug delivery systems that leverage ion transport for diagnostics and therapeutic applications. Additionally, nanoionics has influenced the fields of nanoelectronics, photonics, and optoelectronics, enabling the creation of nanoscale devices and systems with enhanced performance and functionality.

Nanoionics represents a rapidly evolving field that continues to push the boundaries of scientific understanding and technological innovation. Its interdisciplinary nature and potential for cross-domain collaboration make it an exciting area of research, with implications ranging from fundamental science to real-world applications. As researchers continue to explore the intricacies of ion transport and its manipulation at the nanoscale, nanoionics is poised to make significant contributions to technological progress and societal advancement.

Nanoionics: Fundamentals and Applications seeks to provide a detailed and up-to-date overview of the field. It explores the behavior and manipulation of ions at the nanoscale, with applications spanning energy storage, electronics, sensors, and biomedical devices. Across the chapters, we cover everything from basic principles to experimental techniques and emerging trends in nanoionics, highlighting the latest advances and innovations. The chapters included in the book are summarized below:

Chapter 1 introduces the fundamentals of nanoionics, focusing on ion behavior in solid-state materials essential for energy storage and conversion in batteries, supercapacitors, and fuel cells. It emphasizes the need for further research into ion migration and redox processes to enhance performance for electric vehicles and portable devices.

Chapter 2 examines how nanoscale ion movement advances energy storage, conversion, and sensing. It discusses Interface-Dominant Materials (IDMs), nanoarchitectonics, biocompatible electronics, and precise ion transport control in biological and artificial systems, along with the potential of nanocrystalline structures.

Chapter 3 outlines methods for synthesizing nanomaterials—chemical, physical, and biological—and details techniques for their characterization. It highlights the integration of nanoionics into resistive switching memories and lithium batteries, along with future directions.

Chapter 4 explores synthesis techniques, classifications, and applications of nanoporous silica. It explains its mechanisms in drug delivery, biosensors, purification, biomarker detection, and ion transport in energy storage.

Chapter 5 reviews advances in biomimetic ion transport systems and their applications in tissue engineering, neuroprosthetics, and energy storage, while addressing scalability and safety challenges.

Chapter 6 focuses on biomedical applications of nanoionics, including diagnostics, drug delivery, and cancer therapy, highlighting the field's growing role in medical innovation.

Chapter 7 discusses the use of ionic and mixed ionic-electronic conductors to improve the energy efficiency and functionality of sensors, transistors, and storage devices. It reviews recent advancements and the potential of nanoionic materials in next-generation electronics.

Chapter 8 addresses key challenges in nanoionics, such as scalability and safety. It explores the interplay of materials and technologies in energy storage, electronics, and Industry 4.0 applications.

Chapter 9 delves into modeling and simulation in nanoionics, highlighting the Finite Element Method (FEM) and memristor evolution. It presents a SPICE-based model, discusses cation- and anion-based resistive switching, and emphasizes the role of selector devices in optimizing ion movement.

Chapter 10 explores the broader technological impact of nanoionics, including improvements in batteries, sensors, and electronics. It discusses commercialization hurdles, regulatory needs, and the economic potential of the field in addressing global energy demands.

Chapter 11 examines ion migration and defects in nanoionic materials, focusing on cation/anion dynamics and their effects on batteries, fuel cells, and sensors. The role of point defects and material degradation is emphasized for improving device performance.

Chapter 12 investigates the intersection of nanofluidics and ion transport, showcasing nanochannel-based systems for ultra-sensitive biomolecule detection. It highlights practical applications in energy, environmental monitoring, and diagnostics, and calls for continued research and collaboration.

Chapter 13 reviews fuel-free bioinspired micro/nanorobots, powered by external stimuli such as magnetic, acoustic, light, or electric fields, which open new possibilities for *in vivo* microrobot applications due to their biocompatibility and sustainability.

We extend our deepest gratitude to Martin Scrivener and the Scrivener Publishing team for their unwavering support in bringing this book to life.

The Editors
July 2025

1

Nanoionics for Energy Storage and Conversion: Materials and Technologies

Nawishta Jabeen[1]*, Adeela Naz[2], Imtiaz Ahmad[2] and Ahmad Hussain[3]

[1]Department of Physics, Fatima Jinnah Women University, Rawalpindi, Pakistan
[2]School of Materials Science and Engineering, Harbin Institute of Technology, Shenzhen City, China
[3]Department of Physics, The University of Lahore, Sargodha Campus, Sargodha, Pakistan

Abstract

In the domain of nano-electronics and energy storage applications, a crucial scientific development has been made due to nanophotonics, moletronics, and nanoionics. Nanoionics emphasizes on the performance of ions at the nanoscale level principally in solid-state materials. Nanoionics has gained importance in recent years in the field of energy storage and conversion specifically for fuel cells, supercapacitors, and batteries, relying on its ion's transportation, high energy and power densities and longer cycle life. They can be utilized as electrode materials or as electrolyte materials for energy storage devices. In nanoionics, the small nanoscale electrode with higher surface area will be beneficial for the storing of ions and will result to the significant enhancement of the energy and power densities. In recent years the focus has been devoted to the nanoionics related to miniaturized electronic and energy applications that are crucial for electric vehicles and portable electronics. This field of research is flourishing and providing the versatile objective oriented opportunities and solutions for the growing demand of advanced energy related to technological issues for numerous industries. Furthermore, there is still a lot of research required to comprehend the mechanism, ion migration, and coupled redox process deeply.

***Keywords*:** Nanoionics, ions transportation, energy storage, electrolytes, electrodes, supercapacitor, battery

**Corresponding author*: Nawishta.jabeen@fjwu.edu.pk

Inamuddin, Tariq Altalhi, Mohammad Luqman and Jorddy Neves Cruz (eds.) Nanoionics: Fundamentals and Applications, (1–24) © 2025 Scrivener Publishing LLC

1.1 Introduction

Fossil fuels, which have limited supplies and will eventually run out globally, provide the majority of today's energy needs. It is anticipated that by 2050, there will be 10 billion people on the planet. So, a civilization based on batteries has the potential to partially solve the trilemma that arises from the global expansion in energy and food production, global economic growth, and environmental protection. Fossil fuels may be phased out much sooner if there remained the shortage of research on renewable devices for the environmental and economic solutions. It takes a synthetic, non-polluting fuel derived from renewable sources to ensure a sustainable energy system. Rechargeable batteries, supercapacitors, and fuel cells are examples of energy storage technologies that may improve electrical properties by reducing length scales to the nanoscale [1, 2]. Large grain boundaries and surface atoms that alter ions farther away and affect bulk properties are features of nanostructured materials that set them apart from polycrystalline ones [3–6]. The development of nanotechnology allows the creation of material structures over a wide range of length scales, allowing for modification of fundamental attributes and the manipulation of inaccessible constants by humans [7, 8]. The enormous success of the nano-electronics area has been largely attributed to the capacity to create multifunctional materials and devices that are suited to particular

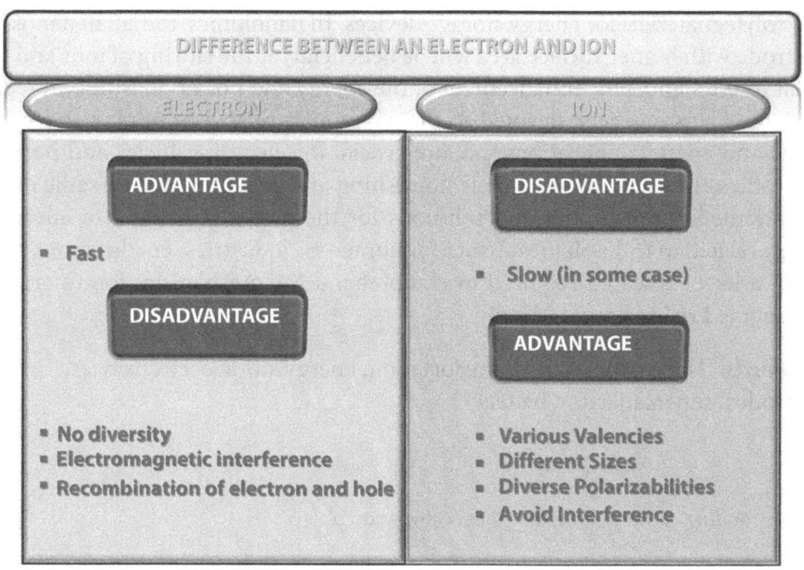

Figure 1.1 Difference between electron and ion.

needs [9]. For many energy storage and conversion applications, including electrochemical capacitors, batteries, fuel cells, and blue energy harvesting, efficient ion storage and quick ion transport in nanoporous electrodes are essential [10]. Interfaces have a substantial impact on ionic transport features, which results in size effects and material density. Nanoionics play a similar role to nano-electronics and semiconductor physics, with the importance of nanostructured arrangements being similar. Stability is crucial for nanostructured arrangements, and metastable interfaces can be durable even at high-temperatures. There is a distinct difference between electrons and ions, influenced by various factors, as shown in Figure 1.1. Their unique behaviors highlight fundamental contrasts in charge, mass, and interaction.

Research on the appliances, mechanisms, properties, methodologies, effects, and applications associated with ion transport in nano-confinement is referred to as "nano"-"ionics" research. It is important to note that the name and concept of nanoionics are not new. The term "nanoionics" and its concept were first used in 1992 [11]. Earlier, Maier and numerous other scientists put in a tremendous amount of work to shape this research area and significantly contributed to it [12–14]. Nanoionics, originally focused on solid-state nanoscale systems, now encompasses ion transport and storage phenomena in liquids. Nanostructured conductors, such as nanocrystalline ceramics, nanoscale films, heterostructures, and nanocomposites,

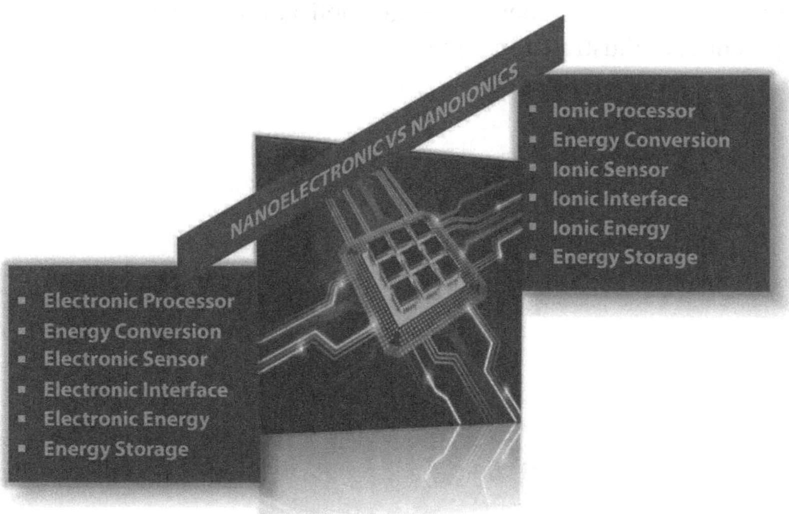

Figure 1.2 Ions demonstrate various exclusive benefits as compared to electrons for energy and information storage applications.

have distinct features and are being explored more and more for energy storage and transfer [15].

For instance, solid-state nanostructured materials should not be the only materials of choice. Similar to this, the use of nanoionics may extend beyond the usual electrochemistry-related areas to include activities like the transfer of energy and information. In the dynamic realm of technological evolution, nanoelectronics and nanoionics diverge in numerous fascinating ways, as vividly illustrated in Figure 1.2.

1.2 Nanoionics for Energy Storage

Already said, the upcoming generation of energy conversion and storage methods are projected to be significantly impacted by nanotechnology and nanostructured materials. A scientific and technological challenge that is unquestionably significant is the search for innovative, affordable, and ecologically acceptable alternatives to the present fuel cells, batteries, and super capacitors utilized in energy applications [16]. To create economically viable hybrid electric vehicles, for instance, it is vital to develop alternative technologies as lithium batteries, which are currently hitting their performance limits. In this chapter, current research into the development and manufacturing of energy devices that utilize nanoionics to enhance their functionality are discussed. Nanoionics have garnered significant attention in the field of energy storage applications, showcasing remarkable potential as illustrated in Figure 1.3.

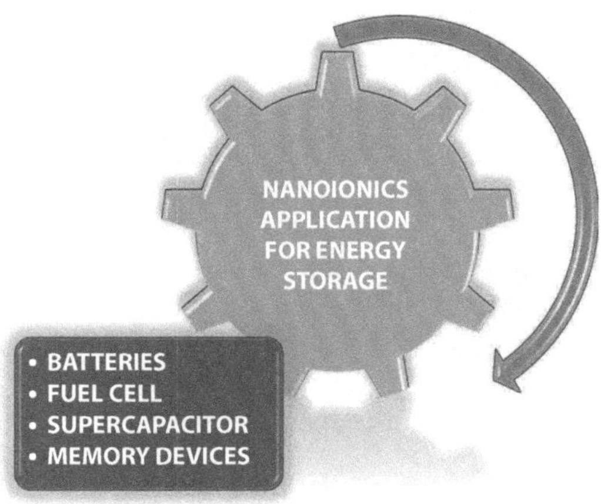

Figure 1.3 Nanomaterials for energy storage applications.

1.2.1 Nanoionics for Batteries

Research on nanoionics is mostly relying onto improvement of the solid-state electrolytes that can conduct ions at the nanoscale. Solid-state electrolytes are of particular interest for next-generation lithium-ion batteries. Since nanoionic materials are better able to withstand the strain of ions intercalation and de-intercalation throughout charge and discharge cycles, it is anticipated that the implementation of nanostructured materials in batteries as electrode may cause in, a lengthier cycle life. Faster charging times may also be achieved using nanostructured electrodes because of their larger electrode/electrolyte contact area. Numerous studies have established that using Si-C-based nanocomposites or glasses containing SnO_2 or different Sn-alloys as anodes enhances lithium battery cycling performance and keeps the batteries from shattering after a lot of cycles. It has also been demonstrated that employing TiO_2 nanowires with just 40–60 nm in diameter, which can tolerate up to $Li_{0.9}ITiO_2$ (306 mAh/g) at 1.5-1.6 V vs Li^+ (1M)/Li, improves capacity retention during cycling [17]. It is important to note that while working with nanoscale particles, electrochemical reaction pathways might be drastically different. Hematite, for instance, exhibits an irreversible phase transition in this situation.

Using large particles (1-2 m) prevents it from being used as an anode material; however, when using nanoparticles (20 nm), it exhibits amazing capabilities for Li-ions insertion [18]. Another way to boost electrode capacities is to utilize porous materials to enhance the electrodes' surface area. V_2O_5 aerogels have revealed to function improved as cathodes in lithium battery compared to polycrystalline non-porous powders of the similar arrangement [19, 20]. An additional example is the noticeably enhanced electrochemical responsiveness of carbon-coated nanoparticles containing phosphor-olivine $LiFePO_4$, as this electrically insulating substance is prevented from forming inactive areas in the bulk form [21]. As a result of realizing that high electronic conductivity and high lithium diffusivity are not requirements for new electroactive materials for lithium batteries, but rather that these requirements can be met by using clever nanoscale designs, the search for new electroactive materials has been greatly accelerated.

1.2.2 Nanoionics for Supercapacitors

Traditional supercapacitors use liquid electrolytes, which can have limitations in terms of stability and performance. Nanostructured electrodes can also improve the performance of supercapacitors. Supercapacitors resemble batteries in certain ways, but they are made to have lower energy densities and longer cycle lives and power (discharge rates). Nanoionics can play a role in designing nanostructured materials that enable reversible Faradaic reactions. Supercapacitors can store more energy when they undergo redox reactions at the electrode-electrolyte interface. The requirements for the electrodes are less stringent than those for batteries because their operation depends on surface charge buildup at the electrode/electrolyte boundary rather than intercalation and de-intercalation of ions from the electrodes. Numerous studies have previously shown that adding nanostructured carbons to supercapacitors can improve their capacitance and cyclability [22–24]. Similar improvements have been made to the lithium batteries electrolyte transport characteristics. Large gains in lithium-ion conductivity in polymer electrolytes have been made possible by the use of nano-composites, [25] particularly when ionic liquids and polymer-in-salt nanostructures [26]. It has long been known that crystalline polymer electrolytes are insulators and that ionic conduction in polymer electrolytes happens only in the amorphous phase directly above glass transition. Nevertheless, recent research has shown that cutting the chain length in the crystalline 6:1 complexes PEO_6:$LiXF_6$ ($X_{1/4}$P, As, Sb), opening up new avenues for the improvement of electrolytes [27].

1.2.3 Nanoionics for Fuel Cell

It is also crucial to use nanostructured materials to increase the ionic conductivity of electrolytes in fuel cell devices [28]. It has been discovered that the presence of tiny amounts of SiO_2 nanoparticles within the membrane helps to store the water produced during the electrochemical process while operating at temperatures between 130 to 150°C with low humidity [29]. It improves the conduction of protons. The application of nanostructured ceramic materials has also proven beneficial for transitional IT-SOFCs (temperature solid oxide fuel cells). Ceria-based nanocrystalline powders and YSZ (SDC, CGO and YDC) can be used to reduce the temperature of the electrolyte and charge transport at the electrode/electrolyte interface is enhanced by nanocrystalline ceria, which displays mixed electronic-ionic conduction [25]. Recently, nanoionics has demonstrated its potential for usage in energy applications in a different scenario: oxygen exchange. Nanoionics is essential for enhancing ion conduction, catalyst activity, and electrode material properties, among other elements of fuel cell functioning. If the conductivity electrolyte is raised by increasing its conductivity or by decreasing its width, it is typically found that electrode polarization losses are the main problem limiting the performance of modern solid oxide fuel cells [30]. The thermally accelerating oxygen reduction process at the cathode exhibits slow kinetics, which is mostly to blame for this, is one of the key obstacles to SOFC operation at lower temperatures. Although it has shown that epitaxial strain is particularly effective in accelerating oxygen exchange, more study is required to establish how it affects surface chemistry and ion exchange. Epitaxial strain favors the creation of oxygen vacancies and increased oxygen mobility in $LaCoO_3$, a well-researched material utilized as a fuel cell cathode, according to first principles calculations. When compared to single cobaltite surfaces, oxygen surface exchange at hetero interfaces $La_{0.6}Sr_{0.4}CoO_3$/(La, Sr)$_2$ CoO_4 is three orders of magnitude higher, as demonstrated by Sase [31]. The local strain effects at the contact provided an explanation for this outcome. Moreover, it has been shown that significant molecular alterations brought on by strain, including Sr enrichment and accelerated oxygen vacancy production, take place at the surface of $La_{0.7}Sr_{0.3}MnO_3$ epitaxial films. When compared to layers that were compressively strained and formed on $LaAIO_3$ substrates, the diffusion tests with isotopically in tensile strained $La_{1-x}Sr_xCoO_3$ were conducted. These findings are explained by decreased activation obstacles for diffusion as well as by the energy required to create oxygen vacancies, which leads to increased oxygen diffusion. More recently, it was demonstrated that the oxygen surface exchange coefficient of single crystal YSZ increases when the dopant content at the surface is optimized by atomic layer deposition of 5-10 nm YSZ layer [32].

The results of this research reveal how nanoionics may improve the performance of electrochemical devices.

1.3 Nanostructured Materials of Transport Behavior

Nanostructuring has a considerable impact on the transport performance of ionic and mixed conductors [33]. For situations where space charges at the contact dominate the electrical conduction, this has been demonstrated pretty well. Considered are two fundamental scenarios, one involving the formation of space charges and the other involving their depletion, and the effects of each on electrical conduction.

1.3.1 Accumulating of Space Charges

In the context of multilayers composed of BaF_2 and CaF_2, Sata *et al.* [6], provided a classic illustration of the formation of space charges and their effect on ionic conductivity. The experimental results were found to be coherently explained by redistributing the segregation of ions of F in the center of CaF_2-BaF_2 interfaces, which resulted in larger contributions from interstitial spaces and vacancies. The small effects that cause the nominal increase in conductivity occur for as long as each layer's thickness is greater than the length of Debye screening. Very short spacing among the charge layers will be expected to cause non trivial size effects.

1.3.2 Space Charges Depletion

Let's look into the acceptor doped $SrTiO_3$, whose electrical properties have been extensively investigated in considering space charge depletion [34]. In this instance, it is found that the positively charged grain boundaries result to reduced densities of oxygen vacancies ($VO^{\bullet\bullet}$) and electron holes ($h^{\bullet}$) in specific zones. For that reason, the stoichiometry at grain border zones differs greatly from that in the bulk. As a consequence of these depletion processes, the bulk and the blocking grain boundary of the depletion layers are responsible for the two semicircles shown in the impedance spectra, one at high frequency and the other at lower frequency. Based on the way that particle size influences various electrochemical potential components, several nanoscale effects have been anticipated too far [35]. The phrase indicates these:

$$\mu_j = \mu_{0j} + RT \ln c + RT \ln \gamma + zje\Phi \tag{1.1}$$

The standard word is the electrical potential of a single charge carrier, c is the concentration, and 0j denotes the chemical free energy used to generate an isolated carrier j. In non-dilute systems, the activity coefficient accounts for the impacts of energetic or entropic excess. We have gone over the effects of energy and transportation. The next section is devoted to scale effects in relation to energy storage, specifically with regard to lithium batteries.

1.4 Nanomaterials for Energy Storage Applications

Researchers are concentrating on ecologically acceptable renewable energy sources like solar cells and wind turbines in the expanding field of energy storage. Future energy storage technologies may benefit from nanomaterials with nanometer-scale dimensions, distinctive optical and electrical characteristics, and huge electrochemically active surfaces. These materials are excellent for ultrathin, flexible, and structural energy storage solutions because to their high inherent strength, electronic, and ionic conductivities, and flexibility. In spite of its large capacity, lithium-ion batteries (LIBs) have a lengthy cycle life, and are recyclable, but have low electrical conductivity, a slow rate of Li^+ diffusion, and a lengthy length of Li^+ diffusion. SnO_2 and TiO_2 nanoparticles, metal oxide nanostructures, are attractive alternatives, while mono-disperse Si quantum dots are appropriate for Si/C nanocomposites in LIBs [36].

1.4.1 Nanoionics Application and Technologies in Fuel Cells

There is a lot of potential for accelerating electrochemical processes by adding electrocatalysts to the inside surface of porous cathodes in solid oxide fuel cells. The atomic layer deposition was employed for dual catalysts to the cathode made of yttria-stabilized zirconia and lanthanum strontium manganite consisting of $(Mn_{0.8}Co_{0.2})_3O_4$ and a trace of the amount of Pt. This material exhibited a 35% increase in solid oxide fuel cells peak power density at 750°C and a 53% drop in polarization resistance when the ideal thickness of the atomic layer deposition layer was applied. Together, the two catalysts work well during the electrochemical reactions to generate nanoionics $(Mn_{0.8}Co_{0.2})_3O_4$ with higher density of grain boundaries and nearby distinct Pt particles that are uniformly split on the lanthanum strontium manganite and yttria-stabilized zirconia [37].

Formation thereby expands the sites of active electrochemical reactions to encompass the whole interior of cathode surface. First time in the realm of solid oxide fuel cells, the current work establishes how electrocatalytic surface nanoionics is created and how quick mass and charge transfer result in a considerable improvement in cell performance.

1.4.2 Nanoionics Application and Technologies in Lithium Batteries

1.4.2.1 Nanocrystalline Electrodes

Lithium (Li) batteries are used in the majority of commercial applications because to their superior cycle performance and over 95% coulombic efficiency. Rocking chair cells are used in these batteries, with $LiCoO_2$ acting as the positive electrode and C serving as the negative electrode. Such a homogeneous storage method has a low energy density (120-150 $Whkg^{-1}$) due to its low storage capacity [38]. The majority of modern high-performance handheld microelectronic devices consume a significant amount of energy. The Tarascon group [39] created a novel method of lithium storage that includes heterogeneous lithium storage to obtain high storage of Li in a number of compounds of transition metal oxide. When $2Li^+$ is added to the transition metal oxide CoO, such a conversion mechanism turns it into a Co/Li_2O nanocomposite. Furthermore, a further 0.6 Li^+ is stored at low potential, bringing the total storage capacity to about 900 $mAhg^{-1}$.

It was demonstrated, however, that only around 70% of Li^+ could be extracted from Li_2O during the initial charging process, meaning that the coulombic efficiency is only roughly 70%. Of the materials that have been studied thus far. In the first cycle, only 70–80% of the lithium is removed; the remaining 30% is permanently lost in Li_2O host matrix. RuO_2 is an electrode material with a coulombic efficiency of around 100% and a notable lithium storage capacity of 11-10 $mAhg^{-1}$ [40]. The whole charge and discharge curve of Ruthenium (IV) oxide vs Li over the voltage range of 0-4.3V possessed the particle size ranges from 30 to 200 nm. The microstructure of a fully lithiated (5.6 Li) Ruthenium (IV) oxide nanocomposite was discovered to have a thick solid electrolyte interface (SEI) coating covering it, and takes pictures for high resolution from transmission electron microscopy (TEM) [41].

The complete re-oxidation, or potential of 4.3V, causes the SEI layer to completely disappear and the extraction of 5.4 Li results in the reform of nanocrystalline Ruthenium (IV) oxide [40]. Stated otherwise, the 5.6 Li insertion causes the original Ruthenium (IV) oxide phase to disappear, and the 5.5 Li extraction results in the formation of the RuO_2 phase with 3-5 nm crystallite sizes. Evidently, even with the cycling restricted to the 0.8-4.3 V range, 99% coulombic efficiency is still attained. This shows that SEI and other processes (explained later) are not influenced by the increase of RuO_2 decomposition and reformation as stated by Equation 1.2 and are extremely reversible:

$$RuO_2 + 4Li^+ + 4e^- \rightleftarrows Ru + Li_2O \tag{1.2}$$

Ru/Li_2O nanocomposite's outstanding transport performance in nanoscale (2–5 nm) phases may be the cause of Li^+'s total extraction. The extraction of Li is made easier by RuO_2, a metallic conductor with a high electronic conductivity and Li^+ diffusion coefficient, as well as good mass and charge transport characteristics and 2-5 nm transport lengths.

1.4.2.2 Shape of the Curve and Cell Voltage

Nano size affects cell voltage by Equation 1.3, with a 1 nm grain size and $\bar{\gamma}$ of 0.1 Jm^{-2} and V of 25 cm^3mole^{-1}. Different charging-discharging curves occur from various storage systems. For electrode materials with reversible conversion reactions, there are three storage mechanisms: homogeneous

Li insertion, flat conversation reaction, and storage due to interfaces. The Gibbs-Kelvin term for the chemical potential of lithium performs an important role in these curves.

$$\mu Li(\bar{r}) = \mu Li(\bar{r} \to \infty) + 2\bar{\gamma}(\bar{r}) \tag{1.3}$$

A disparity in that term may result in a sloping discharge charge curve even in circumstances where there is no variation macroscopically. Lithium in the nanocrystalline electrode has a constant value, a change in the grain size distribution results in a change in composition [42].

1.4.2.3 Low Potential Extra Storage of Lithium

It should be noted that the sloping voltage zones' equivalent Li-storage capacities in RuO_2 are roughly 330 and 120 $mAhg^{-1}$, respectively, during the first discharge and charge processes, if the voltage is controlled within the range of 0-1.2 V. After the initial discharge in 45 min at the rate of C/5, a very reversible behavior of Lithium storage with a capacity of 120 $mAhg^{-1}$ is observed. Moreover, limited potential storage is achievable even with great rate performance. For instance, Ru/Li_2O can be kept at a rate of 4C for 70 $mAhg^{-1}$ [43]. Thus, the storage behavior is compatible with a capacitive behavior with the best high rate of performance and cycle performance and within 0-1.2 V voltage window.

1.4.2.4 Interfacial Lithium Storage: Phenomenological Model

We examine geometry, such as a heterostructure: Li_2O/metal, where metal is indicated by the dotted lines at x = 0 and L, to understand how particle size affects a composite's ability to store lithium made up pure ionic conductor and pure electronic. The theory of this structure is that it functions similarly to a supercapacitor, with the storage capacity being determined by the interfacial gap L. To enable the use of straightforward space charge model, we assume a dilute defect scenario.

In this essay, we have underlined the importance of nanoionics for both basic science and technology applications. It is demonstrated that the enormous theoretical capacities and great coulombic efficacy of the nanocrystalline electrodes. The unique interfacial lithium storage mechanism that nano crystallinity has produced has been used to explain the unusually low potential storage. Additionally, a novel class of soft matter electrolytes called "soggy sand electrolytes" is discussed, which serves as an example of how interfacial effects are applied in liquid systems.

1.4.3 Nanoionics Application and Technologies in Supercapacitors

Nanosystems of electron conductors and/or dielectrics make up the majority of the objects studied in a wide range of fundamental and practical fields in the science and technology of nanomaterials and nanostructures. Additionally, solid electrolytes (SE), superionic conductors (SIC), and fast ion transfer (FIT) materials exist. FIT-containing nano systems are nanoionics-related items. In the article "A step towards nanoionics" [11], the criterion for the concept and the name "nanoionics" were presented.

$$P = \lambda / L, \tag{1.4}$$

A definition of nanoionics includes (1) study objects (FIT nanostructures), (2) study subject (fit in the nanoscale range: features, effects of phenomena, process and applications), (3) study methodology (fit-related material and nanoobject design), and (4) scale criterion 1/L [44]. Introduced for FIT nano systems within the framework of the "Research on Priority Area Nanoionics" project, L denotes the size of the nanodevice and typical length of FIT processes realized, nanoionics is developing quickly in Europe, the United States, and Japan. The development of novel materials for fuel cells and lithium is a major area of study.

Recently, research on switches having both quantum electron conductivity and FIT has emerged. These memristors and switches, or resistors with memory, are referred to as a specific type of P = 1/L, ionic devices (ITRS) for semiconductors by International Technology Roadmap (ITR). FIT-based nanodevice manufacturing was first proposed in [11]. This chapter goal is to review the advancement of both theoretical and practical nanoionics.

1.4.3.1 Novel Nanoionic Phenomena, Effects, and Physicochemical Nano Systems

Designing nanodevice topologies is a one-step technique using direct electron beam lithography. It was found that in superionic conductor sheets, solid electrolytes with high Ag^+ ionic conductivity, $RbAg_4I_5$ and $CsAg_4B_{3-x}I_{2+x}$, were sensitive to electron irradiation. As a result, the first FIT nanostructures could be produced via direct electron beam lithography [11]. The study structures can be viewed as prototypes of the first tri-electrode nano-devices with FIT. The scheme of the experiment is presented.

As device sizes decrease, there is an increasing need for materials with large concentrations of dopants optically and magnetically active impurities in the fields of optoelectronics, photonics, and spintronics. Important characteristics of silver and copper halides (Ag, Cu) include their non-hydroscopic nature, doping with rare-earth and transition elements, high refraction indices, and transparency.

1.4.3.2 Ionic Conductors Classification: Innovative Superionic Conductors

A proposal to reclassify solid ionic conductors was made in [44]. AdSIC is a new class of conductors with nearly FIT-optimal crystal structure and record-high ion transport properties. For compounds with modest i, the double electric layer's (DEL) thickness determines by the value of i. If DEL is not available, alternative standards ought to be utilized.

1.4.3.3 AdSIC/EC Heterojunctions for Ion-Electron Mechanisms

Since it has been demonstrated [44] that Fermi level alignment for electron carriers in nano system of AdSIC proceeds deprived of considerable zone twisting, AdSIC/EC (metal) heterojunction contact potentials (V) meet the requirements. The potential limit value is (E_C - AdSIC)/e

$$V << kBT/e \tag{1.5}$$

There is a production of V with the potential limit, for instance, when the electron work function of EC is greater compared to that of AdSIC (E_C > AdSIC). As a result, the charge of mobile ions on AdSIC's surface has opposite sign. Concentration of heteroboundary mobile ions to control both the thickness of a feature-rich layer and electric field depth; because electric field penetrate in AdSIC due to its depth, these qualities are both of order Q (charge screening length). This layer has an electric capacity per unit area. Equating the ion charge's surface density (where is it) using C = 1/V.

1.4.3.4 Creation of Nanoionic Supercapacitors: Models and Methods

To maintain potential barriers for ions on surfaces at equal level to the bulk AdSIC of the material is the aim of AdSIC nanoionics. To achieve high frequency capacity features in nanoionic supercapacitors (NSC), coherent

AdSIC/EC heterojunctions must be created. IMT RAS produced the first high-capacity AdSIC/EC heterostructures between 2004 and 2006. Thin films are open, dynamic systems with various behavioral characteristics, and understanding internal order parameters and external parameters is crucial for controlling these systems [45].

1.4.3.5 AdSIC-Based Devices

AdSIC-based nanoionic devices such as supercapacitors have been established for a very long time. Because the AdSIC/EC interface transition layers disrupt the FIT conditions, typical designs have low-capacity densities and maximum recharging frequencies (10 microF/cm^2; 10-100 Hz). A unique class of impulse energy and charge storage devices called NSC was presented as a solution to this issue. High mobile ion densities on AdSIC/EC heterojunctions are responsible for the high frequency-capacity or energy power properties of NSC. A film capacitor heterostructure of the form M/$CsAg_4Br_{3-x}I_{2+x}$/Ag, was utilized to put the concept of an interface AdSIC/EC design into practice, where M is a metal alloy.

Heterostructures were fabricated using the superionic compound $CsAg_4Br_{3-x}I_{2+x}$, resulting in high conductivity and reversible changes in electrode capacity. AdSIC-based capacitors store energy in a 0.5 nm thick layer, making them more effective in charge conservation under radiation circumstances than traditional ones. These tools can be applied to areas affected by radioactive waste and space research [46].

1.4.3.6 Deep-Sub-Voltage Nanoelectronics as Impulse Storage Capacitors in

Research on energy storage and impulse charge capacitors in nanoelectronics focuses on substitute devices such as high-capacity film capacitors and field effect transistors. Nevertheless, there are still issues with high-capacity devices; it demands the development of high impulse and capacity storage devices, the size of microns.

Experimental 3D electrode nanostructure film capacitors have a capacitance of 10 μF/cm^2, but in advance portable devices with thick surface mounts, the size of capacitors with C > 0.01 F are huge. The relationship between supply voltage V_{dd} and the breakdown electric field Fmax, dielectric constant k, or densities of surface and volume capacitance in capacitors

$$V_{dd}\, F_{\max}\, (k\varepsilon_0/\rho C)^{1/2} \approx F_{\max} = k\varepsilon_0/\delta C, \tag{1.6}$$

The reduction of voltage-dependent current in micro and nanoelectronics is a long-term trend, with ICs becoming sub-voltage. Large-scale manufacture of integrated circuits (Ics) with V_{dd} = 0.5 V is anticipated in 2016, in addition to the 2016 timeline for the manufacturing of application specific integrated circuits with V_{dd} 0.5 V. It is suggested that deep-sub-voltage nanoelectronics (DSVN) mirror human brain activity and have the potential to advance Russian national development.

1.4.3.7 *Micron Size Supercapacitors Based Advanced Superionic Conductors*

Ferroelectric ceramics, ZrO_2, and HfO_2 are examples of dielectrics with significant k that cannot have a capacity density greater than 15 F/cm^2, and multilayer ferroelectric capacitors have a Cmax of 0.01 μF. Conventional micron capacitors do not meet minimum values for applications [35].

$$kF_{max} < 2 \text{ x}10^9 \text{ V/cm}, \tag{1.7}$$

The limiting charge density Qmax on the crystallographic planes of small-index ionic crystals is 1.5 10-4 C/cm^2, which is used to deduce this. Since the maximum energy of the electric field, Fmax, is 2 109 V/cm (k = 1), the value of Qmax must not be exceeded.

$$\rho_E = \varepsilon_0 k\, F^2 \max /2 \tag{1.8}$$

Chemical compounds have a higher enthalpy of formation, limiting energy storage. States caused by metals arise in atomically ordered heterojunction band gaps. AdSIC requires higher k values due to potential relief and external field. Current improvements focus on component density and AdSIC/EC heterojunctions [47].

1.4.4 Nanoionics Application and Technologies in Novel Memory Devices

The ability to achieve improved ionic conductivity at the nanoscale is crucial for energy storage and conversion. The term "electrolytes" also refers to dielectric materials that, as previously indicated, may transport ions at the nanoscale under an applied voltage in addition to more conventional materials. Nevertheless, the uses of nanoionics go far beyond only converting and storing energy. Nanoionics devices have been developed extensively and have had significant success to realize the functions that are not possible in conventional devices. Nanostructured ionic conductors and ion transport are the key components of these devices. For instance, extensive study has been done on nanoscale memristor, a new hot topic in science that may change the information sector [48].

1.4.4.1 Resistive-Switching Memories of Nanoionics

Two-terminal sandwich of metal, insulator, and metal architectures, the cation or anion migration and redox reactions result in nanoionics-based resistive memory. The device consists of an active electrode, an insulator, and an inert counter-electrode.

$$O \leftrightarrow 1/2O_2 + V\H{o} + 2e' \qquad (1.9)$$

Cations move to the counter-electrode, creating metal filaments that grow and form low resistive states. Resistive switching materials include oxides like TiO_{2-x}, doped Ta_2O_5, and ITO (Figure 1.4). It was confirmed by the redox reaction that oxygen is also produced Yang *et al.*, oxygen vacancies form filaments in $Pt/TiO_2/Pt$ devices, created by different electrodes.

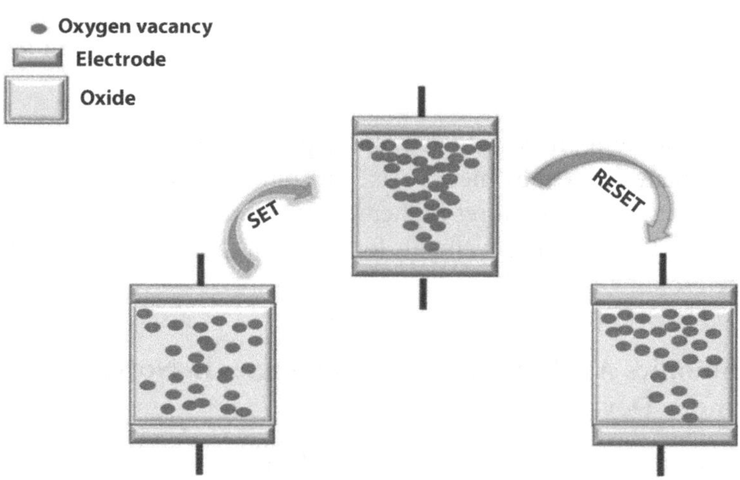

Figure 1.4 Drawing of resistive switching process dependent on vacancy of oxygen movement.

Resistive switching in TiO_x, HfO_x, and TiO_x devices is caused by nanoscale cation movement. Non-filamentary conduction scales with device area for neuromorphic applications [49].

1.4.4.2 Memristors for Non-Volatile Memories (NVM)

Music players, digital cameras, or computers all utilize memory technology like flash and DRAM. Flash memories, however, have slower endurance, longer write/erase times, and higher power requirements. When device size reduces, physical degradations like charge leakage and cell-cell coupling occur, which leads to problems with data retention. One potential option for future non-volatile memory is the reactive random access memory (RRAM), which has the benefits of non-volatility, rapid reaction rates, lower power utilization, and straightforward stack ability. RRAM devices can achieve high scalability and miniaturization to nanometer scales. To improve integration density and solve scaling issues, crossbar designs, vertical RRAM structures, and nanoscale memory devices have been created. However, crosstalk between adjacent cells can cause undesirable misreading of switching states. In an effort to increase density and streamline the production process, nanoscale memory systems without additional nonlinearity components have been developed [50].

1.4.4.3 Memristors for Artificial Synapses

Numerous synapses and the billions of neutrons that make up the human brain are essential to the biological processes that occur there. Information technology may undergo a revolution; thanks to neuromorphic computing, which has higher efficiency, parallelism, fault tolerance, and low power requirements. Nanoionics-based memristors have been researched for implementing high densities of synaptic connections and simulating biological processes. Three synaptic mechanisms involved in memory and learning are spike-timing-dependent plasticity (STDP) and the second mechanisms is long-term potentiation (LTP) and the third mechanism is short-term plasticity (STP), all rely on variations in synaptic weight. The fundamental learning rule of Hebbian learning, STDP, is applied to implement these features. Different input pulse repetition times have allowed Ag_2S synapses to successfully mimic STP and LTP synaptic functions. Ag as an active electrolyte and Ag^2S as an ionic conductor can be used to produce biological synaptic plasticity. Heterosynaptic plasticity is achieved using modulatory stimuli such as Ag thin films put on SiO_2, Pt electrodes, and SiO_2 as an electrolyte. These stimuli have the ability to push Ag ions into filaments and pull them out of them, significantly influencing synaptic function [51].

1.4.4.4 Recognition of LTP and STP in Oxide Memristors

To achieve long-term and short-term memory in metal oxide-based memristors, numerous investigations have been carried out. Similar to STP, these devices become more conductive and quickly degrade due to oxygen vacancies diffusion. Engineering of pulse characteristics, such as duration, number, repetition interval, and amplitude, is necessary to induce LTP. Repeated stimulation can induce LTP in biological systems, which is distinguished by an increase in retention time and pulse amplitude. The oxidation of MoS_2 produces an ultrathin layer of MoO_x, which allows memristors-based on MoO_x and MoS_2 to exhibit high on and off ratios at low working voltages. Additionally, STP and LTP can be applied to artificial synapses by changing the input frequency, the number of pulses, and the pulse amplitude [52].

1.4.4.5 Realization of STDP in Oxide Memristors

Through the management of overlapping programming pulses, many memristors have effectively implemented STDP. A conductive filament is made between the top electrode and TaO_x in a $Pd/Ta_2O_5/TaO_x/Pd$ memristor, for instance, to enable resistive switching. On the top and bottom electrodes, respectively, voltages are applied as prespikes and postspikes. Conductance varies at brief time intervals (Dt) when two pulses overlap. Still, simple, non-overlapping pulses are needed to mimic the synaptic plasticity observed in biological systems. For a biorealistic simulation of synaptic plasticity, Kim *et al.* presented a second order memristor. Such a state variable of second order and the local temperature of memristors (T), which depends on the input pulse, is added by this second-order memristor to change conductance in addition to voltage. This allows long-term conductance modulation and synaptic plasticity to be achieved with simple pulses. The Pt/WOx/W gadget uses the mobility of oxygen vacancies as a second-order state variable [53].

1.5 Prospects and Outlook: Why Nanoionics?

Nanoelectronics, a crucial scientific development process, complements nanophotonics, moletronics, and nanoionics. Nanoionics offers potential for developing ionic information processors due to their insensibility to electric field noise and lightweight nature. Lighter-mass carriers are preferred for smaller devices, requiring less energy and faster movement. Ions offer efficient information processors with smell and color, offering nonlinear conduction, memory-effect transistors, low power consumption, and insensibility to electric field noise [54].

1.5.1 Future of Nanoionic Devices

The Russian Ministry of Industry aims to develop a prospective electron component basis for radio electronic system development, while the US emphasizes the importance of an innovative constituent based on nanoelectronics. In hybrid energy sources for spacecraft, nanoionic supercapacitors have the tendency to produce best wE product values. Patenting laboratory-tested constructions and technologies with large-scale applications could advance work advancement by record-high wE product values can be produced using nanoionic supercapacitors in hybrid energy sources for spacecraft. Nanoionic devices with quantum electron transport and classical ion motion can utilize kinetic inductance for novel hybrid devices.

1.6 Conclusions

Nanoionics is a field focusing on nanostructured materials that enhance ionic conductivity in solid electrolytes, improving energy generation, conversion, storage, and battery performance. It may be used in non-volatile memories and systems with metal-insulator-metal ion-migration-induced resistive switching.

Transition metal cations trap electrons, enhancing electronic conductivity and metallic behavior. The MIM system determines the electroforming conditions, and the process is controlled by total charge. Bipolar resistive switching establishes intermediate resistance levels for memristive behavior and multibit storage. The mechanism, ion migration, and associated redox process require more investigation.

Interfacial oxidation states can be dynamically controlled using nanoionics devices; it makes it possible to regulate the electrical characteristics of an interface in an electric field. For instance, voltage in bilayer of Co and metal-oxide can be modulated by oxygen migration in/out of the interface. This could result in a reduction in magnetic anisotropy energy and an upsurge in the coercive field. Gadget reaction times can be significantly accelerated by temperature. By researching oxides for improved ionic conductivity, like yttria-stabilized zirconia (YSZ), creating nanostructures, performance and functionality can be further improved. If the process causing higher ion diffusivity in ionic conducting nanostructured materials is comprehended, then better materials and nanostructures can be used to develop future nanoionic devices.

References

1. Schoonman, J., Nanostructured materials in solid state ionics. *Solid State Ionics*, 135, 1–4, 5–19, 2000.
2. Jabeen, N., Hassan, N.U., Bokhari, A., Khan, M.F., Eldin, S.M., Arifeen, W.U., Bahajjaj, A.A.A., High performance δ-Bi2O3 nanosheets transformed Bi2S3 nanoflakes interconnected nanosheets as negative electrode for supercapacitor applications. *Fuel*, *347*, 128392, 2023.
3. Jamnik, J. and Maier, J., Transport across Boundary Layers in Ionic Crystals Part I: General Formalism and Conception. *Ber. Bunsenges. Phys. Chem.*, 101, 1, 23–40, 1997.
4. Puin, W., Rodewald, S., Ramlau, R., Heitjans, P., Maier, J., Local and overall ionic conductivity in nanocrystalline CaF2. *Solid State Ionics*, 131, 1–2, 159–164, 2000.
5. Jabeen, N., Hussain, A., Elsaeedy, H.I., Rahman, A.U., Tarique, M., Unique hierarchical architecture of SnO2 hexagonal interconnected nanolayered arrays as negative electrode for high performance asymmetric supercapacitors. *Mater. Chem. Phys.*, *303*, 127796, 2023.
6. Sata, N., Eberman, K., Eberl, K., Maier, J., Mesoscopic fast ion conduction in nanometre-scale planar heterostructures. *Nature*, 408, 6815, 946–949, 2000.
7. Jabeen, N., Hussain, A., Ali, J., Metal Phosphates/Phosphonates for Supercapacitor Applications, in: *Metal Phosphates and Phosphonates: Fundamental to Advanced Emerging Applications*, pp. 245–266, Springer International Publishing, Cham, 2023.
8. Whitesides, G.M., Nanoscience, nanotechnology, and chemistry. *Small*, 1, 2, 172–179, 2005.
9. Chau, R., Doyle, B., Datta, S., Kavalieros, J., Zhang, K., Integrated nanoelectronics for the future. *Nat. Mater.*, 6, 11, 810–812, 2007.
10. Jabeen, N., Hussain, A., Khan, Z.U., Khedher, N.B., Alturki, M., Khan, M.I., Ultrastable and high-performance 3D core-shell NiCo2S4@ Bi2O3 nanorod arrays as anode for LiS batteries. *J. Energy Storage*, *72*, 108519, 2023.
11. Despotuli, A.L. and Nikolaichik, V.I., A step towards nanoionics. *Solid State Ionics*, 60, 4, 275–278, 1993.
12. Kern, K. and Maier, J., Nanoionics and nanoelectronics. *Adv. Mater.*, 21, 25-26, 2569–2569, 2009.

13. Habasaki, J., Leon, C., Ngai, K.L., Dynamics of glassy, crystalline and liquid ionic conductors. *Top. Appl. Phys.*, 132, 355–4105, 2017.
14. Schoonman, J., Nanoionics. *Solid State Ionics*, 157, 1-4, 319–326, 2003.
15. Hussain, A., Jabeen, N., ul Hassan, N., Rasheed, S., Idrees, A., Eldin, S.M., Javed, M.S., Structural, piezoelectric and ferroelectric analysis of 0.96Bi0.5Na0.5TiO3-0.06BaTiO3: xwt% MnO2 ceramics for high-tech applications. *Ceram. Int.*, 49, 17, 27848–27854, 2023.
16. Arico, A.S., Bruce, P., Scrosati, B., Tarascon, J.-M., Van Schalkwijk, W., Nanostructured materials for advanced energy conversion and storage devices. *Nat. Mater.*, 4, 5, 366–377, 2005.
17. Armstrong, A.R., Armstrong, G., Canales, J., García, R., Bruce, P.G., Lithium-ion intercalation into TiO2-B nanowires. *Adv. Mater.*, 17, 7, 862–865, 2005.
18. Mustafa, Z., Jabeen, N., Hassan, N.U., Jelani, M., Muhammad, A.Q., Sehar, F., Bajaber, M.A., Realization of room temperature exchange bias effect in Co/NiO bilayer via all-solid-state Li-ion redox capacitor. *Ceram. Int.*, *49*, 2, 2115–2120, 2023.
19. Dong, W., Rolison, D.R., Dunna, B., Electrochemical properties of high surface area vanadium oxide aerogels. *Electrochem. Solid-State Lett.*, 3, 10, 457, 2000.
20. Le, D.B., Passerini, S., Guo, J., Ressler, J., Owens, B.B., Smyrl, W.H., High surface area V 2 O 5 aerogel intercalation electrodes. *J. Electrochem. Soc.*, 143, 7, 2099, 1996.
21. Huang, H., Yin, S.-C., Nazar, L.F.S., Approaching theoretical capacity of LiFePO4 at room temperature at high rates. *Electrochem. Solid-State Lett.*, 4, 10, A170, 2001.
22. Nelson, P.A. and Owen, J.R., A high-performance supercapacitor/battery hybrid incorporating templated mesoporous electrodes. *J. Electrochem. Soc.*, 150, 10, A1313, 2003.
23. Wang, J., Zhang, S.Q., Guo, Y.Z., Shen, J., Attia, S.M., Zhou, B., Zheng, G.Z., Gui, Y.S., Morphological effects on the electrical and electrochemical properties of carbon aerogels. *J. Electrochem. Soc.*, 148, 6, D75, 2001.
24. Niu, C., Sichel, E.K., Hoch, R., Moy, D., Tennent, H., High power electrochemical capacitors based on carbon nanotube electrodes. *Appl. Phys. Lett.*, 70, 11, 1480–1482, 1997.
25. Zan, F., Jabeen, N., Xiong, W., Hussain, A., Wang, Y., Xia, H., SnO2/Fe2O3 hybrid nanofibers as high performance anodes for lithium-ion batteries. *Nanotechnology*, *31*, 18, 185402, 2020.
26. Howlett, P.C., MacFarlane, D.R., Hollenkamp, A.F., High lithium metal cycling efficiency in a room-temperature ionic liquid. *Electrochem. Solid-State Lett.*, 7, 5, A97, 2004.
27. Stoeva, Z., Martin-Litas, I., Staunton, E., Andreev, Y.G., Bruce, P.G., Ionic conductivity in the crystalline polymer electrolytes PEO6: LiXF6, X= P, As, Sb. *J. Am. Chem. Soc.*, 125, 15, 4619–4626, 2003.

28. Ali, J., Li, Y., Song, J., Zhang, M., Zhou, G., Xu, J., Liu, F., A Mixed PC61BM-COi8DFIC Based Electron Transport Material for Inverted High-performance Perovskite Solar Cells. *ChemNanoMat, 9*, 1, e202200377, 2023.
29. Watanabe, M., Uchida, H., Seki, Y., Emori, M., Stonehart, P., Self-humidifying polymer electrolyte membranes for fuel cells. *J. Electrochem. Soc.*, 143, 12, 3847, 1996.
30. Jabeen, N., Zaidi, A., Hussain, A., Hassan, N.U., Ali, J., Ahmed, F., Helal, M.H., Single-and Multilayered Perovskite Thin Films for Photovoltaic Applications. *Nanomaterials, 12*, 18, 3208, 2022.
31. Sase, M., Yashiro, K., Sato, K., Mizusaki, J., Kawada, T., Sakai, N., Yamaji, K., Horita, T., Yokokawa, H., Enhancement of oxygen exchange at the hetero interface of (La, Sr) CoO3/(La, Sr) 2CoO4 in composite ceramics. *Solid State Ionics*, 178, 35-36, 1843–1852, 2008.
32. Chao, C.-C., Park, J.S., Tian, X., Shim, J.H., Gür, T.M., Prinz, F.B., Enhanced oxygen exchange on surface-engineered yttria-stabilized zirconia. *ACS Nano*, 7, 3, 2186–2191, 2013.
33. Maier, J., Ionic conduction in space charge regions. *Prog. Solid State Chem.*, 23, 3, 171–263, 1995.
34. De Souza, R.A., Fleig, J., Maier, J., Kienzle, O., Zhang, Z., Sigle, W., Rühle, M., Electrical and structural characterization of a low-angle tilt grain boundary in iron-doped strontium titanate. *J. Am. Ceram. Soc.*, 86, 6, 922–928, 2003.
35. Maier, J., Nanoionics: ion transport and electrochemical storage in confined systems. *Nat. Mater.*, 4, 11, 805–815, 2005.
36. Das, H., Pathak, B., Khanam, S., Kalita, P.K., Datta, P., Nanomaterials for next generation energy storage applications. *MRS Commun.*, 12, 3, 285–294, 2022.
37. Shen, M., Ai, F., Ma, H., Xu, H., Zhang, Y., Progress and prospects of reversible solid oxide fuel cell materials. *Iscience, 24*, 12, 103464, 2021.
38. Khan, M.A.R., Development of nanocomposites for energy storage devices, The University of Texas at El Paso, 2012.
39. Debart, A., Dupont, L., Poizot, P., Leriche, J.B., Tarascon, J.M., A transmission electron microscopy study of the reactivity mechanism of tailor-made CuO particles toward lithium. *J. Electrochem. Soc.*, 148, 11, A1266, 2001.
40. Balaya, P., Li, H., Kienle, L., Maier, J., Fully reversible homogeneous and heterogeneous Li storage in RuO2 with high capacity. *Adv. Funct. Mater.*, 13, 8, 621–625, 2003.
41. Balaya, P., Bhattacharyya, A.J., Jamnik, J., Zhukovskii, Y.F., Kotomin, E.A., Maier, J., Nano-ionics in the context of lithium batteries. *J. Power Sources, 159*, 1, 171–178, 2006.
42. Jamnik, J. and Maier, J., Nanocrystallinity effects in lithium battery materials Aspects of nano-ionics. Part IV. *PCCP*, 5, 23, 5215–5220, 2003.
43. Zhukovskii, Y.F., Balaya, P., Kotomin, E.A., Maier, J., Evidence for interfacial-storage anomaly in nanocomposites for lithium batteries from first-principles simulations. *Phys. Rev. Lett.*, 96, 5, 058302, 2006.

44. Despotuli, A.L., Andreeva, A.V., Rambabu, B., Nanoionics of advanced superionic conductors. *Ionics*, 11, 306–3145, 2005.
45. Bauer, U., Yao, L., Tan, A.J., Agrawal, P., Emori, S., Tuller, H.L., Van Dijken, S., Beach, G.S.D., Magneto-ionic control of interfacial magnetism. *Nat. Mater.*, 14, 2, 174–181, 2015.
46. Despotuli, A.L., Zagorodnev, V.N., Lichkova, N.V., Minenkova, N.A., New High Conductive CsAg4Br3–xI2+ x (0.25< x< 1) Solid Electrolytes. *Fiz. Tverd. Tela (Leningrad)*, 31, 9, 242–244, 1989.
47. Krohns, S., Lunkenheimer, P., Kant, C., Pronin, A.V., Brom, H.B., Nugroho, A.A., Diantoro, M., Loidl, A., Colossal dielectric constant up to gigahertz at room temperature. *Appl. Phys. Lett.*, 94, 12, 122903, 2009.
48. Strukov, D.B., Snider, G.S., Stewart, D.R., Stanley Williams, R., The missing memristor found. *Nature*, 453, 7191, 80–83, 2008.
49. Yang, J.J., Miao, F., Pickett, M.D., Ohlberg, D.A.A., Stewart, D.R., Lau, C.N., Stanley Williams, R., The mechanism of electroforming of metal oxide memristive switches. *Nanotechnology*, 20, 21, 215201, 2009.
50. Kim, G.H., Lee, J.H., Ahn, Y., Jeon, W., Song, S.J., Seok, J.Y., Yoon, J.H., Yoon, K.J., Park, T.J., Hwang, C.S., 32× 32 crossbar array resistive memory composed of a stacked Schottky diode and unipolar resistive memory. *Adv. Funct. Mater.*, 23, 11, 1440–1449, 2013.
51. Jeong, D.S., Kim, I., Ziegler, M., Kohlstedt, H., Towards artificial neurons and synapses: a materials point of view. *RSC Adv.*, 3, 10, 3169–3183, 2013.
52. Wang, Z.Q., Xu, H.Y., Li, X.H., Yu, H., Liu, Y.C., Zhu, X.J., Synaptic learning and memory functions achieved using oxygen ion migration/diffusion in an amorphous InGaZnO memristor. *Adv. Funct. Mater.*, 22, 13, 2759–2765, 2012.
53. Du, C., Ma, W., Chang, T., Sheridan, P., Lu, W.D., Biorealistic implementation of synaptic functions with oxide memristors through internal ionic dynamics. *Adv. Funct. Mater.*, 25, 27, 4290–4299, 2015.
54. Hodgkin, A.L. and Huxley, A.F., This Week's Citation Classic. *J. Physiol.*, 117, 500–544, 1952.

2

Fundamentals of Nanoionics and their Applications

Mustafa Aamir Hussain[1], Shruti Mishra[1], Nisha V. Bora[2] and Leena V. Bora[1*]

[1]*Chemical Engineering Department, Institute of Technology, Nirma University, Ahmedabad, Gujarat, India*
[2]*Mechanical Engineering Department, L.D. College of Engineering, Ahmedabad, Gujarat, India*

Abstract
Nanoionics, the study of nanoscale ion transport, could improve energy storage and conversion, sensing, and AI. This chapter explores the intriguing prospects of this rapidly evolving sector, highlighting major progress and future directions. Interface-Dominant Materials (IDMs) shine in innovative solid-state power devices, showcasing excellent transport properties. High-energy batteries, supercapacitors, and fuel cells from IDMs provide a sustainable future. Nanoarchitectonics explores atom-based devices, pushing the limits. Building with individual atoms enables unprecedented control over electrical and ionic pathways, suggesting computing and sensing advancements. Inspired by biological systems, biological nanoionics combine nature and technology. Replicating complicated biological structures and ion transport systems creates biocompatible, efficient electronics. In enhanced sensing and medication delivery, artificial nanoionics mimic natural ion channels to control ion transport precisely. Customizing artificial nanochannels enables tailored treatment and ultrasensitive biosensors. This chapter explores nanochannels' remarkable electrochemical energy storage potential, especially in nanocrystalline formations. These materials improve ionic conductivity and stability, enabling next-generation batteries and supercapacitors. It also covers contemporary nanoionics advancements and emphasizes their importance in shaping energy, healthcare, and technology. Nanoionics is enabling fresh discoveries in interface-dominant materials and bio-inspired structures.

**Corresponding author*: leena.bora@nirmauni.ac.in

Inamuddin, Tariq Altalhi, Mohammad Luqman and Jorddy Neves Cruz (eds.) Nanoionics: Fundamentals and Applications, (25–50) © 2025 Scrivener Publishing LLC

***Keywords*:** Nanoionics, microtechnology, solid oxide fluid cell (SOFC), nanoarchitectonics, nanocrystalline

2.1 Introduction

The word Nanoionics was first introduced in 1992 by Despotuli *et al.* [1], which means the phenomenon of transportation of nano-sized ions. It is the study of effects, properties, mechanisms, etc. of these movements that are an indirect effect of the migration of ions across heterogeneous barriers. For a better understanding of nanoionics, it is important to study both solid and liquid states of nanoionics, effects such as size effects, quantum effects and enhanced ion conductivity will prevail in ion transport at nanoscale, as compared to its bulk material. One of the crucial features of nanoionics is the capability to control ion transport at the nanoscale level. It utilizes nanostructured materials with adapted parcels, such as severance size, connectivity, and face chemistry. Nanoionic materials' high face-to-volume ratio is one of their key advantages. This implies that many ions can be stored on the material's face, increasing the density of the energy. Additionally, high ionic conductivity materials, required for quick charging and discharging rates, can be created using nanoionics [2]. It is possible to create new nanoionic materials with performance that is optimized for a given operation by managing ion transport.

Nanoionics, a field within solid-state physics, originated from material science, where quantum mechanics was employed as an analytical instrument to investigate the properties of crystal materials. Solid-state ionics have advanced through multiple research avenues, encompassing the study of transport laws, structural analysis, entropy, energy storage, and energy transformation. The investigations primarily examine the equilibrium and non-equilibrium properties of disarranged ionic materials. These studies offer valuable insights into the development of environmentally-friendly smart materials and efficient energy conversion and storage devices. Notable technological applications encompass cutting-edge solid-state battery systems, chemical sensors, and fuel cells. The emergence of nanoionics is a consequence of the increasing significance of crystal imperfections and the quest for an analytical framework to comprehend them. The field of material science has progressed from the examination of perfectly ordered crystals to the observation of point disorder and the investigation of crystals with structural disorder. It has also involved the creation of ion-conducting glasses and polymer electrolytes, as well as the construction of nano-sized systems and thin films [3].

The concept of nanoionics is not confined to just solid-state applications; this chapter includes its liquid phase, biological, energy storage and information transfer applications. Pore sizes are an important factor for nanoionic diffusion; they are negligible in bulk systems but interfacial effects come into the picture when these effects are observed at nanoscale [4]. Interfacial forces or effects refer to surface tension and cohesive forces. It is an adhesive force between a liquid phase of one substance with either a solid or liquid. These interfacial forces increase with decrease in pore sizes. The dependency of ion transport on these factors such as H-bonding, van der Waals forces and ion-dipole interaction play an important role in the manufacturing of batteries, biological phenomenon, biochemistry, microfluids, conductors, etc.

2.2 Applications

2.2.1 Employment of Interface - Dominant Materials (IDMs) in Novel Solid State Power Devices

The employment of IDMs in new solid-state power devices represents a promising frontier in material science and energy technology. These materials, often nanoscale, play a crucial role at the interfaces within solid-state devices, affecting their performance, efficiency, and durability.

New attempts have been made to implement real device systems employing nanoionics generalizations, but two restrictions still exist: the usage of substrates grounded on ceramic single chargers and the narrow area capabilities of conventional growth styles [5]. The incorporation of nanoionics generalities should count on wide range area deposits of pottery on dependable substrates attuned with existing microtechnology in order to make a significant effect on the MEMS (micro-electromechanical systems) assiduity. The downsizing required to combine power sources of solid-state with other micro/nanodevices would be made possible by the application of silicon substrate in combination with accessible thin film deposit methods [6]. Additionally, the potential for constructing substantial infrastructures for high specific energy, power and faultless system integration is being investigated.

To put the phenomena found at the interfaces of single-resistant substrates and various semiconductors, functional nano-finagled flicks are being integrated into silicon technology. On typical Si substrates, certain exploration groups have demonstrated the ability to construct epitaxial functional flicks by using oxide buffer layers. In order to create high-quality, free-standing flick structures that may be transmitted from the master substrate to any support, further sacrificial sub caste absorbers are being investigated [7].

To provide homogenous conditions over greater regions, the adoption of a technically relevant substrate like silicon must be coupled with the expansion of batch-product deposit techniques. At exploratory institutes, large-area PLD (Pulsed Laser Deposition) systems are available, and artificial scripts have been implemented using techniques like mail-to-mail or roll-to-roll lading [8]. The primary obstacle in ALD (Atomic Layer Deposition) is finding new strategies to raise the deposit rates for accessories. With the introduction of Spatial ALD (SALD), deposit speed can be significantly increased without the need for a high vacuum, which makes the process more difficult and expensive to gauge up. New solid-state energy microdevices built on ceramic flips with dominant nanoengineered interfaces may benefit from the use of nanoionics [9].

2.2.1.1 Micro Solid Oxide Fuel Cells (μSOFC)

Fuel cells, also known as electrochemical devices, are capable of direct conversion of chemical energy from electrical energy. A fuel cell's fundamental physical composition is an electrolyte layer in interaction with a permeable anode and cathode on both sides. A diagrammatic representation of a fuel

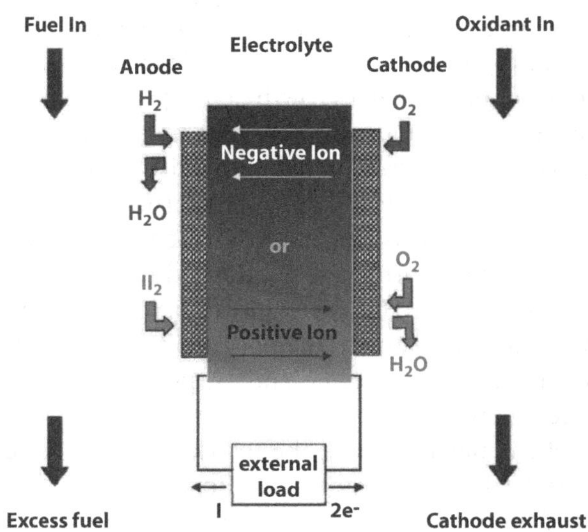

Figure 2.1 Diagram of a single fuel cell. The electrolyte permits the passage of either positive or negative ions across it, in this typical scenario [16].

cell is shown in Figure 2.1, along with the directions of ion-conduction movement through the cell and the reactant and product gases.

SOFCs through oxygen ion conducting ceramics in the electrolyte and composite electrodes benefit more from nanoionics. SOFCs use hydrocarbons from natural gas and convert chemical to electrical energy more efficiently than other fuel cells. Many nanocrystalline interface manipulations are being used to improve the conductivity of oxygen ion conductors like doped zirconia and ceria used in SOFCs [10].

The advancement of μSOFCs has been facilitated through the reduction in size of ceramic materials to slim films and their subsequent incorporation into Si technology [11]. Lowering the inner resistance of standard SOFC systems can be attained by decreasing the regularity of ceramic electrolyte materials. This reduction in consistency enables operation at lower temperatures [12]. Conventional silicon technology has facilitated the production of ceramic free-standing membranes with low thermal mass, enabling rapid and energy-efficient start-up procedures. Experiments have been conducted on cells connected to yttria-stabilized zirconia (YSZ) electrolyte membranes that possess symmetrical metallic electrodes [13, 14]. These cells have been operated at temperatures as low as 350°C, a notable reduction in temperature compared to the usual operating temperatures of SOFCs. Remarkably, these cells have demonstrated power outputs exceeding 1 Wcm^{-2}.

The development of solar energy storehouse technology has been slow due to two factors viz., low internal resistance of current materials used as electrolytes [12], and the rapid declination of metallic-grounded active electrodes at reduced temperatures [15]. To overcome these limitations, a new advance is demanded, with nanoionics generalities anticipated to play a crucial part in reducing operating temperatures and easing system encapsulation and integration with other CMOS (Complementary Metal-Oxide-Semiconductor) grounded factors.

Electrolytes with advanced oxygen ion conductivity are in demand, and an approach could be to substitute YSZ with other ceramic materials [16]. Still, this would bear conforming microfabrication processes. Taking advantage of nanoionics, generalities have high eventuality due to their superior performances. Exemplifications include conductivity improvement via strain or space-charge goods and columnar structures that enhance oxide ion conductivity through perpendicular heterointerfaces [17].

Substituting metallic electrodes with ceramic bones would ensure long-term cell operation, barring the unfavorable effect of metallic film declination. Still, thin film ceramic electrodes are confined to operate at fairly elevated temperatures (T> 700 °C) [18]. Saranya *et al.* [19], strategized for enhancing ORR (oxygen reduction reactions) rates in these electrodes as a necessity, similar to enhancing oxygen objectification and prolixity in $La_{0.8}Sr_{0.2}MnO_{3+\delta}$ (LSM) and $La_{0.8}Sr_{0.2}(Mn_{1-x}Co_x)_{0.85}O_{3\pm\delta}$ (LSMC) grain boundaries. Combining this strategy with the purposeful development of functional flicks with perpendicularly aligned grains can significantly decrease the workable temperature of μSOFC cathodes.

2.2.1.2 *Ion Gated Thermoelectrics*

The efficiency of thermoelectric bias is highly impacted by charge carrier mobility and absorption. The most popular method for increasing carrier concentration is chemical doping, although it can have a negative impact on the performance of materials due to flaws and dopant isolation, particularly in nanostructured materials. An additional path based on nanoionic products is utilized to get around this problem.

An effective method is to use an electrolyte as a gate in place of thin film paddings. At the point of contact with the semiconductor, ions in the electrolyte rearrange and form electrochemical double layers (EDLs) when a voltage is applied. Because of the proximity of the ionic and electronic charges, this results in a significant interface capacitance. Ionic gating can be utilized in silicon microfabrication to create thermoelectric transistors, which are similar to conventional electrical transistors. This opens up opportunities for novel operations, such as intelligent detectors and electronic mortal-organic interfaces, like practical clothing and materials for medical monitoring.

A novel design for a thermocouple device that controls both the p-type Cu_2O and n-type ZnO was used to from an electric double–layer capacitor structure by applying a potential in the common gate by Takayanagi *et al.* [20]. Under specific circumstances, this enhanced the device's joining power by a factor of more than ten. It also created an implicit modification in the gate when open to heat by using an electrolyte with a high Seebeck measure. An exterior heat-gated organic transistor that complies with electrolyte-gated transistors and an ionic thermoelectric super capacitor (ITESC) was created by Zhao *et al.* [21]. The Soret effect was found to cause the ionic thermoelectric leg to produce a high open voltage when subjected to a temperature gradient, which gates the transistor. The drain current varied as a result of the temperature differential. The researchers were able to adjust the transistor affair current by a minimum of two orders of magnitude with induction of temperature changes by providing an operating voltage that varied in the same order as the thermal voltage variation of the ITESC.

2.2.1.3 Solid Oxide Photoelectrochemical Cells (SOPECs)

Renewable energy sources are being sought to meet the rising energy demand due to climate change and fossil fuel autarchy. Despite efforts in chemistry, physics, and material science, collecting and storing solar energy is difficult. This requires finding and testing different light conversion and storage systems. A key method uses light to drive electrochemical reactions to store reaction products like H_2 and O_2 from water splitting. For decades, photoelectrochemical systems using aqueous electrolytes have been studied, with TiO_2, Fe_2O_3, and other transition metal oxides preferred for water splitting. Light's effect on inorganic solid electrolyte cells, such as oxide ion conductors, has been little studied. This is true despite extensive knowledge of ion oxides and mixed conductors used in electrolysis cells and SOFCs [22].

SOPECs are developed by combining photovoltaics and renewable energy/electrolysis cells. Historically, SOPEC fabrication was grounded on two distinct cell systems, with only single cells being liquid-grounded. Still, these approaches aren't ideal due to volumetric constraints and high voltage drops [23].

Figure 2.2 shows oxygen ions flowing from cathode to the anode of an O^- SOFC due to the oxygen chemical potential gradient in the electrolyte. The O_2 on the cathode side needs to enter the electrolyte, exit the lattice on the anode side, and react with fuel in order to guarantee constant movement of O^{2-} across the electrolyte. When H_2 or hydrocarbon reacts with O^{2-} supplied by the electrolyte at the anode, electrons, H_2O, and CO_2 are formed. This process is known as the cathodic oxygen reduction reaction.

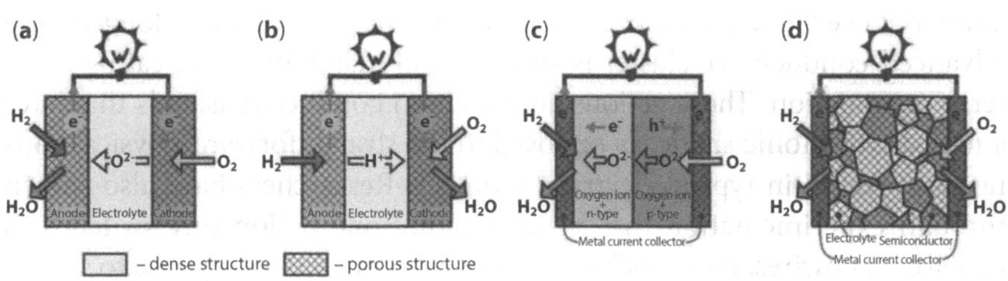

Figure 2.2 Operation schemes of the following fuel cells: (a) double-layer fuel cells, (b) proton-conducting electrolyte SOFC, (c) oxygen-ion conducting electrolyte SOFC, and (d) single-layer fuel cells [57].

2.2.2 Nanoarchitectonics for Atom-Based Devices

Nanoarchitectonics can be used to create new types of transistors and solar cells, while nanoionics can be used to develop new batteries and sensors. Atomic switches, a type of electrical switching system, are created by regulating atoms and molecules at the nanoscale. These switches, which are designed with infinitesimal conductance, differ from conventional switches in macroscopic scales [24]. The structure of the switch involves several infinitesimal position events, similar to electrochemical responses, prolixity, and rush at the nano position gap of two electrodes (Ag_2S and Pt) reported by Ariga [25], demonstrated the fabrication and operation of neuromorphic networks using atomic switches. He showed that the networks can learn from input signals and generate output signals that are similar to the brain's electrical activity. Application of a negative bias to the Pt electrodes gives rise to the formation of Ag clusters, which can be switched on by a rush of several small or tiny amounts of them. Again, a positive bias causes the oxidation of the Ag snippet to Ag ion, leading to the disbandment of the switch, which turns the device off. To achieve more efficient and adaptable advancements in material research with nanoscale structural features, a post-nanotechnology concept of unified methodology is needed. As demonstrated in Figure 2.3, this task can be attributed to the developing field of nanoarchitectonics, a universal concept for creating functional materials from nanoscale building units.

Atomic switches grounded on Ag_2S exhibit clear distinctions between their switching mechanisms and memory capabilities compared to conventional electrical switches. Compared to artificial electrical switches, such prejudices may have implicit capacities that are more similar to natural synapses [26]. Researchers have used low-frequency beats with a 20-alternate interval to display bio-like capabilities, short-term malleability, and permanent potentiation. But even with a shorter 2-alternate interval period, stable, long-lasting advanced-conductance cluster islands are still maintained, indicating long-term potentiation. These actions are based on conductive islands that have a nanoarchitectonic shape, as opposed to the straightforward physical connections found in typical electrical switches. Researchers have also shown through experimentation that by connecting one billion tiny switches to tabletop nanowires, it is possible to achieve a cooperative response to stimulant inputs that mimics nanoarchitected neuromorphic bias [27].

Krishnan *et al.* [28], investigated the fabrication of gapless-type switches using polymer-grounded atomic switches with poly (ethylene oxide) (PEO) across electrodes. In three different types of switches with Ag/Ag-PEO/Pt, Ag/PEO/Pt, and Pt/Ag-PEO/Pt nanoarchitecture, they approximated the

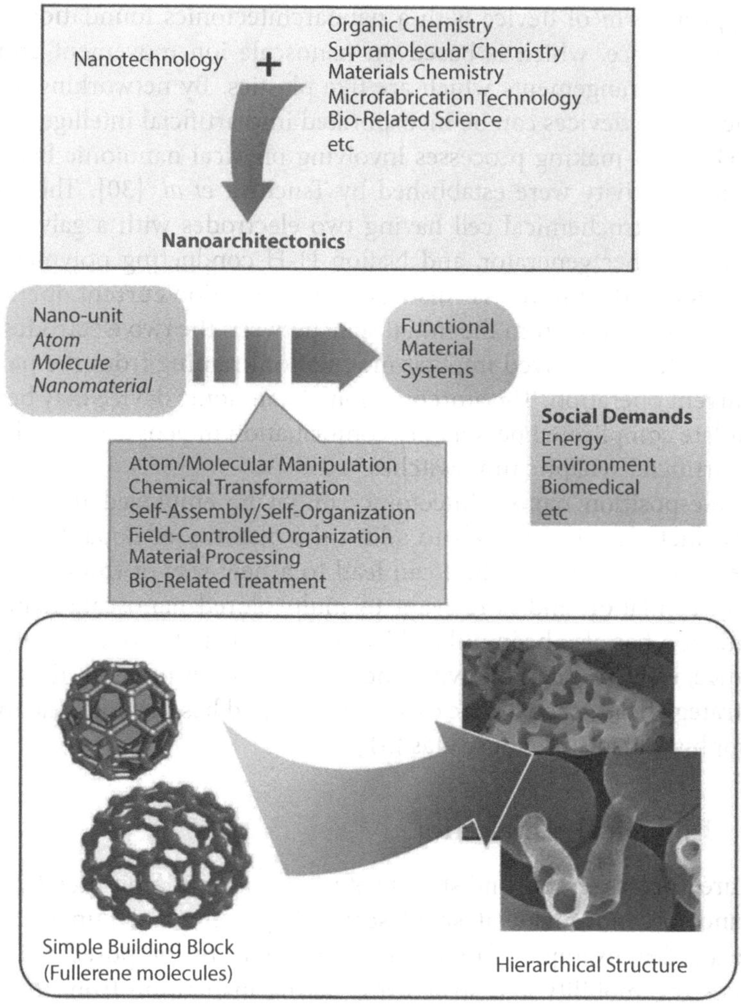

Figure 2.3 By emphasizing the design and construction of nanostructures, nanoarchitectonics is a logical approach that integrates nanotechnology and nanoionics research fields, whereas nanoionics concentrates on the properties and transport of ions in nanomaterials [58].

developing exploit of conductive fibers within the PEO matrix. They identified a number of elements that influence switching processes, including the tone-assembly of polymer flicks into lamellar nanoarchitecture, the addition of Ag swabs to the polymer matrix, and the role of Ag electrodes in actual growth. Krishnan *et al.* [29], also investigated the advanced functionality of the polymer-grounded micro switch nanoarchitecture of the Ag/ PEO/ Pt device pattern.

A typical form of device with a nanoarchitectonics foundation is the nanoionic device, which is based on nanoscale ion movement and has structural rearrangements which are like plastics. By networking atomic switches, these devices can be incorporated into artificial intelligence systems. Decision-making processes involving physical nanoionic bias with dynamic adaptivity were established by Tsuchiya *et al.* [30]. They made use of an electrochemical cell having two electrodes with a galvanostat, arbitrary number generator, and Nafion H-H-conducting polymer electrolyte. Due to the uneven H^+ distribution, palpitation current operations generated deviations from the initial state between the two electrodes. The nanoionic device received implicit modulation training from the palpitation current operation. The aforementioned nanoionic device may be used to simulate complicated personality conformation in people by combining it with artificial synaptic tiny switches.

Atomic-position nanoarchitectonics might be employed in materials with distinct electrical parcels to adjust the quantity of dopant atoms in solid electrolytes. This method can lead to a new way of thinking about material conflation and is relevant to multicolored nanoscale materials. This method has also been utilized by experimenters to control magnetoresistance, electrical conductivity, and magnetization in Fe_3O_4 thin films. This strategy beats electrostatic carrier doping and has the potential to create a sophisticated storehouse bias [31].

2.2.3 Biological Nanoionics

In nature, the processing and storage of information is arbitrated by using fluid and ion movement at small scale. Organisms and human beings can be an ideal example of ionic circuits as they are composed of ions of various sizes, mobility and structures. Taking inspiration from the ionic structure found in the organisms, the new age artificial intelligence is constructed in a nanoionic way that includes sensory devices, human machine interface, artificial neural networks etc. This type of construction possesses key advantages such as lower usage of energy, and power requirements as compared to the conventional processor.

In the field of chemical energy conversion and storage, the concept of ions as charge exporters has drawn a lot of attention. Ions have a wide variety of applications because of their polarizability, size, and valencies. A developing technology that seamlessly connects artificial intelligence systems with biological systems is nanoionics, which is built on multi-ionic exporters, much like nanoelectronics. Nanoionics can mimic biological functioning directly.

In single cells, ion flow is responsible for pH balance, processing, action potential and signaling. Xu and Lavan [32], studied an electric eel, a freshwater fish, which generates 600 V and a current of 1 Ampere when it comes across predators. Na^+ and K^+ permeation into membrane channels generate the potential. This energy transport relies only on accurate ion transport through nanopores. All these ion transportation and information transfer depends on precise transfer of ions through pores and protein. Protein ion channels are generally divided into two categories: ion channels and passive conduits. Ions flow due to differences in concentration and electric potential difference.

Ion channels generally allow the transfer of only one type of ions which is why it is also known as ion filters. In some cases, ion channels allow the transfer of multiple ions such as Ca^{2+} and Zn^{2+}. Meanwhile, different ion fluxes play distinct roles. Ca^{2+} ion concentrations usually transmit information, but ionic currents in Na^+ or K^+ fluxes act as physical signals.

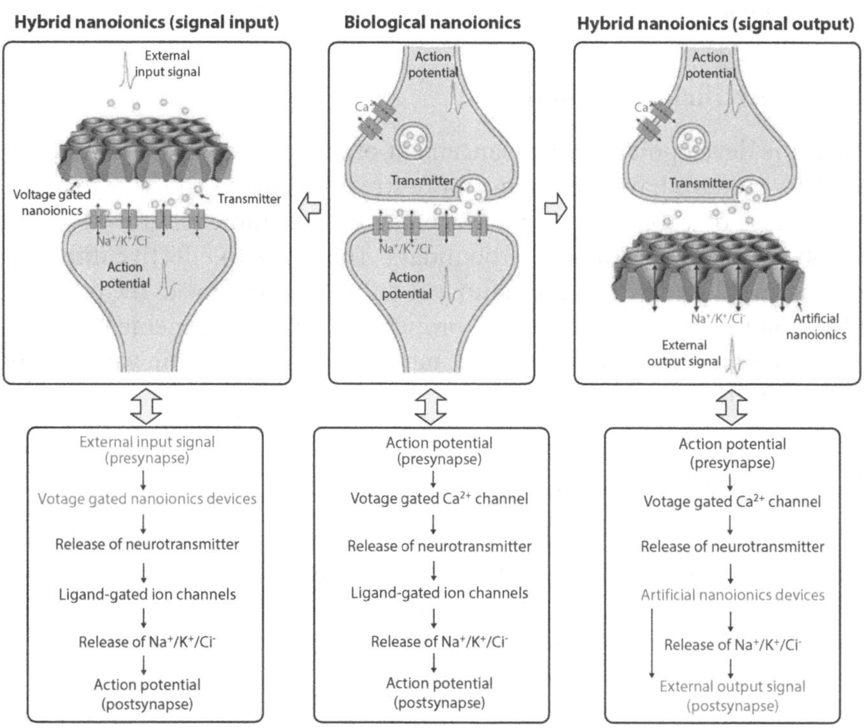

Figure 2.4 The figure shows how bioinspired hybrid nanoionics and neural networks generate biological action potentials. Action potentials in biological systems are a result of ion transport by neurotransmitters. External input signals can be read as pre-synapse action potentials or written as post-synapse action potentials in hybrid nanoionics [59].

Membrane potentials in inhibitory synapses and skeletal muscle fibers can be sustained by Cl^- ion flow ionic currents [33].

These biological nanoionics energy systems and information processors, which are related to biological ion transport, offer an excellent model for resolving current energy- and information-related contemporary issues.

Figure 2.4 shows, two hybrid nanoionics systems combining artificial and biological elements which are suggested for signal input and output. In the first system, artificial nanoionics devices are triggered by external stimuli viz., pressure, light, heat, electricity, etc., which release neurotransmitters. An action potential is produced post-synapse when biological nanopores are activated by neurotransmitters, allowing Na^+ or K^+ ions to pass through the transmembrane. As a result, a hybrid nanoionics system that combines biological and artificial elements is produced that performs like the biological one. In the second scenario, the "read" and "write" models of biological signals are matched by artificial signals transduced from biological signals. Human-machine "two-way" communication can be achieved with an ionic device that combines these features.

2.2.4 Artificial Nanoionics

With the development and advancement of nanotechnology, artificial ion delivery systems have grown in a more streamlined and cohesive manner. In the case of a solid - solid interface, there is a significant growth in the number of atoms in the phase boundary. This leads to effective diffusion of ions and the growth of the volume of diffusion flux. Due to the presence of nanoions, the thermodynamic properties of nanoconfinement objects are also improved. The presence of nanoions also affects the variation of energy at the interface.

2.2.4.1 Liquid Nanoionics

It refers to the transportation of ions in liquid in the nanoscale system, which includes nanotubes, nanopores, and nanotubes. In a liquid phase, the ions exist either as solvents or solvated ions. Some of the factors affecting ionic diffusion are conductivity of the pore, pore connectivity, charge dynamics and ionic concentration [34].

2.2.4.1.1 Transport of Ion in Nanoconfinement

For materials of different sizes, the electrostatic attractive forces are different, as most of these forces weaken for distances more than 100 nm.

Interaction of forces with various intensities helps in determining the movement of ions through nanochannels. Distribution of these ions in the influence of electrical charges leads to the formation of electrical double layer. This thickness is known as Debye length. When the nanospace size lies in the Debye length, the ions will be able to experience the electrostatic force from the surface of the adjacent wall. Ion-ion interactions are different in cases of conducting and non-conducting confinement. In non-conducting confinement, the ions follow Coulomb's law, on the other hand, in conducting confinement, the energy due to interaction reduces exponentially because of the screening of ions. The movement of ions can be classified into three categories; when the potential of charge is zero, the ions move very slowly inside the pores, when the charging proceeds, the diffusion of ions increases; once the pores become charged the diffusion of ions slows down [35].

2.2.4.1.2 Ion Transport Phenomenon in Large Pore Size (2–100 nm)

Ion transport properties in larger nanopores (2-100 nm) are influenced by an external force, electrostatic force, wettability, and steric effects. Ion transport in biological nanopores is methodical and compliant, with properties such as ionic gating, rectification, and pumps.

A nanofluidic module can function as a gate for the flow of ions; thanks to ionic gating, which opens in response to stimuli similar to voltage-gated physiological nanopores. Qian *et al.* [36], used metal-organic framework nanochannels modified with azobenzene (Azo) to develop an active ionic gate with an on/off ratio as high as 17.8. When heat or UV radiation is applied, the popular photo responsive organic group azo experiences reversible isomerization among trans and cis, which causes an unusual open-to-close channel shifting. In addition to these and many other external stimuli, the ionic gate may additionally be triggered by electricity, pH, thermal energy, magnetism, and particular molecules and ions.

Biological nanopores have a feature called ionic rectification that allows ion flow only in one direction. The efficiency of the ionic system is improved by a high rectification ratio of more than 500. An asymmetric funnel-shaped ionic rectification system was developed by Xiao *et al.* [37], with cylindrical and conical segments filled with various molecules to produce a surface charge distribution and synergistic asymmetric structure. A nanochannelled dynamic curvature framework with a reversed rectification switch was developed by Hou *et al.* [38], to enable flexible ion transport. Under a range of electric fields, pH levels, and salt concentrations, the membrane can display aberrant ionic transport behaviors, including negative differential resistance, massive ionic current improvements, and adjustable ionic current rectification. It also functions as a nanochannel-curvature-adjustable smart ionic diode. It was suggested that a theoretical model be used to clarify these processes. Effective drug delivery or cell commination research shows that the nanofluidic system with ionic rectification may carry multiple charged drugs or cell messengers unilaterally without diffusive leakage.

Ion pumps transport ions across variations in concentration by using energy. Ion pumping moves protons in biological systems from low to high concentrations, creating a voltage difference that activates an ATPase and causes it to produce ATP [39]. In artificial systems, photo-responsive particles are placed into porous or lipid membranes to achieve ion-pumping [40]. Furthermore, photochemical, photoelectric, and photothermal effects are employed in light-driven, nanofluidic ion transport systems to produce an asymmetric surface charge distribution. Ion transport and photoinduced transporters enable direct ion pumping in some semiconductor nanostructures. To pump ions against a 5000-concentration gradient in a carbon nitride nanotube membrane, Xiao *et al.* [41], successfully created a gradient surface charge density using a light source.

2.2.5 Utilization of Nanochannels for Electrochemical Energy Storage

Because of their ability to control ionic transport, nanochannels hold great potential for electrochemical storage of energy and conversion electronics. Lithium Sulfur Batteries, or LSBs, were first introduced in the early 1960s and have since been regarded as essential electronic components. To meet growing power demands, LSBs must have high cycling lifetime, power density, and energy density. Issues with ionic transport in current LSB systems are impeding their development. In lithium-ion Batteries (LIBs), a portion of the cathode reaction's byproducts disintegrate in the electrolyte and travels via the divider to the anode, where they react to produce low battery coulombic efficiency and irreversible capacity loss.

2.2.5.1 Lithium-Ion Batteries (LIB)

LIBs need Li^+ ions to quickly switch between cathode and anode. If not, the anode will become overloaded with electrons, decreasing the voltage and capacity of the high-current discharge. Excessive anion transport may lead to low Li^+ transference. Anion transfer that is not under control can shorten battery life by producing side reactions, Joule heat, and battery polarization. Therefore, high transference number and Li^+ conductivity are necessary for LIB ionic transport. Negatively charged nanochannels in separators or electrode materials can only theoretically permit rapid Li^+ transmission because of strong electrostatic interactions. This will improve battery electrochemical performance by resolving some problems brought on by uncontrollably high anion transport and achieving high Li^+ conductivity.

One or more electrochemical cells, each consisting of two electrodes enclosed in an electrolyte and spaced apart by a separator over a predetermined distance, make up a battery. By looking at Figure 2.5, one can gain a more intuitive understanding of the makeup and operation of a Li-ion battery.

High ionic transport and ion selectivity are characteristics of surface-charged 2D nanofluidic structures that make them perfect for LIBs. Yan *et al.* [42], arranged ultrathin Co_3O_4 nanosheets with surface functionality. Lower structural stress, a shorter Li^+ diffusion distance, and exclusive and quick Li^+ transport were all features of the special structure. Capacity is enhanced in 2D nanofluidic channels by capacitive surface storage and negatively charged functional groups. In LIBs and SIBs (Sodium-ion batteries), the composite film demonstrated rapid Li^+ dispersal kinetics and good rate performance. Negatively charged RGO and CNTs penetrate the

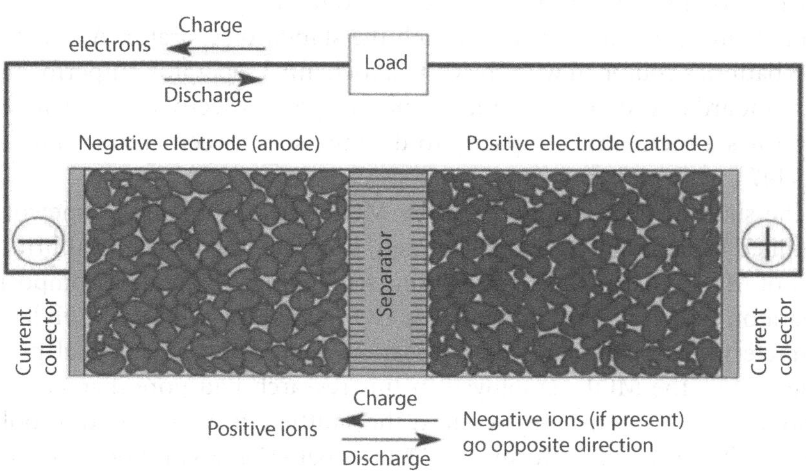

Figure 2.5 Interior structure of a (Li-ion) battery's electrochemical cell [60].

composite film preferentially against Li^+, enabling fast diffusion, an electrolyte ion-PTCDA molecule reaction, and a highly conductive pathway.

2.2.5.2 Lithium Sulfur Batteries

In recent years, Li-S batteries have gained attraction for their high theoretical specific capacity (1675 mA h g^{-1}) and energy density (2600 Wh kg^{-1}) [43]. The shuttle effect limits the use of normal Li–S batteries. The shuttle effect of polysulfides diffusing to the anode and reacting with lithium metal can cause decrease in capacity and low coulombic efficiency [44]. Many solutions have been proposed [45], including nanochannel-based cation selectivity to inhibit polysulfide anion transport and improve Li–S battery coulombic efficiency and cycling lifetime.

Zhang *et al.* [46], created a novel ion selective Li–S membrane layer using multifunctional Nafion. Routine membranes allowed polysulfides to diffuse between cathode and anode. Polysulfides could be contained on the cathode by ion selective membranes. Coulombic interactions allow Li^+ to hop through the membrane nanochannels but not to polysulfide anions. Membranes that electrostatically shield polysulfide anions could be advantageous for Li-S batteries. Since the dense 2D nanochannels of ultrathin graphene oxide (GO) membranes had better permeability than Nafion membranes, GO membranes were used to create separators. GO separators with a negative charge effectively block anion transport. Polysulfide diffusion was found to be inhibited in the double-liquid cell divided by the GO membrane after 16 hours. In eight hours, polysulfides darkened the right chamber as they spread through the standard separator. As a result, Li–S batteries equipped with the GO nanochannel separator outperformed the standard one, demonstrating membrane perm selectivity, a metric that quantifies a membrane's capacity to distinguish between anions and cations [47].

The shuttle effect is suppressed by MOF (Metal Organic Framework) modified separators with nanochannels; however, the mechanical brittleness of MOFs makes their fabrication challenging. To create a composite separator, Bai *et al.* [48], combined a GO membrane with an MOF layer, characterized by simplicity of fabrication and strong mechanical properties. Since the MOFs employed in this research had pore diameters of approximately 0.9 nm, they reduced the shuttle effect compared to polysulfides. The cycling performance of Li–S batteries was found to be significantly better at 1 Coulomb than it was with the original GO separator.

2.2.5.3 Lithium Organic Batteries (LOB)

Due to their lightness, structural flexibility, and chemical diversity, organic materials show great promise as a lithium battery substitute for transition metal oxide cathodes. Lately, organic cathodes and metallic lithium anodes have been used to create LOBs. There are many options available for organic cathode redox processes and structures due to their diversity. The benefits of LOB include low molecular weight, good recyclability, economical, and high theoretical specific capacity. LOB experience undesired diffusion, specifically, the discharging products close to the cathode, and electroactive organic species move to the anode made of lithium metal and react with them, causing rapid capacity decay, self-discharge, and low coulombic efficiency.

Song *et al.* [49], proposed two main solutions: using a solid ion conductor, ceramic, or polymer electrolyte to suppress the shuttle effect and improving the electrode structure and framework to refine the material's stagnation during the charge/discharge process. To prevent transmembrane anion transport, they designed polypropylene/Nafion/polypropylene (PNP) separators with nanochannels. The PNP separator stopped the movement of dispersed electroactive anions and had great Li^+ perm selectivity because of coulombic interactions and nanochannel size confinement. Diffusion tests revealed that the PNP separator (0.5 mg cm^{-2} Nafion loading) outperformed the polypropylene (PP) one in blocking Iminodianthraquinone (IDAQ) – anions, which are used to identify the type of anion used. For Li–S batteries as well as new dissolved electroactive species batteries, sandwich-type separators show promise; however, they are not suitable for LOB.

By coating GO and Super P (or graphene) onto a celgard separator (a microporous membrane), Pan *et al.* [50], produced a cation selective nanochannel separator for LOB. The majority of organic molecules, discharge products, and intermediates in carbonyl organic cathode materials dissolve in the electrolyte. In this separator, organic anions were rejected while Li^+ was transported selectively; thanks to the reconstruction of GO sheets to create negatively charged nanochannels. As the upper current collector, Super P (or graphene) additionally supplied dissolved electroactive species reactivation sites. As opposed to the pristine separator, the nanochannel membrane improved the performance of the LOB by preventing electroactive anions from diffusing to the anode.

2.2.6 Nanocrystalline Structures

There are a few techniques that are used for the fabrication of nanocrystalline solids. These techniques include: Ionized gas condensation (IGC) (a process used to produce nanoparticles or thin films by condensing vaporized materials in the presence of a plasma or an ionized gas), Sol-gel, mechanical attrition and spray pyrolysis. In spray pyrolysis, a chemical precursor is distributed as aerosol drops. These droplets are carried to the hot zone and are reduced to oxide particles. This method is employed in the formation of oxides like ZnO, ZrO_2 and Al_2O_3 [51].

2.2.6.1 Sol-Gel (Chemical Deposition Method)

$LiNbO_3$'s excessive lithium-ion dispersal in ball-milled samples tends to be intriguing because it is a poor ionic conductor. Nb K-edge X-ray absorption spectroscopy (XAS) studies of sol-gel-prepared nanocrystalline $LiNbO_3$ from alkoxides and ball-milled $LiNbO_3$ were carried out by Chadwick *et al.* [52]. Given the size of the particles and the likelihood that the average coordination number (CN) will not decrease, the sol-gel sample (21 nm particles) exhibits a change that is comparable to bulk material. Samples

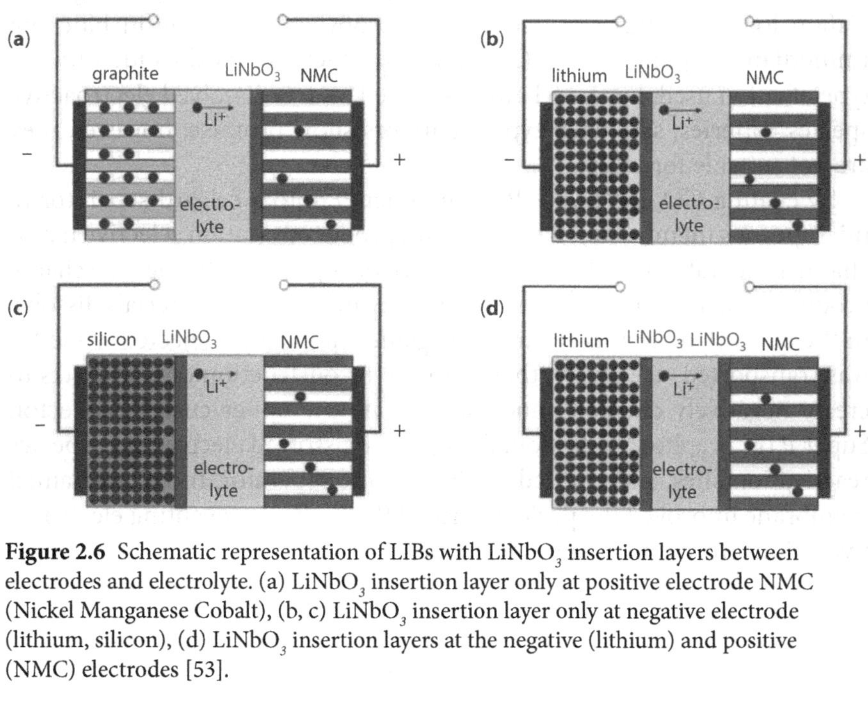

Figure 2.6 Schematic representation of LIBs with $LiNbO_3$ insertion layers between electrodes and electrolyte. (a) $LiNbO_3$ insertion layer only at positive electrode NMC (Nickel Manganese Cobalt), (b, c) $LiNbO_3$ insertion layer only at negative electrode (lithium, silicon), (d) $LiNbO_3$ insertion layers at the negative (lithium) and positive (NMC) electrodes [53].

that were ball-milled (particle size of 23 nm) showed fewer peaks overall, including the Nb–O correlation. This is because there is 50% amorphous material in the ball-milled sample. Amorphous material was also observed in the ball-milled sample by transmission electron microscopy (TEM) measurements. Schematized cross-sections through the different types of LIBs covered in the work carried out by Erwin Hüger *et al.* [53], are shown in Figure 2.6, where $LiNbO_3$ insertion layers are positioned between the electrolyte and the positive and negative electrodes, respectively. It should be noted that studying LIBs with $LiNbO_3$ insertion layers at both electrolyte surfaces, that is, the interfaces between the electrolyte and the positive electrode and the negative electrode would be one area of future research.

This method has been used for the synthesis of ceramics and oxides as it offers control and alteration on a molecular level. The procedure of a sol-gel process is as follows: a metal alkoxide $(MOR)_x$ is subjected to hydrolysis which results in the replacement of OR group with OH group (known as formation of sol). These small particles are condensed to form an interconnected network known as gel. After the drying of this gel, the final product can either be an oxide or a hydroxide; calcination under high temperature is the last step towards the formation of these oxides.

2.2.6.2 *Microstructure Investigation*

The reason for the various properties of nanocrystalline material lies in its microstructure. Microstructures or transportation of nanoions in some early studies is mentioned as a 'liquid-like' state as referred to in section 2.4.1. This structure would address the possibility of diffusion of nanoions in the nanocrystalline material. Its structure and questions related to it such as its nature, in terms of atomic order, its level, etc. have always attracted the attention of researchers. To answer such questions, two great possibilities have been measured. One extreme is that there is widespread disarray in the structure of nanocrystalline frame that is several atoms in thickness. Another possibility is that the structure has a normal boundary as found in bulk material.

Backscattering causes oscillation by reflecting waves into the outgoing wave. Immersion has two implicit issues. When the two swells are out of phase, destructive hindrance occurs, while when they are in phase, formative hindrance occurs. As incident photoelectron energy rises, so will the emitted photoelectron energy. Due to the fixed distance between the target and its near titles, surge relations can be in-phase or out-of-phase, resulting in oscillations. These oscillations' intensity depends on the number of backscattering neighbors and the EXAFS Debye-Waller effect.

2.2.6.3 *Storage of Hydrogen*

Due to the pore walls' attractive potential, which raises the density of the material adsorbed, molecular pores have the capacity to absorb significant volumes of gases. Pederson and Broughton [54], studied that the capillarity effect of large-diameter, multi-walled carbon nanotubes, with diameters of a few nanometers, has been observed for low-surface-tension liquids.

The storage of hydrogen is generally done in high pressure vessels in the form of liquid and is adsorbed as hydrides. The problem with storage of liquid hydrogen is weight and space, some part of hydrogen is adsorbed by the metallic container under atmospheric conditions of temperature and pressure whereas desorption takes place at higher temperatures. Hydrogen embrittlement is another difficulty that leads to the deterioration of metallic containers. A convenient solution to all these problems was attained by creating nanostructured hydrides using high energy ball milling.

Carbon nanotubes have gained interest in hydrogen storage due to their capability to increase weight gain through responses with alkali essence dopants on different carbon forms. While alkali- essence doped carbon nanotubes can adsorb hydrogen in dry hydrogen, their medium of hydrogen adsorption remains unclear [55]. Yang's [56], results show nearly 2 hydrogens can be adsorbed, but the medium remains unclear.

2.3 Future Perspective

Nanoionics has potential applications in various fields, such as medicine, energy harvesting, data storage and nanoelectronics. Future applications of nanoionics include:

- Fabrication of biocompatible and targeted nanomaterials for disease detection and therapy: For example, nanoionics can be used to design a smart drug delivery system that can liberate drugs in response to specific stimuli or signals.
- High-performance, low-cost nanomaterials for energy conversion and storage. Using nanoionics can be used to improve the performance and stability of solar cells, fuel cells, batteries and supercapacitors.
- Research on novel and functional nanomaterials for data processing and communication: For example, nanoionics can be used to design new low-power, high-speed memory devices, logic gates, sensors, and transistors.

- Nanoarchitectonics, the design and fabrication of nanostructured materials and devices: By controlling the size, shape, and structure of nanostructures, scientists are able to create materials with fixed ionic conductivity and other properties such as nanoarchitectured electrodes with high surface fast-ion-diffusion processes that can be used to power batteries and electrochemical devices that showcase improved performance.

2.4 Conclusion

We have explored the exciting frontiers of nanoionics, delving into the various materials, design principles, and applications that underpin this rapidly evolving field. From the tailored functionality of interface-dominant materials to the atomic-level precision of nanoarchitectonics, nanoionics' potential to revolutionize various industries is undeniable. The highlighted areas, which include biological nanoionics, artificial nanoionics, nanochannel-based energy storage, and nanocrystalline structures, are only a sampling of the vast possibilities available in this field. Biological nanoionics has enormous potential for understanding and mimicking natural ion transport processes, paving the way for advances in bioelectronics and bioinspired devices. Artificial nanoionics, on the other hand, enables the development of entirely new functionalities and materials with tailored ionic properties, thereby pushing the boundaries of traditional device design. The use of nanochannels for electrochemical energy storage offers a compelling opportunity to develop high-performance batteries and supercapacitors, addressing the growing demand for sustainable and efficient energy solutions. Nanocrystalline structures, with their distinct ionic transport properties, provide additional opportunities for improving device performance and investigating new applications.

References

1. Despotuli, A. and Nikolaichik, V., A step towards nanoionics. *Solid State Ion*, 60, 4, 275–278, Apr. 1993, doi: 10.1016/0167-2738(93)90005-N.
2. Kilner, J.A., Brian Steele's Contributions to Solid State Ionics. *Proc. Electrochem. Soc.*, 2005-07, 13–19, Jan. 2005, doi: 10.1149/200507.0013PV.
3. Sepúlveda, P., Muga, I., Sainz, N., Rojas, R.G., Ossandón, S., Nanoionics from a quantum mechanics point of view: Mathematical modeling and numerical

simulation. *Comput. Methods Appl. Mech. Eng.*, 407, 115926, Mar. 2023, doi: 10.1016/j.cma.2023.115926.

4. Sparreboom, W., van den Berg, A., Eijkel, J.C.T., Principles and applications of nanofluidic transport. *Nat. Nanotechnol.*, 4, 11, 713–720, Nov. 2009, doi: 10.1038/nnano.2009.332.
5. F. Baiutti, F. Wrobel, G. Christiani and G. Logvenov, 3 - Oxide molecular beam epitaxy of complex oxide heterointerfaces, in: *Metal Oxide-Based Thin Film Structures*, pp. 53–78, Amsterdam , Elsevier, 2018.
6. Lu, D., Baek, D.J., Hong, S.S., Kourkoutis, L.F., Hikita, Y., Hwang, H.Y., Synthesis of freestanding single-crystal perovskite films and heterostructures by etching of sacrificial water-soluble layers. *Nat. Mater.*, 15, 12, 1255–1260, Dec. 2016, doi: 10.1038/nmat4749.
7. Spreitzer, M., Egoavil, R., Verbeeck, J., Blank, D.H.A., Rijnders, G., Pulsed laser deposition of SrTiO3 on a H-terminated Si substrate. *J. Mater. Chem. C. Mater.*, 1, 34, 5216, 2013, doi: 10.1039/c3tc30913d.
8. Greer, J.A., History and current status of commercial pulsed laser deposition equipment. *J. Phys. D. Appl. Phys.*, 47, 3, 034005, Jan. 2014, doi: 10.1088/0022-3727/47/3/034005.
9. Muñoz-Rojas, D. and MacManus-Driscoll, J., Spatial atmospheric atomic layer deposition: a new laboratory and industrial tool for low-cost photovoltaics. *Mater. Horiz.*, 1, 3, 314–320, 2014, doi: 10.1039/C3MH00136A.
10. Liu, M., Lynch, M.E., Blinn, K., Alamgir, F.M., Choi, Y., Rational SOFC material design: new advances and tools. *Mater. Today*, 14, 11, 534–546, Nov. 2011, doi: 10.1016/S1369-7021(11)70279-6.
11. Bieberle-Hütter, A., *et al.*, A micro-solid oxide fuel cell system as battery replacement. *J. Power Sources*, 177, 1, 123–130, Feb. 2008, doi: 10.1016/j.jpowsour.2007.10.092.
12. Garbayo, I., *et al.*, Electrical characterization of thermomechanically stable YSZ membranes for micro solid oxide fuel cells applications. *Solid State Ion*, 181, 5–7, 322–331, Mar. 2010, doi: 10.1016/j.ssi.2009.12.019.
13. Kerman, K., Lai, B.-K., Ramanathan, S., Nanoscale Compositionally Graded Thin-Film Electrolyte Membranes for Low-Temperature Solid Oxide Fuel Cells. *Adv. Energy Mater.*, 2, 6, 656–661, Jun. 2012, doi: 10.1002/aenm.201100751.
14. Fan, Z., An, J., Iancu, A., Prinz, F.B., Thickness effects of yttria-doped ceria interlayers on solid oxide fuel cells. *J. Power Sources*, 218, 187–191, Nov. 2012, doi: 10.1016/j.jpowsour.2012.06.103.
15. Beckel, D., *et al.*, Thin films for micro solid oxide fuel cells. *J. Power Sources*, 173, 1, 325–345, Nov. 2007, doi: 10.1016/j.jpowsour.2007.04.070.
16. Tarancón, A., Strategies for Lowering Solid Oxide Fuel Cells Operating Temperature. *Energies (Basel)*, 2, 4, 1130–1150, Nov. 2009, doi: 10.3390/en20401130.

17. Yang, S.M., *et al.*, Strongly enhanced oxygen ion transport through samarium-doped CeO2 nanopillars in nanocomposite films. *Nat. Commun.*, 6, 1, 8588, Oct. 2015, doi: 10.1038/ncomms9588.
18. Garbayo, I., Pla, D., Morata, A., Fonseca, L., Sabaté, N., Tarancón, A., Full ceramic micro solid oxide fuel cells: towards more reliable MEMS power generators operating at high temperatures. *Energy Environ. Sci.*, 7, 11, 3617–3629, 2014, doi: 10.1039/C4EE00748D.
19. Saranya, A.M., *et al.*, Unveiling the Outstanding Oxygen Mass Transport Properties of Mn-Rich Perovskites in Grain Boundary-Dominated $La_{0.8}Sr_{0.2}(Mn_{1-x}Co_x)_{0.85}O3_{\pm\delta}$ Nanostructures. *Chem. Mater.*, 30, 16, 5621–5629, Aug. 2018, doi: 10.1021/acs.chemmater.8b01771.
20. Takayanagi, R., Fujii, T., Asamitsu, A., Simultaneous control of thermoelectric properties in p- and n-type materials by electric double-layer gating: New design for thermoelectric device. *Appl. Phys. Express*, 8, 5, 051101, May 2015, doi: 10.7567/APEX.8.051101.
21. Zhao, D., Fabiano, S., Berggren, M., Crispin, X., Ionic thermoelectric gating organic transistors. *Nat. Commun.*, 8, 1, 14214, Jan. 2017, doi: 10.1038/ncomms14214.
22. Walter, M.G., *et al.*, Solar Water Splitting Cells. *Chem. Rev.*, 110, 11, 6446–6473, Nov. 2010, doi: 10.1021/cr1002326.
23. Licht, S., Wang, B., Ghosh, S., Ayub, H., Jiang, D., Ganley, J., A New Solar Carbon Capture Process: Solar Thermal Electrochemical Photo (STEP) Carbon Capture. *J. Phys. Chem. Lett.*, 1, 15, 2363–2368, Aug. 2010, doi: 10.1021/jz100829s.
24. Hasegawa, T., Terabe, K., Tsuruoka, T., Aono, M., Atomic Switch: Atom/Ion Movement Controlled Devices for Beyond Von-Neumann Computers. *Adv. Mater.*, 24, 2, 252–267, Jan. 2012, doi: 10.1002/adma.201102597.
25. Ariga, K., Nanoarchitectonics: what's coming next after nanotechnology? *Nanoscale Horiz.*, 6, 5, 364–378, 2021, doi: 10.1039/D0NH00680G.
26. Ohno, T., Hasegawa, T., Tsuruoka, T., Terabe, K., Gimzewski, J.K., Aono, M., Short-term plasticity and long-term potentiation mimicked in single inorganic synapses. *Nat. Mater.*, 10, 8, 591–595, Aug. 2011, doi: 10.1038/nmat3054.
27. Avizienis, A.V., *et al.*, Neuromorphic Atomic Switch Networks. *PLoS One*, 7, 8, e42772, Aug. 2012, doi: 10.1371/journal.pone.0042772.
28. Krishnan, K., Tsuruoka, T., Mannequin, C., Aono, M., Mechanism for Conducting Filament Growth in Self-Assembled Polymer Thin Films for Redox-Based Atomic Switches. *Adv. Mater.*, 28, 4, 640–648, Jan. 2016, doi: 10.1002/adma.201504202.
29. Krishnan, K., Muruganathan, M., Tsuruoka, T., Mizuta, H., Aono, M., Highly Reproducible and Regulated Conductance Quantization in a Polymer-Based Atomic Switch. *Adv. Funct. Mater.*, 27, 10, 1605104, Mar. 2017, doi: 10.1002/adfm.201605104.

30. Tsuchiya, T., *et al.*, Ionic decision-maker created as novel, solid-state devices. *Sci. Adv.*, 4, eaau2057, 2018. DOI:10.1126/sciadv.aau2057
31. Tsuchiya, T., *et al.*, *In Situ* Tuning of Magnetization and Magnetoresistance in Fe_3O_4 Thin Film Achieved with All-Solid-State Redox Device. *ACS Nano*, 10, 1, 1655–1661, Jan. 2016, doi: 10.1021/acsnano.5b07374.
32. Xu, J. and Lavan, D.A., Designing artificial cells to harness the biological ion concentration gradient. *Nat. Nanotechnol.*, 3, 11, 666–670, Nov. 2008, doi: 10.1038/nnano.2008.274.
33. Doyle, D.A., *et al.*, The Structure of the Potassium Channel: Molecular Basis of K^+ Conduction and Selectivity. *Science (1979)*, 280, 5360, 69–77, Apr. 1998, doi: 10.1126/science.280.5360.69.
34. Kondrat, S., Vasilyev, O.A., Kornyshev, A.A., Feeling Your Neighbors across the Walls: How Interpore Ionic Interactions Affect Capacitive Energy Storage. *J. Phys. Chem. Lett.*, 10, 16, 4523–4527, Aug. 2019, doi: 10.1021/acs.jpclett.9b01623.
35. He, Y., *et al.*, Importance of Ion Packing on the Dynamics of Ionic Liquids during Micropore Charging. *J. Phys. Chem. Lett.*, 7, 1, 36–42, Jan. 2016, doi: 10.1021/acs.jpclett.5b02378.
36. Qian, T., *et al.*, Efficient Gating of Ion Transport in Three-Dimensional Metal–Organic Framework Sub-Nanochannels with Confined Light-Responsive Azobenzene Molecules. *Angew. Chem. Int. Ed.*, 59, 31, 13051–13056, Jul. 2020, doi: 10.1002/anie.202004657.
37. Xiao, K., *et al.*, Enhanced Stability and Controllability of an Ionic Diode Based on Funnel-Shaped Nanochannels with an Extended Critical Region. *Adv. Mater.*, 28, 17, 3345–3350, May 2016, doi: 10.1002/adma.201505842.
38. Wang, M., Meng, H., Wang, D, Yin, Y., Stroeve, P., Zhang, Y., Sheng, Z., Chen, B., Zhan, K., Hou, X., Dynamic Curvature Nanochannel-Based Membrane with Anomalous Ionic Transport Behaviors and Reversible Rectification Switch. *Adv. Mater.*, 31, 11, e1805130, 2019 Mar. doi: 10.1002/adma.201805130. Epub 2019 Jan 11. PMID: 30633407.
39. Gadsby, D.C., Ion channels versus ion pumps: the principal difference, in principle. *Nat. Rev. Mol. Cell Biol.*, 10, 5, 344–352, May 2009, doi: 10.1038/nrm2668.
40. Bhosale, S., *et al.*, Photoproduction of Proton Gradients with π-Stacked Fluorophore Scaffolds in Lipid Bilayers. *Science (1979)*, 313, 5783, 84–86, Jul. 2006, doi: 10.1126/science.1126524.
41. Xiao, K., *et al.*, Artificial light-driven ion pump for photoelectric energy conversion. *Nat. Commun.*, 10, 1, 74, Jan. 2019, doi: 10.1038/s41467-018-08029-5.
42. Yan, C., Lv, C., Zhu, Y., Chen, G., Sun J., Yu, G., Engineering 2D Nanofluidic Li-Ion Transport Channels for Superior Electrochemical Energy Storage. *Adv. Mater.*, 29, 46, 2017 Dec. doi: 10.1002/adma.201703909. Epub 2017 Oct 17. PMID: 29044794.
43. Geng, Z., *et al.*, Improved Electrochemical Performance of Biomass-Derived Nanoporous Carbon/Sulfur Composites Cathode for Lithium-Sulfur Batteries

by Nitrogen Doping. *Electrochim. Acta*, 202, 131–139, Jun. 2016, doi: 10.1016/j.electacta.2016.03.176.
44. Hou, T., Xu, W., Chen, X., Peng, H., Huang, J., Zhang, Q., Lithium Bond Chemistry in Lithium–Sulfur Batteries. *Angew. Chem. Int. Ed.*, 56, 28, 8178–8182, Jul. 2017, doi: 10.1002/anie.201704324.
45. Xiao, L., *et al.*, A Soft Approach to Encapsulate Sulfur: Polyaniline Nanotubes for Lithium-Sulfur Batteries with Long Cycle Life. *Adv. Mater.*, 24, 9, 1176–1181, Mar. 2012, doi: 10.1002/adma.201103392.
46. Huang, J.-Q., Zhang, Q., Peng, H.-J., Liu, X.-Y., Qian, W.-Z., Wei, F., Ionic shield for polysulfides towards highly-stable lithium–sulfur batteries. *Energy Environ. Sci.*, 7, 1, 347–353, 2014, doi: 10.1039/C3EE42223B.
47. Huang, J.-Q., Zhuang, T.-Z., Zhang, Q., Peng, H.-J., Chen, C.-M., Wei, F., Permselective Graphene Oxide Membrane for Highly Stable and Anti-Self-Discharge Lithium–Sulfur Batteries. *ACS Nano*, 9, 3, 3002–3011, Mar. 2015, doi: 10.1021/nn507178a.
48. Bai, S., Liu, X., Zhu, K., Wu, S., Zhou, H., Metal–organic framework-based separator for lithium–sulfur batteries. *Nat. Energy*, 1, 7, 16094, Jun. 2016, doi: 10.1038/nenergy.2016.94.
49. Song, Z., Qian, Y., Otani, M., Zhou.,Stable Li–Organic Batteries with Nafion-Based Sandwich-Type Separators. *Adv. Energy Mater.*, 6, 1501780, 2016. doi: 10.1002/aenm.201501780.
50. Pan, Y., Hao, J., Zhu, X., Zhou, Y., Chou, S.-L., Ion selective separators based on graphene oxide for stabilizing lithium organic batteries. *Inorg. Chem. Front.*, 5, 8, 1869–1875, 2018, doi: 10.1039/C8QI00374B.
51. Messing, G.L., Zhang, S.-C., Jayanthi, G.V., Ceramic Powder Synthesis by Spray Pyrolysis. *J. Am. Ceram. Soc.*, 76, 11, 2707–2726, Nov. 1993, doi: 10.1111/j.1151-2916.1993.tb04007.x.
52. Chadwick, A.V., Pooley, M.J., Savin, S.L.P., Lithium ion transport and microstructure in nanocrystalline lithium niobate. *Phys. Status Solidi (C)*, 2, 1, 302–305, Jan. 2005, doi: 10.1002/pssc.200460170.
53. Hüger, E., Riedel, L., Zhu, J., Stahn, J., Heitjans, P., Schmidt, H., Lithium Niobate for Fast Cycling in Li-ion Batteries: Review and New Experimental Results. *Batteries*, 9, 5, 244, Apr. 2023, doi: 10.3390/batteries9050244.
54. Pederson, M.R. and Broughton, J.Q., Nanocapillarity in fullerene tubules. *Phys. Rev. Lett.*, 69, 18, 2689–2692, Nov. 1992, doi: 10.1103/PhysRevLett.69.2689.
55. Chen, P., Wu, X., Lin, J., Tan, K.L., High H_2 Uptake by Alkali-Doped Carbon Nanotubes Under Ambient Pressure and Moderate Temperatures. *Science (1979)*, 285, 5424, 91–93, Jul. 1999, doi: 10.1126/science.285.5424.91.
56. Yang, R.T., Hydrogen storage by alkali-doped carbon nanotubes–revisited. *Carbon N Y*, 38, 4, 623–626, 2000, doi: 10.1016/S0008-6223(99)00273-0.
57. Kuterbekov, K.A., *et al.*, Classification of Solid Oxide Fuel Cells. *Nanomaterials*, 12, 7, 1059, Mar. 2022, doi: 10.3390/nano12071059.

58. Maji, S., Shrestha, L.K., Ariga, K., Nanoarchitectonics for Hierarchical Fullerene Nanomaterials. *Nanomaterials*, 11, 8, 2146, Aug. 2021, doi: 10.3390/nano11082146.
59. Zhang, J., Liu, W., Dai, J., Xiao K., Nanoionics from Biological to Artificial Systems: An Alternative Beyond Nanoelectronics. *Adv. Sci.*, 9, 2200534, 2022. https://doi.org/10.1002/advs.202200534.
60. Al-Gabalawy, M., Hosny, N.S., Hussien, S.A., Lithium-Ion Battery Modeling Including Degradation Based on Single-Particle Approximations. *Batteries*, 6, 3, 37, Jul. 2020, doi: 10.3390/batteries6030037.

3

Nanomaterials for Nanoionics Applications: Synthesis, Characterization and Device Integration

Amita[1], A.S. Mathur[2*] and B.P. Singh[2]

[1]*Department of Physics, RBS College, Agra (U.P.), India*
[2]*Department of Physics, Institute of Basic Science, Dr. Bhimrao Ambedkar University, Agra (U.P.), India*

Abstract
Recent advances in technology have demonstrated that nanotechnology and nanoscience evolution is crucial. Nanotechnology involves the manipulation of materials and devices at the nanoscale through processes such as development, synthesis, characterization, and modification of size and shape. Ion transfer in a nanoconfined environment is the subject of a novel field of study known as "nanoionics." The properties of nanostructured materials are critical to the fundamental understanding of nanoionics research as well as the investigation of possible applications. In this context, the current chapter describes various techniques for the synthesis and characterization of nanomaterials. Additionally highlighted is the device integration of nanoionics in the fields of lithium batteries and resistive switching memories.

Keywords: Nanoionics, nanomaterials, synthesis, characterization, resistive switching memories, lithium batteries

3.1 Introduction

"There is Plenty of Room at the Bottom"
This was the title of the talk by renowned physicist Richard Feynman in 1959, at the annual meeting of the American Physical Society. It was for the

**Corresponding author*: arpitswarupmathur@gmail.com

Inamuddin, Tariq Altalhi, Mohammad Luqman and Jorddy Neves Cruz (eds.) Nanoionics: Fundamentals and Applications, (51–78) © 2025 Scrivener Publishing LLC

first time that the concept of nanosized particles was introduced. This was the time when a computer needed an entire room and Feynman conceptualized the idea of looking at matter on the atomic scale. Nanotechnology is a branch of technology where people manufacture materials on nanoscale and use it for various applications. It means they use atoms or group of atoms to form matter that exhibits properties (physical, chemical, biological) very different from the parent substance due to the quantum effects. This allows us to manipulate and hence control the properties of matter. This miniaturized matter can then be used for specific purposes. We can thus keep this nano particulate matter, distinct from the known four states of matter i.e. solid, liquid, gas and plasma state. Today, the scope of nanomaterials has increased so much in every walk of life, be it medical science, information technology, environment, transportation, cosmetics and many more. New researches on quantum dots and short pulses can lead to new insights into the field. In the year 1993, A. L. Despotuli and V. I. Nikolaichik [1] coined the term nanoionics in analogy to microionics, which was used to fabricate and study the sensors, batteries and new devices of the size of a few microns where ionic conduction takes place. These microionic devices were quite compatible with the semiconductor devices. Nanoionics is, thus, a branch of nanoscience that deals with the movement of ions in yet smaller devices of the size of a few nanometres (of the order of Debye radius). The extent of the Debye radius is directly influenced by both the ionic conductivity and the concentration of mobile ions. The greater the conductivity and concentration of ionic conductors in a material, the lesser is the Debye length. Ionic conduction in solid and liquid electrolytes was first studied by Faraday in 1834 [2], while studying the electrolysis, he observed that the laws of electrolysis hold good for the ionic solids as well. He observed that the movement of ions played a role in generating electric current within an electrolyte, as they have the ability to exchange charges with an electrode during the electrolysis process. He took heated solid electrolytes, silver sulfide (Ag_2S) and lead fluoride (PbF_2), for his experiment and found that the conductivity of ions was a function of temperature and the conductivity power was greatly enhanced upon heating. His theories upon the ionic conduction gave rise to two branches of science namely, electrochemistry and solid-state ionics.

Frenkel in 1926, attributed the conduction of ions in crystals to the point defects in ionic crystals. He suggested that even in thermodynamic equilibrium condition, sometimes, a few numbers of cations move from their regular position in the crystal lattice to some interstitial position and thereby giving rise to a number of ions free to move inside the crystal and thus, ionic conduction in the crystalline material. Such a vacancy-interstitial pair is called Frenkel defect [3]. Later the work of Frenkel was extended by Wagner and Schottky; they considered the pairs of cation and anion lattice vacancies within the crystal. This defect is observed mostly in materials having almost similar size of cations and anions. Typical examples exhibiting the Schottky effect are alkali halides [4]. The glassy and polymeric materials when used as electrolytes suggested another conduction mechanism [5]. A decoupling index R is defined where $1<R<10^{13}$. For glassy electrolytes, $R{\sim}10^{8}$, the decoupling index provides information on decoupling of fast ion conducting modes and the amorphous matrix. In case of polymer electrolytes, a value of R close to 1 and the presence of nearly complete structural disorder are essential requirements for achieving rapid ion conduction in polymers.

In addition to lattice imperfections, the interaction between the mobile ions and the host framework is another essential requirement for achieving rapid ion conduction. The form of this interaction is yet unknown; some local phonons may provide an insight into the phenomenon. It is evident that the weak affinity between the framework and mobile ions is a certainty, or it can be stated that it is imperative for the mobile ions to loosen their bonds with the host framework to enable rapid ion conduction. The interfaces of crystalline solids play a pivotal role in governing ion mobility within nanoscale electrochemical devices. Thus, the distance between two neighboring interfaces and interfacial density is another control parameter for the designing of these devices.

This chapter presents the different paths for synthesizing nanomaterials and discusses various methods for their characterization. Device integration of nanoionics in resistive switching memories and lithium batteries has also been reviewed in detail in this chapter.

3.2 Synthesis of Nanomaterials

Three different approaches can be employed for the synthesis of nanoparticles. These methods are outlined as follows:

(1) Chemical
(2) Physical
(3) Biological

3.2.1 Chemical Route of Synthesis of Nanomaterials

Various bottom-up approaches for synthesis are employed in the synthesis of nanoparticles, usually in gas or liquid phases. The generation of pure nanoparticles with regulated particle sizes is ensured by this method. Several methods fall under the umbrella of bottom-up techniques. The correct choice of the method depends on factors such as nanomaterial type, desired particle size, method simplicity, and nanocomposite properties. Choosing a method for nanoparticle syntheses, such as co-precipitation, sol-gel, solvothermal, hydrothermal, sonochemical, microemulsion, pyrolysis, vapor deposition, microwave-assisted, ion-exchange, intercalation, and reflux, is contingent on these considerations.

Co-precipitation method

This is a straightforward and commonly employed technique for synthesizing various types of nanoparticles. It produces homogenous nanoparticles through precipitation in an aqueous media [6–12]. The co-precipitation technique involves the combination of water-soluble salts, typically containing divalent and trivalent metal ions, with the soluble salts often containing trivalent metal ions. A chemical reaction takes place between these salts, leading to the precipitation of at least one water-insoluble salt. Depending on the reaction conditions and the reducing agent used, the solution requires constant stirring, either with or without additional heat [13–15]. This process often results in particles with lower crystallinity, though this can be increased by adding thermal energy [16]. In order to maintain the necessary pH, typical reducing agents like sodium hydroxide and ammonia solution are added during the entire process, which is carried out in an alkaline medium. The salt ratio, solution pH, reaction temperature, and base type all have an impact on the size of the nanoparticles. The separation and purification of the solvent can be achieved through filtration or centrifugation, followed by drying. Additionally, this method allows for the preparation of doped ferrites by incorporating different rare earth metals. This technique makes it easier to produce biocompatible nanoparticles, which are essential for use in biomedical applications where small nanoparticles are required. Processes including nucleation, growth, coarsening (Ostwald ripening), agglomeration, and stabilization are involved in the production of nanoparticles. The smallest constituent particles of the emerging thermodynamic phase manifest during nucleation, and a crucial factor in this phenomenon is the level of supersaturation. Coarsening, also called Ostwald ripening, is the result of larger particles devouring smaller ones during the development process so as to reduce the surface energy. Agglomeration can also take place to lower the surface energy. Particles expanding outside of the nanoscale could result from uncontrolled coarsening and aggregation. Agents that cap or stabilize growth are essential in avoiding overgrowth. Through the chemisorption of charged species, these agents attach to the surface of the nanoparticles, causing electrostatic (van der Waals) repulsion. Stable nanoparticles are produced by strong repulsive forces, whereas coagulation is caused by weak forces [17]. Numerous reducing agents exist, including hydrazine dichloride ($N_2H_4.2HCl$), hydrazine hydrate ($N_2H_4.H_2O$), hydrogen (H_2), and metal borohydride ($NaBH_4$, $LiBH_4$). The reduction reaction in this case will be:

$$M^{n+} + ne^- \rightarrow M^0 M^{n+} + ne^- \rightarrow M^0 (M^{n+} M^{n+} \text{ is a metal})$$

Additionally, an oxidation reaction takes place simultaneously and is represented as follows:

$$X^m - ne^- \rightarrow X^{m-n} X^m - ne^- \rightarrow X^{m-n} (X^m X^m \text{ is a reducing agent})$$

Although the final product gets collected in a solid state, the entire reaction takes place in a liquid state. Thus, another name for this phenomenon is reactive crystallization.

Sol-gel method

The sol-gel technique was originally devised for the low-temperature synthesis of glass and ceramic materials. In this approach, the metal alkoxide solution undergoes hydrolysis with water or alcohols in the presence of an acid or base, followed by polycondensation of the resulting mixture. Polycondensation increases the viscosity of the liquid phase and removes water molecules from it, converting it into a gel phase. The transition from the gel phase to the powder phase occurs once all the water molecules have undergone condensation. To provide fine crystalline properties to this powder, further heat is required [18–22]. This process works well for creating hybrids, oxides, and composites from inorganic and organic materials. The main benefit of the sol-gel technology, which is based on inorganic polymerization processes, is the simplicity of the process. However, composite formation compromises the method's purity, requiring post-treatment to purify the material. In Figure 3.1, an overview of different sol-gel coating processes is presented [23–31].

Hydrothermal method

Solutions are subjected to extreme temperature and pressure conditions in this method. This method's main benefit is its capacity to create superior crystals while maintaining composition control. Salts of transition metals

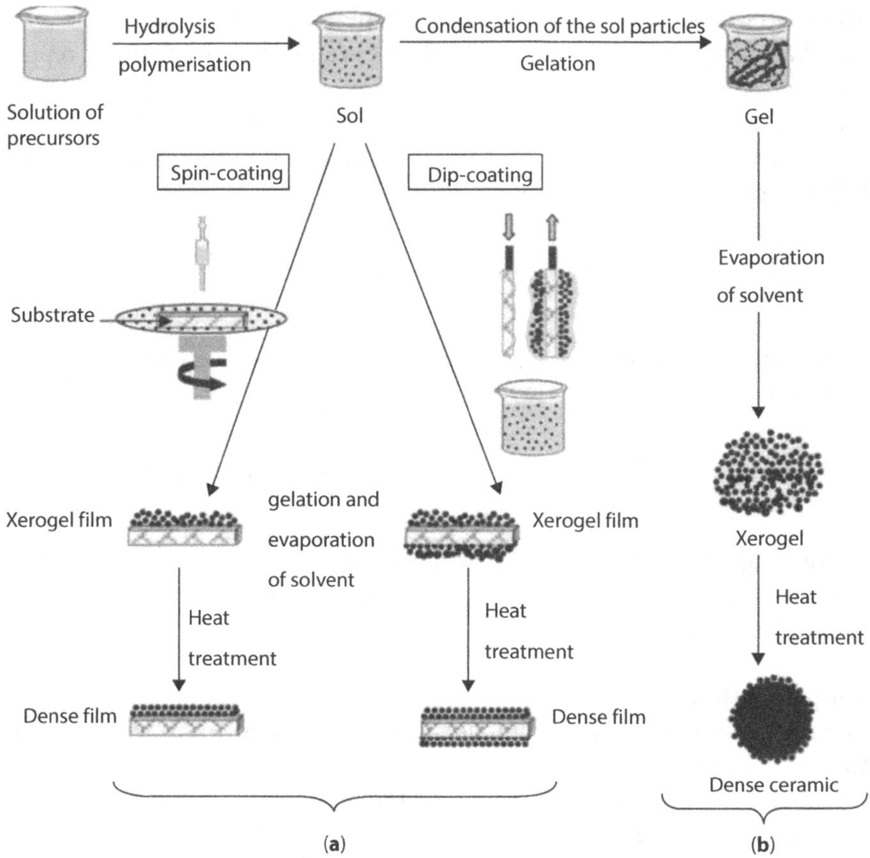

Figure 3.1 Schematic representation of sample synthesis using the sol-gel method: (a) films produced from colloidal sol and (b) powder produced from gel [119–123]. (Adapted from [23]).

that are divalent and trivalent are combined in a 1:2 mol ratio [32–36]. An organic solvent is combined while being constantly stirred to produce a homogenous solution. The solutions are then added to a closed container, which is referred to as an autoclave (bomb). The solvents rise over their boiling points due to an autogenous increase in pressure that happens during heating. Depending on the kind of nanoparticle being manufactured, temperature and time adjustments must be made [37]. This technique is especially useful for producing small nanoparticles, which makes it useful in the biological sciences. As an illustration, copper ferrites, which are frequently encountered magnetic nanoparticles synthesized through hydrothermal methods, underwent cytotoxicity evaluation using the MTT test. The reaction of cells was subsequently examined in human breast cancer cells (MCF-7) [38].

This method enables the controlled creation of nanoparticles with desired size, shape, and surface chemistry by carefully choosing solvents and changing variables including pressure, temperature, aging time, pH, reactant concentration, and reaction time. Because of the technique's high degree of homogeneity, post-annealing procedures are not necessary. Researchers find it more attractive than traditional methods because it enables the synthesis of nanoparticles with minimal size in the desired crystalline phase at comparatively lower temperatures. Numerous doped metals [39], oxides [40], single crystals, zeolites [41], selenides, and sulphides have all been prepared successfully using the hydrothermal approach.

Sonochemical method

The sonochemical process is a particularly quick and safe method. It involves employing ultrasonic radiation to generate cavities, or bubbles, in a liquid medium. The medium is penetrated by ultrasonic energy, which raises the energy mass inside the bubble to pressures of 20 MPa and temperatures of almost 5000 K. The matter inside and around the bubbles is chemically excited as a result of the bubbles' autogenous collapse. Metals like CoS_2 [42], oxides [43], alloys [44], and selenides like CdSe and ZnSe [45] can all be synthesized with this approach.

Microwave-assisted synthesis

Initiated in the 1950s, the microwave-assisted approach has gained general recognition in the last 20 years. By employing electromagnetic radiation (EMR), this approach directly transfers microwave radiations to materials, heating them with movable electric charges. Thermal energy is produced *via* this process from electromagnetic energy [46]. Temperatures between 100 and 200°C are achieved by using frequencies in the range of 1 to 2.5 GHz. More laborious reactions can be finished in a matter of minutes because of the method's lower reaction time requirement [47, 48]. Narrow size-distributed, small-sized particles can be produced using microwave-assisted synthesis. This method can be applied for the preparation of diverse materials, including ferrites, oxides, selenides, and colloidal metals [49]. Fievet *et al.*'s microwave polyol technique represents an improved version of microwave synthesis. In this technique, microwaves are used to produce metals in the colloid state by heating metal precursors in the presence of polyalcohol.

3.2.2 Physical Route of Synthesis of Nanomaterials

The two main categories of physical procedures are "top-down" and "bottom-up" techniques as outlined in Figure 3.2. Using mechanical milling procedures, bigger materials are reduced to tiny particles in the "top-down" approach. But one significant drawback of this approach is that, it is very difficult to get the right shape and size of the particle [50]. Furthermore, samples made by milling may have different magnetic properties than ordinary particles of the same size. Defects in the lattice parameters that emerged during the milling process are the cause of this deviation [51]. On the other hand, the "bottom-up" approach creates nanoparticles by letting smaller ions contract in a liquid or gaseous phase, which subsequently combines chemically to produce larger materials.

Laser evaporation method

One feasible bottom-up technique for producing magnetic nano powders is laser evaporation. In this process, raw metal oxides, employed as the initial ingredients for synthesis, are evaporated using a laser. As a result, the steep temperature gradient helps to promote fast condensation and

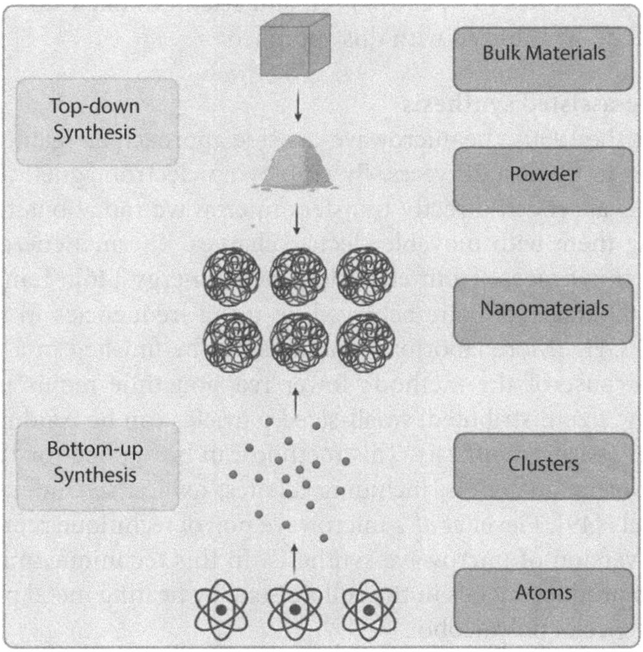

Figure 3.2 Top-down versus bottom-up synthesis approaches. (Adapted from [124]).

nucleation, which produces nanoparticles outside the evaporation zone [52, 53]. The size and magnetic phase of the particles can be altered by precisely controlling the laser intensity and the composition of the gas within the evaporation chamber [54]. In Figure 3.3, the diagram illustrates the process of nanoparticle formation through laser ablation.

RF plasma method

The RF Plasma method is another physical technique that necessitates elevated temperatures. By surrounding the evacuated system with high-voltage radiofrequency coils, this method elevates the metal temperature beyond its evaporation point. Subsequently, helium gas is introduced into the system, and the coils in that region generate high temperatures.

Metal vapor nucleation takes place on helium gas atoms, diffusing into the cooler collection rod to generate nanoparticles [3]. Figure 3.4 depicts the schematic process of nanoparticle synthesis through RF thermal plasma.

Thermal decomposition or thermolysis

Heat-induced decomposition is an endothermic reaction that leads to the breaking of chemical bonds within a compound, resulting in the formation of new, smaller bonds. 13 nm-sized monodispersed nanoparticles were effectively created by Park *et al.* [55]. Monodispersed nanoparticles are especially helpful in the field of biomedicine for treating cancer. According to Kelly *et al.*, coordination compounds and metallocene are proposed as optimal precursors for achieving monodispersed nanoparticles through thermal decomposition [56, 57]. Before undergoing thermal degradation, coordination compounds require the application of stabilizing and

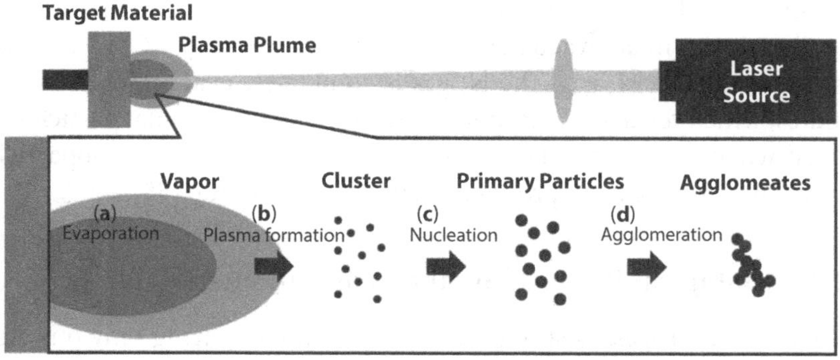

Figure 3.3 Schematic of particle generation procedure in laser ablation method. (Adapted from [125]).

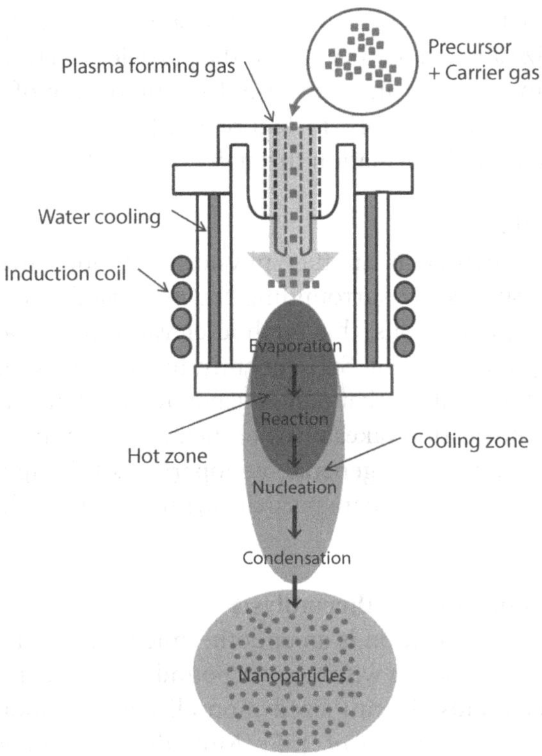

Figure 3.4 Schematic process of utilizing RF thermal plasma for nanoparticle synthesis. (Adapted with permission from [126]).

capping agents. Modifying these elements, such as the concentration of precursors, solvents, or the reaction time, enables the alteration of size and shape. The stabilizing agent may occasionally act as a capping agent as well. Lithium azide, LiN_3, for instance, can produce tiny lithium particles. It is heated to 400°C in an evacuated quartz tube. At about 370°C, LiN_3 breaks down and releases N_2 gas. The N_2 gas is completely evacuated after a few minutes, which results in a decline in pressure. Metal colloidal particles are created when the remaining lithium atoms come together. Nanoparticles smaller than 5 nm can be produced using this technique [58].

3.2.3 Biological Route of Synthesis of Nanomaterials

The biological process is straightforward, usually requiring only one step, and is environmentally beneficial. In this case, we can produce the nanomaterials using both diverse plant parts and microbes [59].

Using microorganisms

The aqueous solution of metal salts can be utilized to manufacture various nanomaterials using different microorganisms such as bacteria, fungus, and algae.

Use of bacteria

Living organisms use proteins as an active ingredient in the biomineralization process that yields nanoparticles. For instance, magnetotactic bacteria produce magnetic particles in anaerobic environments, such as the seafloor. They create nanoscale magnetic iron oxide crystals using magnetosomes, which are protein-coated structures. This allows them to utilize the magnetosomes as a compass to help them find their desired home [60, 61]. It is possible to create homogenous particles *in vitro* that have a core diameter of 20–45 nm [62–64]. In spite of this, magnetosomes have advantageous magnetic properties for use in medicine, including the treatment of hyperthermia [65, 66]. He *et al.*, have developed an alternative method that utilizes photosynthetic bacteria, such as rhodopseudomonas capsulata, to generate extracellular gold nanoparticles measuring between 10 and 20 nm. The transformation of gold ions into gold nanoparticles is predominantly reliant on the bacterial enzyme nicotinamide adenine dinucleotide hydride (NADH)-dependent reductase. They found that the morphology and form of the nanoparticles are affected by the pH of the growing media [67]. Schluter *et al.* utilized pseudomonas cells sourced from an alpine environment to externally generate palladium nanoparticles [68].

Use of fungi

Silver nanoparticles outside the cell were produced through the fusarium oxysporum fungus. The enduring stability of these nanoparticles is attributed to the enzymatic activity of NADH-reductase. It is worth noting that fungal cells demonstrate a higher level of protein production compared to bacterial cells [69]. T. Reesei finds extensive application in various industries, including food, textiles, animal feed, medical, and paper industries.

Use of algae

Singaravelu *et al.*, outlined a method for the extracellular production of gold nanoparticles using Sargassum wightii algae. Amazingly, after just 12 hours of incubation, a 95% output rate was reached [70]. In spite of this, nothing is known about the synthesis of nanoparticles from algae. It's crucial to remember that some fungi, bacteria, and algae can be harmful, therefore, safety precautions must be used during this procedure.

Use of biological templates

The biological mechanism may create nanomaterials inside of organisms by using biological templates as vital tools. These templates, such as DNA and proteins, facilitate the production of unique and sophisticated nanostructures. This approach allows for the design of biosensors [71], bioNEMS, and bioelectronics systems [72] using nanoparticles. A crucial part of the creation of nanocomposite materials is played by proteins. An example of such a protein is ferritin, an internal iron-storage protein found in both prokaryotes and eukaryotes. Ferritin stores iron oxide and releases it gradually, serving as a buffer in humans to regulate iron overload and deficiency. The iron oxide core of ferritin is encased in a protein shell. Iron oxide can be used to selectively dissolve the core to create apoferritin, a protein that allows iron oxide or other desirable nanoparticles to fill the empty core without causing damage to the surrounding protein. As a result of this technique, an inorganic nanocomposite protein is formed. Gold nanoparticles were produced in horse spleen ferritin by Fan *et al.* [73]. Wu *et al.*, synthesized radioactive nanoparticles containing yttrium phosphate within apoferritin, which were subsequently coupled with biotin [74]. Nanoparticle assembly also heavily relies on DNA templates, particularly on plasmids. Spin-coating cadmium perchlorate and plasmid DNA can result in the formation of a conjugate comprising 5–10 nm CdS DNA nanoparticles. Plasmids, closed circular DNA molecules found in many bacterial species, are utilized in this process.

Use of different plant parts

The synthesis of nanoparticles has been achieved using plants and plant extracts, wherein the phytochemicals present in the plants act as reducing agents for metal nanoparticles. Phytochemicals like flavones, organic acids, and quinones serve as effective natural reducing agents in the nanoparticle synthesis process. The biomass of the plants pelargonium graveolens (geranium) and medicago sativa (alfalfa) has been used to create gold nanoparticles of various forms [75, 76]. Nanoparticles composed of bimetallic Au, Ag, and bimetallic Au core-Ag shell have been produced from the leaves of azadirachta indica, commonly known as neem. In this synthesis, the plant's terpenoids and/or sugars function as reducing agents [77]. Gold nanotriangle preparation has been made possible by aloe vera leaf extract [78]. Furthermore, several plants such as helianthus annuus (sunflower) [79] and brassica juncea (Indian mustard) [80] have been employed for the synthesis of nanoparticles made of nickel, silver, zinc, cobalt, and copper.

3.3 Characterization of Nanomaterials

Nanoparticles exhibit diverse physicochemical properties, with variations in size leading to distinct characteristics at the nanoscale. To assess these properties, characterization of nanoparticles is essential and can be achieved using various instruments. UV spectrophotometers, Fourier Transform Infrared (FT-IR) Spectroscopy, Atomic Force Microscopy (AFM), Scanning Electron Microscopy (SEM), Transmission Electron Microscopy (TEM), Vibrating Sample Magnetometers (VSM), Magnetic Force Microscopy (MFM), Energy Dispersive X-ray Spectroscopy (EDS), Superconducting Quantum Interference Devices (SQUID), Mossbauer Spectroscopy (MS), Electron Paramagnetic Resonance (EPR), X-ray Photoelectron Spectroscopy (XPS), and Thermo Gravimetric Analysis (TGA) are a few of the frequently used instruments for nanoparticle characterization [81].

3.3.1 Surface Morphology, Surface Area, Size and Shape of Nanoparticles

The unique physicochemical properties of nanoparticles are significantly influenced by their size and shape. AFM, TEM, and FESEM are a few examples of the techniques that can be used to analyze surface morphology. These techniques yield pictures that reveal information about the morphology of nanoparticles, such as whether they are rod-shaped, spherical, or porous. It is also possible to compute a nanoparticle's diameter. In comparison to SEM, TEM offers more details regarding the morphological, crystallographic, and compositional characteristics of nanoparticles. Using an electron beam to interact with sample atoms, SEM generates a range of signals that reveal information about the sample's surface topography and composition. Samples must therefore have an electrically conductive surface, or ultrathin electrically conducting coatings which must be applied to nonconductive samples. The main tools used to determine the size of nanoparticles are HRTEM, FESEM, and XRD, although TEM is unique in that it can acquire high-resolution images. It makes it possible to retrieve precise and in-depth data on the properties of nanoparticles, such as their lattice spacing, electron phase shift, aggregation state, and crystal orientation. TEM facilitates the determination of nanoparticle size, and the Scherrer equation can be employed in conjunction with XRD spectroscopy to aid in size determination. Sharp XRD peaks are advantageous in sizing crystalline nanoparticles, but broad XRD peaks pose challenges in determining the size of non-crystalline nanoparticles, especially when compared to TEM. Mean particle size and size distribution determination are facilitated by dynamic light scattering (DLS), Mossbauer spectroscopy, and photon correlation spectroscopy. For figuring out a nanoparticle's surface area, the Brunauer Emmet Teller (BET) approach works well [82, 83].

3.3.2 Analysis of Elemental and Mineral Composition

Utilizing EDS alongside SEM and TEM instruments facilitates the determination of both elemental composition and surface morphology analysis. Elemental content can be determined through methods such as Atomic Absorption Spectroscopy (AAA) and Inductively Coupled Plasma Mass Spectroscopy (ICP-MS). In the case of AAA spectroscopy, solid nanoparticles cannot be directly utilized; they must be dissolved in suitable acids or bases before analysis. XRD is employed to ascertain the mineral composition in aggregated crystalline nanoparticles and can calculate percentage crystallinity and unit cell dimensions [84]. X-ray Photoelectron Spectroscopy (XPS) can also furnish information on elemental composition.

3.3.3 Structures and Bonds in Nanoparticles

Many approaches can be used to gather information about the structure and bonding properties, including widely used methods such as XPS, FT-IR, Raman Spectroscopy (RS), TGA and X-ray absorption spectroscopy (XAS). Metal-oxygen bonds are verified by FT-IR and XPS, with the latter also offering information on the surface arrangement of nanoparticles. It keeps track of information on elemental composition, binding energy of materials, and oxidation states. Compounds' spinel lattice and structure are ascertained *via* the use of Raman spectroscopy. Many details regarding the necessary elements' oxidation states, nearby atoms, bond length, coordination number, and electrical configuration can be found using XAS [85, 86].

3.3.4 Magnetic Properties of Nanoparticles

SQUID, VSM, and EPR are useful tools for analyzing a nanoparticle's magnetic characteristics. Free radicals and paramagnetic centers can both be found using the EPR approach. With its exceptional sensitivity, the SQUID instrument can examine a wide range of materials, including crystals, thin films, powders, liquids, and gases. At a constant applied external magnetic field, SQUID and VSM may detect residual magnetization, magnetic saturations, and other characteristics [87, 88]. Many details are available from Mossbauer Spectroscopy, including covalence, electronegativity, oxidation and spin states, bonding, structural and magnetic properties, and more.

3.4 Device Integration of Nanoionics

3.4.1 Resistive Switching Memories

Recent advancements in memory theory encompass a diverse array of techniques, including spin-based memories like magneto resistive random-access memories (MRAM) and related concepts where resistance switching depends on a magnetic field. Ongoing research explores Phase-Change RAM (PCRAM), involving a phase transition from the amorphous to the crystalline state in the switching material orchestrated by thermal processes. Resistance Switching RAM (RRAM) refers to a distinct form of resistive switching phenomenon based on the electrically driven modification of resistance in a metal-insulator-metal (MIM) memory cell.

In Metal-Insulator-Metal (MIM), the 'M' denotes any moderately efficient electron conductor, typically discernible on both sides, while the 'I' signifies an insulator, often a material that conducts ions. The initial pristine sample usually undergoes an electroforming process, such as a current-limited electric breakdown, establishing the system for subsequent switching between a conductive ON state and a less conductive OFF state. As the research suggests, the switching class has a major impact on the mechanism and requirements for this first phase. If the switching process operates irrespective of the polarity of the current and voltage signal, it is termed as symmetric (or unipolar) switching.

In unipolar switching, a threshold voltage is responsible for transitioning a system from its high-resistance state (OFF) to the low-resistance state (ON). The current is restricted by the compliance current of the control circuit. Below the set voltage and with higher current, the device resets to the off state. Without compliance current, PCRAMs display unipolar switching. In contrast, bipolar or antisymmetric characteristics involve transitioning between an ON state at one voltage polarity and an OFF state at the reversed voltage polarity. To observe bipolar switching behavior, the system's structure must exhibit some asymmetry, such as distinct electrode materials or voltage polarity during the initial electroforming step. State readings in unipolar and bipolar characteristics are made at low voltages that have no effect on the state.

In MIM systems, the resistive switching mechanisms frequently combine chemical and/or physical effects. A conventional resistive switching based on a thermal effect exhibits a unipolar characteristic. The resistive switching is influenced by electronic charge injection and/or charge displacement effects. An illustration of this is the charge-trap model [89], where charges undergo Fowler-Nordheim tunnelling at high electric fields, subsequently becoming trapped at sites such as defects or metal nanoparticles within the insulator. This process modifies the resistance and electrostatic barrier characteristics of the MIM structure, rendering it akin to the gate-channel resistance of a Flash field-effect transistor (FET). As an example, trapping sites can be provided by gold nanoclusters embedded in polymeric [90, 91], or inorganic insulator films [92]. It is thought that at different metal/semiconducting perovskite interfaces, trapping at interface states affects the nearby Schottky barrier in a modified model [93, 94]. ZnSe–Ge heterojunctions have been shown to operate *via* a similar method [20].

An alternative model that could be considered is the insulator–metal transition (IMT) model, wherein perovskite-type oxides like SrTiO3:Cr [95] and (Pr,Ca)MnO3 [96–98] undergo an IMT due to electronic charge injection acting similarly to doping. Bipolar switching has recently been added to a general model by Rozenberg *et al.* [99, 100]. Esaki [101] put up a ferroelectricity-based model, which Kohlstedt *et al.* [102, 103] theoretically explained. MIM systems represent the convergence of nanoelectronics [104] and nanoionics [105], where bipolar resistive switching is governed by ionic transport and electrochemical redox reactions, constituting the fundamental mechanism.

3.4.2 Lithium Batteries

An extensively researched strategy involves modifying interface spacing at the nanoscale and interfacial design to manipulate electrical transport characteristics. As a result, in the fields of physics and materials science, nanoelectronics has gained significant traction. The significance of ionic transport characteristics has also been acknowledged; these qualities are pertinent to the chemical kinetics of solids and electrochemical devices. Numerous studies and theoretical analyses conducted during the previous 20 years have demonstrated the significant influence that interfaces, and their spacing have on the characteristics of ionic transport [106].

Typically, selecting the appropriate materials required two main strategies: (a) selecting the best bulk structures; and (b) optimizing composition through doping. Intercalation compounds with advantageous storage capacities and Li-mobilities are notable examples of the first method

[6, 7, 59]. The feasibility of using several transition metal oxides (e.g., Co, Cu, Ni, and Fe) as anodes was prominently shown by Poizot *et al.* [107]. $LiFePO_4$ can function as a cathode material even if its electronic conductivity is low, as demonstrated by Padhi *et al.* [108]. Beaulieu *et al.* [109], reported on the storage of Li at the grain boundaries of $SnMn_3C$, which is normally regarded as "inactive". Grugeon and colleagues [110], documented the storing process inside the metastable organic layer that developed on the redox-active electrode. While the nano-crystalline state yields intriguing and advantageous outcomes, there are drawbacks due to the additional energy introduced by nature and the stability issues associated with nano-crystallization.

The charging and discharging rates are influenced by factors such as electrolyte transport, the transport of Li^+ ions and electrons within the composite electrodes, surface incorporation of Li^+ into electrochemically active particles, and transport within them in coordination with the electrochemical reaction. The practical application of electrochemical energy storage, shown by nonaqueous $Li–O_2$ batteries, is restricted by the charging and discharging speed. Nanoionics, which are materials that have been nanostructured to improve ion-transport kinetics, offer a possible remedy. Heitjans *et al.* [111, 112], noticed a significant enhancement in Li-ion diffusivity achieved through the establishment of a mixed system consisting of nanocrystalline ion conductors and nanocrystalline insulators. Confirmation of this was attained through spin-lattice relaxation measurements and Li nuclear magnetic resonance line-shape studies [113]. Additionally, it was demonstrated that ion transit was facilitated by strain and defects induced by high-energy ball milling. Notably, due to interface energy-level bending, rapid ions were found to reside in the interfacial regions, while slower ions were located within the grains [112]. Li *et al.*'s recent study [114], described how atomically close solid-solid contact may be achieved by creating epitaxial surfaces, which significantly increases ion conductivity. These examples highlight how important nanoionics are to the conversion and storage of energy.

3.5 Summary and Future Prospects

In the last decade, nanotechnology has emerged as a highly captivating field of research, propelled by diverse applications and growing demand, leading to the development of new techniques for producing high-quality nanomaterials [1]. Operating at the intersection of physics, chemistry, materials science, and various engineering disciplines, nanotechnology finds applications across a wide array of scientific and technological domains [105, 113, 115, 116]. One specific area within nanotechnology, termed "nanoionics," focuses on studying the mechanisms, phenomena, properties, effects, methods, and applications related to ion transport in nano confinement [1]. Although not a recent concept, the term "nanoionics" was introduced in 1992 [1], and subsequent efforts by Maier and other scientists have significantly shaped and contributed to this research topic [105, 113, 115, 116]. Nanoionics has broadened its usefulness by studying ion-transport and storage phenomena in liquids, in addition to its original focus on ion transport in all-solid-state nanoscale systems. Nanomaterials, typically on the scale of 10^{-9} meters, exhibit distinct physicochemical properties compared to bulk materials, influenced by their size and shape [1]. Nanomaterials with critical sizes (<100 nm) display unique and captivating properties, leading to increased interest [1]. As the particle size decreases and the surface area increases, nanomaterials exhibit different characteristics compared to bulk materials. Surface atoms are more mobile than inside atoms because they have a lower coordination number. Nanomaterials are used in sensors, adsorbents, functional coatings, catalysis, and nanoelectronics by utilizing these surface characteristics [1]. This chapter presents various approaches for synthesizing different types of nanomaterials, outlines methods for their characterization, and explores device integration into nanoionics-based applications [1]. While solid-state nanoionics has found broad applications in areas such as solid electrolyte batteries [117], the full potential of resistive switching effects, particularly ion-migration-based redox effects, requires further exploration [117]. The manipulation of nanoscale structures, characterized by a high density of interfaces different from bulk materials, can enhance the electrical performance of batteries. For example, great coulombic efficiency and high theoretical capacity are both present in nanocrystalline electrodes [118]. Nanoionics research, although still in its early stages, holds unlimited potential. Notwithstanding current obstacles, the swift advancement in nanoscience and the promising advantages indicate that a comprehensive investigation into nanoionics is warranted in the near future [117]. By fully using nanoionics, active devices that are dominated by nanoelectronics might be revolutionized and new types of artificial intelligence with potentially transformative power could be made possible.

References

1. Despotuli, A.L. and Nikolaichik, V.I., A step towards nanoionics. *Solid State Ionics*, 60, 275, 1993.

2. Funke, K., Solid State Ionics: From Michael Faraday to green energy - The European dimension. *Sci. Technol. Adv. Mater.*, 14, 43502, 2013.
3. Ferrari, M., Introduction to Nanotechnology. Charles P. Poole, Jr., Frank J. Owens. Hoboken, NJ: John Wiley & Sons, 2003, 400 pp., $79.95, hardcover. ISBN 0-471-07935-9. *Clin. Chem.*, 50, 981, 2004.
4. Faux, I.D. and Lidiard, A.B., The Volumes of Formation of Schottky Defects in Ionic Crystals. *Z. Naturforsch. - Sect. A J. Phys. Sci.*, 26, 62, 1971.
5. Angell, C.A., Fast ion motion in glassy and amorphous materials. *Solid State Ionics*, 9–10, 3, 1983.
6. Drbohlavova, J., Hrdy, R., Adam, V., Kizek, R., Schneeweiss, O., Hubalek, J., Preparation and properties of various magnetic nanoparticles. *Sensors*, 9, 2352, 2009.
7. Zhao, S., Guo, J., Li, W., Guo, H., You, B., Fabrication of cobalt aluminate nanopigments by coprecipitation method in threonine waterborne solution. *Dyes Pigm.*, 151, 130, 2018.
8. Rahimi, R., Maleki, A., Maleki, S., Morsali, A., Rahimi, M.J., Synthesis and characterization of magnetic dichromate hybrid nanomaterials with triphenylphosphine surface modified iron oxide nanoparticles (Fe 3O4@SiO2@ PPh3@Cr2O 72-). *Solid State Sci.*, 28, 9, 2014.
9. Sanchez-Martinez, A., Ceballos-Sanchez, O., Koop-Santa, C., López-Mena, E.R., Orozco-Guareño, E., García-Guaderrama, M., N-doped TiO2 nanoparticles obtained by a facile coprecipitation method at low temperature. *Ceram. Int.*, 44, 5273, 2018.
10. Zhou, X. and Xiang, C., Comparative Analysis of Solar-to-Fuel Conversion Efficiency: A Direct, One-Step Electrochemical CO2 Reduction Reactor versus a Two-Step, Cascade Electrochemical CO2 Reduction Reactor. *ACS Energy Lett.*, 3, 1892, 2018.
11. Maleki, A., Rahimi, R., Maleki, S., Hamidi, N., Synthesis and characterization of magnetic bromochromate hybrid nanomaterials with triphenylphosphine surface-modified iron oxide nanoparticles and their catalytic application in multicomponent reactions. *RSC Adv.*, 4, 29765, 2014.
12. Amiri, S. and Shokrollahi, H., Magnetic and structural properties of RE doped Co-ferrite (RE=Nd, Eu, and Gd) nano-particles synthesized by co-precipitation. *J. Magn. Magn. Mater.*, 345, 18, 2013.
13. El Ghandoor, H., Zidan, H.M., Khalil, M.M.H., Ismail, M.I.M., Synthesis and some physical properties of magnetite (Fe 3O 4) nanoparticles. *Int. J. Electrochem. Sci.*, 7, 5734, 2012.
14. Sarfraz, M.H., Zubair, M., Aslam, B., Ashraf, A., Siddique, M.H., Hayat, S., Cruz, J.N., Muzammil, S., Khurshid, M., Sarfraz, M.F., Hashem, A., Dawoud, T.M., Avila-Quezada, G.D., Abd_Allah, E.F., Comparative analysis of phyto-fabricated chitosan, copper oxide, and chitosan-based CuO nanoparticles: antibacterial potential against Acinetobacter baumannii isolates and anticancer activity against HepG2 cell lines. *Front. Microbiol.*, 14, 1188743, 2023.

15. Zhang, J., Liu, W., Dai, J., Xiao, K., Nanoionics from Biological to Artificial Systems: An Alternative Beyond Nanoelectronics. *Adv. Sci.*, 9, e2200534, 2022.
16. Dong, H., Du, S.R., Zheng, X.Y., Lyu, G.M., Sun, L.D., Li, L.D., Zhang, P.Z., Zhang, C., Yan, C.H., Lanthanide Nanoparticles: From Design toward Bioimaging and Therapy. *Chem. Rev.*, 115, 10725, 2015.
17. Dubey, A., Singh, A., Narang, R., Saxena, M., Gupta, M., Modeling and Simulation of Junctionless Double Gate Radiation Sensitive FET (RADFET) Dosimeter. *IEEE Trans. Nanotechnol.*, 17, 49, 2018.
18. Yu, S., Jing, W., Tang, M., Xu, T., Yin, W., Kang, B., Fabrication of Nd:YAG transparent ceramics using powders synthesized by citrate sol-gel method. *J. Alloys Compd.*, 772, 751, 2019.
19. Mulens, V., Morales, M.P., Barber, D.F., Development of Magnetic Nanoparticles for Cancer Gene Therapy: A Comprehensive Review. *Int. Sch. Res. Notices*, 2013, 646284, 14, 2013.
20. Hao, S., Lin, T., Ning, S., Qi, Y., Deng, Z., Wang, Y., Research on cracking of SiO2 nanofilms prepared by the sol-gel method. *Mater. Sci. Semicond. Process.*, 91, 181, 2019.
21. Sriondee, M., Dungsuwan, W., Thountom, S., Synthesis and characterization of Bi0.5(Na1-xKx)0.5TiO3 powders by sol–gel combustion method with glycine fuel. *Ceram. Int.*, 44, S168, 2018.
22. Xu, H., Xin, L., Liu, L., Pang, D., Jiao, Y., Cong, R., Yu, W., Large area MoS2/Si heterojunction-based solar cell through sol-gel method. *Mater. Lett.*, 238, 13, 2019.
23. Bokov, D., Turki Jalil, A., Chupradit, S., Suksatan, W., Javed Ansari, M., Shewael, I.H., Valiev, G.H., Kianfar, E., Nanomaterial by Sol-Gel Method: Synthesis and Application. *Adv. Mater. Sci. Eng.*, 2021, 5102014, 21, 2021.
24. Arora, A., Nanomaterials: methods of generation. In *Nanotechnology*, pp. 11–22, CRC Press, 2020.
25. Parkash, V., Gaur, A., Agnihotri, R., Nanoparticles from Fungal Resources: Importance and Applications, in: *Nanotechnol. Life Sci.*, p. 1, 2020.
26. Kianfar, E., Comparison and assessment of zeolite catalysts performance dimethyl ether and light olefins production through methanol: A review. *Rev. Inorg. Chem.*, 39, 157, 2019.
27. Kianfar, E., Salimi, M., Pirouzfar, V., Koohestani, B., Synthesis and modification of zeolite ZSM-5 catalyst with solutions of calcium carbonate (CaCO3) and sodium carbonate (Na2CO3) for methanol to gasoline conversion. *Int. J. Chem. React. Eng.*, 16, 7, 20170229, 2018.
28. Peerakiatkhajorn, P., Chawengkijwanich, C., Onreabroy, W., Chiarakorn, S., Novel photocatalytic Ag/TiO2 thin film on polyvinyl chloride for gaseous BTEX treatment. *Mater. Sci. Forum*, 712, 133, 2012.
29. Yadav, S. and Jaiswar, G., Review on Undoped/Doped TiO2 Nanomaterial; Synthesis and Photocatalytic and Antimicrobial Activity. *J. Chin. Chem. Soc.*, 64, 103, 2017.

30. Hu, B., Pan, H., Yu, L.G., Wang, L.G., Zhao, L., Zhang, Z.J., Preparation and properties of TiO2 nanoparticles *in situ* surface-modified with salicylic acid. *Wuji Cailiao Xuebao/J. Inorg. Mater.*, 26, 1181, 2011.
31. Zhan, H., Xiong, Z., Cheng, C., Liang, Q., Liu, J.Z., Li, D., Solvation-Involved Nanoionics: New Opportunities from 2D Nanomaterial Laminar Membranes. *Adv. Mater.*, 32, e1904562, 2020.
32. Zhang, Z., Wang, L., Xu, X., Dong, Y., Zhang, L., Development of a validated HPLC method for the determination of tenofovir disoproxil fumarate using a green enrichment process. *Anal. Methods*, 7, 6290, 2015.
33. Qiu, W., Feng, X., Zhang, H., Huang, H., Synthesis and luminescence properties of CaB 4 O 7 :Eu 3+ *via* two-step hydrothermal method. *Optik (Stuttg)*, 182, 1039, 2019.
34. Obreja, P., Cristea, D., Dinescu, A., Romanițan, C., Influence of surface substrates on the properties of ZnO nanowires synthesized by hydrothermal method. *Appl. Surf. Sci.*, 463, 1117, 2019.
35. Li, M., Liu, X., Xu, T., Nie, Y., Li, H., Zhang, C., Synthesis and characterization of nanosized MnZn ferrites *via* a modified hydrothermal method. *J. Magn. Magn. Mater.*, 439, 228, 2017.
36. Wongpratat, U., Maensiri, S., Swatsitang, E., EXAFS study of cations distribution dependence of magnetic properties in Co1-xZnxFe2O4 nanoparticles prepared by hydrothermal method. *Microelectron. Eng.*, 146, 68, 2015.
37. Zhang, J., Song, J.M., Niu, H.L., Mao, C.J., Zhang, S.Y., Shen, Y.H., ZnFe2O4 nanoparticles: Synthesis, characterization, and enhanced gas sensing property for acetone. *Sens. Actuators, B*, 221, 55, 2015.
38. Ahamed, M., Akhtar, M.J., Alhadlaq, H.A., Alshamsan, A., Copper ferrite nanoparticle-induced cytotoxicity and oxidative stress in human breast cancer MCF-7 cells. *Colloids Surf., B*, 142, 46, 2016.
39. Mehraz, S., Kongsong, P., Taleb, A., Dokhane, N., Sikong, L., Large scale and facile synthesis of Sn doped TiO2 aggregates using hydrothermal synthesis. *Sol. Energy Mater. Sol. Cells*, 189, 254, 2019.
40. Wang, T., Tang, T., Gao, Y., Chen, Q., Zhang, Z., Bian, H., Hydrothermal preparation of Ag-TiO2-reduced graphene oxide ternary microspheres structure composite for enhancing photocatalytic activity. *Physica E*, 112, 128, 2019.
41. Xia, S., Chen, Y., Xu, H., Lv, D., Yu, J., Wang, P., Synthesis EMT-type zeolite by microwave and hydrothermal heating. *Micropor. Mesopor. Mater.*, 278, 54, 2019.
42. Muradov, M.B., Balayeva, O.O., Azizov, A.A., Maharramov, A.M., Qahramanli, L.R., Eyvazova, G.M., Aghamaliyev, Z.A., Synthesis and characterization of cobalt sulfide nanoparticles by sonochemical method. *Infrared Phys. Technol.*, 89, 255, 2018.
43. Santibenchakul, S., Sirijaturaporn, P., Mekprasart, W., Pechrapa, W., Ga-doped ZnO nanoparticles synthesized by sonochemical-assisted process. *Mater. Today Proc.*, 5, 13865, 2018.

44. Okoli, C.U., Kuttiyiel, K.A., Cole, J., McCutchen, J., Tawfik, H., Adzic, R.R., Mahajan, D., Solvent effect in sonochemical synthesis of metal-alloy nanoparticles for use as electrocatalysts. *Ultrason. Sonochem.*, 41, 427, 2018.
45. Panahi-Kalamuei, M., Mousavi-Kamazani, M., Salavati-Niasari, M., Hosseinpour-Mashkani, S.M., A simple sonochemical approach for synthesis of selenium nanostructures and investigation of its light harvesting application. *Ultrason. Sonochem.*, 23, 246, 2015.
46. Hassanjani-Roshan, A., Vaezi, M.R., Shokuhfar, A., Rajabali, Z., Synthesis of iron oxide nanoparticles *via* sonochemical method and their characterization. *Particuology*, 9, 95, 2011.
47. Gonzalez-Moragas, L., Yu, S.M., Murillo-Cremaes, N., Laromaine, A., Roig, A., Scale-up synthesis of iron oxide nanoparticles by microwave-assisted thermal decomposition. *Chem. Eng. J.*, 281, 87, 2015.
48. Wu, W., Wu, Z., Yu, T., Jiang, C., Kim, W.S., Recent progress on magnetic iron oxide nanoparticles: Synthesis, surface functional strategies and biomedical applications. *Sci. Technol. Adv. Mater.*, 16, 23501, 2015.
49. Seidlerová, J., Životský, O., Drobíková, K., Mamulová Kutláková, K., Modification of microwave assisted preparation of FexOy nanoparticles. *Mater. Today Proc.*, 5, S52, 2018.
50. Thakore, S., Rathore, P.S., Jadeja, R.N., Thounaojam, M., Devkar, R.V., Sunflower oil mediated biomimetic synthesis and cytotoxicity of monodisperse hexagonal silver nanoparticles. *Mater. Sci. Eng., C*, 44, 209, 2014.
51. Dutz, S., Hergt, R., Mürbe, J., Müller, R., Zeisberger, M., Andrä, W., Töpfer, J., Bellemann, M.E., Hysteresis losses of magnetic nanoparticle powders in the single domain size range. *J. Magn. Magn. Mater.*, 308, 305, 2007.
52. Kurland, H.D., Grabow, J., Staupendahl, G., Andrä, W., Dutz, S., Bellemann, M.E., Magnetic iron oxide nanopowders produced by CO2 laser evaporation. *J. Magn. Magn. Mater.*, 311, 73, 2007.
53. Kurland, H.D., Grabow, J., Staupendahl, G., Müller, F.A., Müller, E., Dutz, S., Bellemann, M.E., Magnetic iron oxide nanopowders produced by CO2 laser evaporation-'*In situ*' coating and particle embedding in a ceramic matrix. *J. Magn. Magn. Mater.*, 321, 1381, 2009.
54. Stötzel, C., Kurland, H.D., Grabow, J., Dutz, S., Müller, E., Sierka, M., Müller, F.A., Control of the crystal phase composition of FexOy nanopowders prepared by CO2 laser vaporization. *Cryst. Growth Des.*, 13, 4868, 2013.
55. Bu, L., Xie, J., Chen, K., Huang, J., Aguilar, Z.P., Wang, A., Sun, K.W., Chua, M.S., So, S., Cheng, Z., Eden, H.S., Shen, B., Chen, X., Assessment and comparison of magnetic nanoparticles as MRI contrast agents in a rodent model of human hepatocellular carcinoma. *Contrast Media Mol. Imaging*, 7, 363, 2012.
56. Shumbula, P.M., Moloto, M.J., Tshikhudo, T.R., Fernandes, M., Dichloro (bis[diphenylthiourea]) cadmium complex as a precursor for HDA-capped CdS nanoparticles and their solubility in water. *S. Afr. J. Sci.*, 106, 7–8, Art.#311, 7, 2010.

57. Kelly, C.H.W. and Lein, M., Choosing the right precursor for thermal decomposition solution-phase synthesis of iron nanoparticles: tunable dissociation energies of ferrocene derivatives. *Phys. Chem. Chem. Phys.*, 18, 32448, 2016.
58. McKay, J., The business of magazine publishing, in: *The Magazines Handbook*, vol. 242, 2018.
59. Roy, K. and Ghosh, C.K., Environmental and Biological Applications of Nanoparticles. *Nanotechnology*, 147, 2017.
60. Wells, B.O., Kannan M. Krishnan: Fundamentals and Applications of Magnetic Materials 1st Edition. *J. Mater. Sci.*, 52, 6905, 2017.
61. Bazylinski, D.A., Garratt-Reed, A.J., Frankel, R.B., Electron microscopic studies of magnetosomes in magnetotactic bacteria. *Microsc. Res. Tech.*, 27, 389, 1994.
62. Faivre, D. and Schüler, D., Magnetotactic bacteria and magnetosomes. *Chem. Rev.*, 108, 4875, 2008.
63. Schüler, D. and Frankel, R.B., Bacterial magnetosomes: Microbiology, biomineralization and biotechnological applications. *Appl. Microbiol. Biotechnol.*, 52, 464, 1999.
64. Timko, M., Molcan, M., Hashim, A., Skumiel, A., Muller, M., Gojzewski, H., Jozefczak, A., Kovac, J., Rajnak, M., Makowski, M., Kopčanský, P., Hyperthermic effect in suspension of magnetosomes prepared by various methods. *IEEE Trans. Magn.*, 49, 250, 2013.
65. Hergt, R., Hiergeist, R., Zeisberger, M., Schüler, D., Heyen, U., Hilger, I., Kaiser, W.A., Magnetic properties of bacterial magnetosomes as potential diagnostic and therapeutic tools. *J. Magn. Magn. Mater.*, 293, 80, 2005.
66. Molcan, M., Gojzewski, H., Skumiel, A., Dutz, S., Kovac, J., Kubovcikova, M., Kopcansky, P., Vekas, L., Timko, M., Energy losses in mechanically modified bacterial magnetosomes. *J. Phys. D. Appl. Phys.*, 49, 365002, 2016.
67. He, S., Guo, Z., Zhang, Y., Zhang, S., Wang, J., Gu, N., Biosynthesis of gold nanoparticles using the bacteria Rhodopseudomonas capsulata. *Mater. Lett.*, 61, 3984, 2007.
68. Schlüter, M., Hentzel, T., Suarez, C., Koch, M., Lorenz, W.G., Böhm, L., Düring, R.A., Koinig, K.A., Bunge, M., Synthesis of novel palladium(0) nanocatalysts by microorganisms from heavy-metal-influenced high-alpine sites for dehalogenation of polychlorinated dioxins. *Chemosphere*, 117, 462, 2014.
69. Ahmad, A., Mukherjee, P., Senapati, S., Mandal, D., Khan, M.I., Kumar, R., Sastry, M., Extracellular biosynthesis of silver nanoparticles using the fungus Fusarium oxysporum. *Colloids Surf., B*, 28, 313, 2003.
70. Singaravelu, G., Arockiamary, J.S., Kumar, V.G., Govindaraju, K., A novel extracellular synthesis of monodisperse gold nanoparticles using marine alga, Sargassum wightii Greville. *Colloids Surf., B*, 57, 97, 2007.
71. Zhang, N.M.Y., Qi, M., Wang, Z., Wang, Z., Chen, M., Li, K., Shum, P., Wei, L., One-step synthesis of cyclodextrin-capped gold nanoparticles for

ultra-sensitive and highly-integrated plasmonic biosensors. *Sens. Actuators, B*, 286, 429, 2019.
72. Zhang, T., Liu, J., Wang, C., Leng, X., Xiao, Y., Fu, L., Synthesis of graphene and related two-dimensional materials for bioelectronics devices. *Biosens. Bioelectron.*, 89, 28, 2017.
73. Fan, R., Chew, S.W., Cheong, V.V., Orner, B.P., Fabrication of gold nanoparticles inside unmodified horse spleen apoferritin. *Small*, 6, 1483, 2010.
74. Wu, H., Wang, J., Wang, Z., Fisher, D.R., Lin, Y., Apoferritin-templated yttrium phosphate nanoparticle conjugates for radioimmunotherapy of cancers. *J. Nanosci. Nanotechnol.*, 8, 2316, 2008.
75. Gardea-Torresdey, J.L., Gomez, E., Peralta-Videa, J.R., Parsons, J.G., Troiani, H., Jose-Yacaman, M., Alfalfa sprouts: A natural source for the synthesis of silver nanoparticles. *Langmuir*, 19, 1357, 2003.
76. Carmo Bastos, M.L., Silva-Silva, J.V., Neves Cruz, J., Palheta da Silva, A.R., Bentaberry-Rosa, A.A., da Costa Ramos, G., de Sousa Siqueira, J.E., Coelho-Ferreira, M.R., Percário, S., Santana Barbosa Marinho, P., Marinho, A.M.R., de Oliveira Bahia, M., Dolabela, M.F., Alkaloid from Geissospermum sericeum Benth. & Hook.f. ex Miers (Apocynaceae) Induce Apoptosis by Caspase Pathway in Human Gastric Cancer Cells. *Pharmaceuticals*, 16, 765, 2023.
77. Shankar, S.S., Rai, A., Ahmad, A., Sastry, M., Rapid synthesis of Au, Ag, and bimetallic Au core-Ag shell nanoparticles using Neem (Azadirachta indica) leaf broth. *J. Colloid Interface Sci.*, 275, 496, 2004.
78. Chandran, S.P., Chaudhary, M., Pasricha, R., Ahmad, A., Sastry, M., Synthesis of gold nanotriangles and silver nanoparticles using Aloe vera plant extract. *Biotechnol. Progr.*, 22, 577, 2006.
79. Banerjee, M., Dey, B., Talukdar, J., Chandra Kalita, M., Production of biodiesel from sunflower oil using highly catalytic bimetallic gold-silver core-shell nanoparticle. *Energy*, 69, 695, 2014.
80. Shekhawat, G.S. and Arya, V., Biological synthesis of Ag nanoparticles through *in vitro* cultures of Brassica juncea C. zern. *Adv. Mater. Res.*, 67, 295, 2009.
81. Galloway, J.M., Talbot, J.E., Critchley, K., Miles, J.J., Bramble, J.P., Developing Biotemplated Data Storage: Room Temperature Biomineralization of L10 CoPt Magnetic Nanoparticles. *Adv. Funct. Mater.*, 25, 4590, 2015.
82. Chekli, L., Bayatsarmadi, B., Sekine, R., Sarkar, B., Shen, A.M., Scheckel, K.G., Skinner, W., Naidu, R., Shon, H.K., Lombi, E., Donner, E., Analytical characterisation of nanoscale zero-valent iron: A methodological review. *Anal. Chim. Acta*, 903, 13, 2016.
83. Gabbasov, R., Polikarpov, M., Cherepanov, V., Chuev, M., Mischenko, I., Lomov, A., Wang, A., Panchenko, V., Mössbauer, magnetization and X-ray diffraction characterization methods for iron oxide nanoparticles. *J. Magn. Magn. Mater.*, 380, 111, 2015.

84. Kubickova, S., Vejpravova, J., Holec, P., Niznansky, D., Correlation of crystal structure and magnetic properties of Co (1-x)NixFe2O4/SiO2 nanocomposites. *J. Magn. Magn. Mater.*, 334, 102, 2013.
85. Gräfe, M., Donner, E., Collins, R.N., Lombi, E., Speciation of metal(loid)s in environmental samples by X-ray absorption spectroscopy: A critical review. *Anal. Chim. Acta*, 822, 1, 2014.
86. Reddy, L.H., Arias, J.L., Nicolas, J., Couvreur, P., Magnetic nanoparticles: Design and characterization, toxicity and biocompatibility, pharmaceutical and biomedical applications. *Chem. Rev.*, 112, 5818, 2012.
87. Krzyminiewski, R., Dobosz, B., Schroeder, G., Kurczewska, J., Focusing of Fe3O4 nanoparticles using a rotating magnetic field in various environments. *Phys. Lett. Sect. A Gen. At. Solid State Phys.*, 382, 3192, 2018.
88. Flanders, P.J. and Graham, C.D., Magnetization measurements using the field gradient of a high-field solenoid. *Rev. Sci. Instrum.*, 50, 1564, 1979.
89. New conduction and reversible memory phenomena in thin insulating films. *Proc. R. Soc. London Ser. A. Math. Phys. Sci.*, 301, 77, 1967.
90. Ouyang, J., Chu, C.W., Szmanda, C.R., Ma, L., Yang, Y., Programmable polymer thin film and non-volatile memory device. *Nat. Mater.*, 3, 918, 2004.
91. Bozano, L.D., Kean, B.W., Beinhoff, M., Carter, K.R., Rice, P.M., Scott, J.C., Organic materials and thin-film structures for cross-point memory cells based on trapping in metallic nanoparticles. *Adv. Funct. Mater.*, 15, 1933, 2005.
92. Guan, W., Long, S., Liu, M., Li, Z., Hu, Y., Liu, Q., Fabrication and charging characteristics of MOS capacitor structure with metal nanocrystals embedded in gate oxide. *J. Phys. D. Appl. Phys.*, 40, 2754, 2007.
93. Sawa, A., Fujii, T., Kawasaki, M., Tokura, Y., Interface resistance switching at a few nanometer thick perovskite manganite active layers. *Appl. Phys. Lett.*, 5, 88, 23, 232112, 5 June 2006.
94. Fujii, T., Kawasaki, M., Sawa, A., Akoh, H., Kawazoe, Y., Tokura, Y., Hysteretic current–voltage characteristics and resistance switching at an epitaxial oxide Schottky junction. *Appl. Phys. Lett.*, 86, 1, 012107, 3 January 2005.
95. Meijer, G.I., Staub, U., Janousch, M., Johnson, S.L., Delley, B., Neisius, T., Valence states of Cr and the insulator-to-metal transition in Cr-doped Sr Ti O 3. *Phys. Rev. B - Condens. Matter Mater. Phys.*, 72, 15, 155102, 2005.
96. Asamitsu, A., Tomioka, Y., Kuwahara, H., Tokura, Y., Current switching of resistive states in magnetoresistive manganites. *Nature*, 388, 50, 1997.
97. de Sousa dos Santos, E.L.V., Cruz, J.N., da Costa, G.V., de Sá, E.M.F., da Silva, A.K.P., Fernandes, C.P., Souto, R.N.P., Essential Oil of Ocimum basilicum against Aedes aegypti and Culex quinquefasciatus: Larvicidal Activity of a Nanoemulsion and *In Silico* Study. *Separations*, 11, 4, 97, 2024.
98. Kim, D.S., Kim, Y.H., Lee, C.E., Kim, Y.T., Colossal electroresistance mechanism in a Au/ Pr 0.7 Ca 0.3 Mn O 3/ Pt sandwich structure: Evidence for a Mott transition. *Phys. Rev. B - Condens. Matter Mater. Phys.*, 74, 17, 174430, 2006.

99. Rozenberg, M.J., Inoue, I.H., Sánchez, M.J., Nonvolatile memory with multilevel switching: a basic model. *Phys. Rev. Lett.*, 92, 17, 178302, 2004.
100. Rozenberg, M.J., Inoue, I.H., Sánchez, M.J., Strong electron correlation effects in nonvolatile electronic memory devices. *Appl. Phys. Lett.*, 88, 1, 2006.
101. Laibowitz, R.B., Esaki, L., Stiles, P.J., Electron transport in Nb-Nb oxide-Bi tunnel junctions. *Phys. Lett. A*, 36, 429, 1971.
102. Kohlstedt, H., Pertsev, N.A., Rodríguez Contreras, J., Waser, R., Theoretical current-voltage characteristics of ferroelectric tunnel junctions. *Phys. Rev. B - Condens. Matter Mater. Phys.*, 72, 12, 125341, 2005.
103. Tsymbal, E.Y. and Kohlstedt, H., Tunneling across a ferroelectric. Science, 313, 5784, 181–183, 2006.
104. Cao, G., Book Review: Nanoelectronics and Information Technology. By Rainer Waser (Ed.). *Adv. Mater.*, 16, 1864, 2004.
105. Maier, J., Nanoionics: Ion transport and electrochemical storage in confined systems. *Nat. Mater.*, 4, 805, 2005.
106. Maier, J., Ionic conduction in space charge regions. *Prog. Solid State Chem.*, 23, 171, 1995.
107. Poizot, P., Laruelle, S., Grugeon, S., Dupont, L., Tarascon, J.M., Nano-sized transition-metal oxides as negative-electrode materials for lithium-ion batteries. *Nature*, 407, 496, 2000.
108. Padhi, A.K., Nanjundaswamy, K.S., Goodenough, J.B., Phospho-olivines as Positive-Electrode Materials for Rechargeable Lithium Batteries. *J. Electrochem. Soc.*, 144, 1188, 1997.
109. Beaulieu, L.Y., Larcher, D., Dunlap, R.A., Dahn, J.R., Reaction of Li with Grain-Boundary Atoms in Nanostructured Compounds. *J. Electrochem. Soc.*, 147, 3206, 2000.
110. Grugeon, S., Laruelle, S., Herrera-Urbina, R., Dupont, L., Poizot, P., Tarascon, J.-M., Particle Size Effects on the Electrochemical Performance of Copper Oxides toward Lithium. *J. Electrochem. Soc.*, 148, A285, 2001.
111. Heitjans, P. and Wilkening, M., Ion dynamics at interfaces: nuclear magnetic resonance studies. *MRS Bull.*, 34, 915, 2009.
112. Dunst, A., Epp, V., Hanzu, I., Freunberger, S.A., Wilkening, M., Short-range Li diffusion vs. long-range ionic conduction in nanocrystalline lithium peroxide Li2O2 - The discharge product in lithium-air batteries. *Energy Environ. Sci.*, 7, 2739, 2014.
113. Kern, K. and Maier, J., Nanoionics and nanoelectronics. *Adv. Mater.*, 21, 2569, 2009.
114. Li, F., Li, J., Zhu, F., Liu, T., Xu, B., Kim, T.H., Kramer, M.J., Ma, C., Zhou, L., Nan, C.W., Atomically Intimate Contact between Solid Electrolytes and Electrodes for Li Batteries. *Matter*, 1, 1001, 2019.
115. Habasaki, J., Leon, C., Ngai, K.L., Molecular dynamics simulations. *Top. Appl. Phys.*, 132, 355, 2017.
116. Schoonman, J., Nanoionics. *Solid State Ionics*, 157, 319, 2003.

117. Wagemaker, M. and Mulder, F.M., Properties and promises of nanosized insertion materials for li-ion batteries. *Acc. Chem. Res.*, 46, 1206, 2013.
118. Balaya, P., Bhattacharyya, A.J., Jamnik, J., Zhukovskii, Y.F., Kotomin, E.A., Maier, J., Nano-ionics in the context of lithium batteries. *J. Power Sources*, 159, 171, 2006.
119. Niederberger, M. and Pinna, N., Metal Oxide Nanoparticles in Organic Solvents: Synthesis, Formation, Assembly and Application (Engineering Materials and Processes), Springer, Berlin, Germnay, 2009.
120. Haruta, M., Nanoparticulate gold catalysts for low-temperature CO oxidation. *J. New Mater. Electrochem. Syst.*, 7, 163, 2004.
121. Liao, Y., Xu, Y., Chan, Y., Semiconductor nanocrystals in sol-gel derived matrices. *Phys. Chem. Chem. Phys.*, 15, 13694, 2013.
122. Owens, G.J., Singh, R.K., Foroutan, F., Alqaysi, M., Han, C.M., Mahapatra, C., Kim, H.W., Knowles, J.C., Sol-gel based materials for biomedical applications. *Prog. Mater Sci.*, 77, 1, 2016.
123. Feinle, A., Elsaesser, M.S., Hüsing, N., Sol-gel synthesis of monolithic materials with hierarchical porosity. *Chem. Soc. Rev.*, 45, 3377, 2016.
124. Huston, M., Debella, M., Dibella, M., Gupta, A., Green synthesis of nanomaterials. *Nanomaterials*, 11, 2130, 2021.
125. Kim, M., Osone, S., Kim, T., Higashi, H., Seto, T., Synthesis of nanoparticles by laser ablation: A review. *Kona Powder Part. J.*, 2017, 80, 2017.
126. Hiragino, Y., Tanaka, T., Takeuchi, H., Takeuchi, A., Lin, J., Yoshida, T., Fujita, Y., Synthesis of nitrogen-doped ZnO nanoparticles by RF thermal plasma. *Solid State Electron.*, 118, 41, 2016.

4

Nano-Porous Silica in Devices and Ion-Based Systems - Unveiling the Design, Fabrication, and Diverse Applications

Rupesh K. Tiwari and Rajendra K. Singh*

Ionic Liquid and Solid-State Ionics Laboratory, Department of Physics, Institute of Science, Banaras Hindu University, Varanasi, India

Abstract

The nano-porous silica nanoparticles are considered as one of the most appealing inorganic materials due to its unique properties such as non-toxicity, stable structure, facile designing and handling, excellent bio-compatibility, bio-degradability, etc. Additionally, they have distinct advantages over other porous organic materials, such as adjustable particle size, facile accessibility and tunability of pores, rigid structure, informal functioning of interior and exterior surface area, and so on. Because of complex surface chemistry and high surface area of the nanoporous amorphous silica, it has variety of applications, such as biomedical applications, water filtering, biosensors, biomarker, environmental applications, energy harvesting as well as energy storage applications, etc. The nanostructured silica and its derivatives have been used extensively as battery anodes and anode protecting layers in the form of nanoporous particles, nanotubes, nanowires, and (micro, meso, and macro) porous silicon films, which can be synthesized using various techniques like soft and hard templating. Due to the aforementioned attractive qualities, porous nanostructured silica-based nanoparticles are the potentially better choice as the electrode as well as coating (surface modification) materials for the electrode materials. They can also be used as nanofiller in solid or gel polymer electrolytes to enhance the electrochemical performance of energy storage devices.

Keywords: Nano-ionics, nano silica, biomarkers, sensors, energy storage, device application

**Corresponding author*: rksingh_17@rediffmail.com; rajendrasingh.bhu@gmail.com

Inamuddin, Tariq Altalhi, Mohammad Luqman and Jorddy Neves Cruz (eds.) Nanoionics: Fundamentals and Applications, (79–96) © 2025 Scrivener Publishing LLC

4.1 Introduction

Nanotechnology, nanoionics, and nanoporous materials have become prominent fields of study, attracting significant scientific attention and fostering innovation in recent years. Nanotechnology is the manipulation of materials at a very small scale, allowing for the development and study of structures and functions with extremely high accuracy. The study of materials at the nanoscale has led to significant advancements in wide range of fields, including health and electronics, as scientists exploit the distinctive characteristics displayed by these materials. Nanoionics is a field that specifically examines and controls the movement of ions at the nanoscale. This is an important factor in the advancement of energy storage devices, sensors, and other systems that utilize nanoionic technology. Scientists seek to optimise the efficiency, dependability, and performance of various technological applications by comprehending and managing ion movement at the nanoscale level. Nano-porous materials, have attracted considerable interest due to their complex architecture and diverse range of applications. These materials possess large surface areas and customizable pore sizes, rendering them highly adaptable in several domains such as catalysis, sensing, and energy storage. Nano-porous materials have distinctive attributes that enable the creation of new materials with improved properties, thereby facilitating the development of state-of-the-art technologies. The collaborative investigation of nanotechnology, nanoionics, and nanoporous materials signifies a revolutionary period in scientific and technological progress, providing new and inventive resolutions to intricate problems and laying the groundwork for groundbreaking developments that have the potential to redefine the scope of several industries.

Nano-porous materials have been receiving a lot of attention over the last few decades because they have intrinsic vacant spaces inside them due to which they have high surface area to volume ratio along with high specific surface area [1]. The main advantage of these materials is that the shape and size of these empty spaces can be modified at the nanometric, molecular, and atomic levels. On the basis of the diameter of the pores, these materials are primarily classified by the International Union of Pure and Applied Chemistry (IUPAC) into three categories: first, microporous, with diameter of less than 2 nm; second, mesoporous, which have a diameter in the range of 2–50 nm; and third, macroporous, which have a diameter that is greater than 50 nm (Figure 4.1) [2–6]. The zeolites, metal-organic frameworks (MOFs), covalent-organic frameworks (COFs), activated carbon, ceramics, aerogels, silicates, pillared materials, non-siliceous materials,

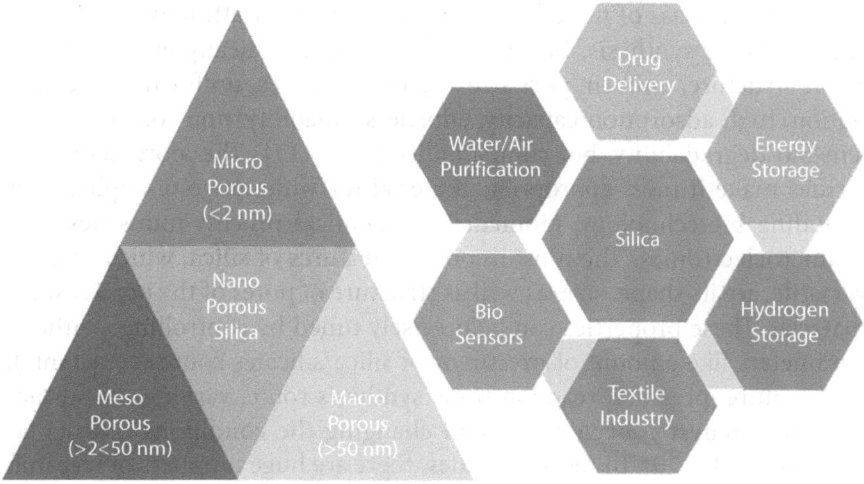

Figure 4.1 Classification of nonporous silica on the basis of porosity and its applications.

various porous-polymers, and inorganic porous-hybrid materials are some well-known examples of natural as well as artificial nano-porous materials. The use of these nano-porous materials is primarily determined by the necessary design at the molecular as well as at atomic levels, which regulate their porosity, nature, and design of pore, and surface area. These porous materials show a lot of promise for use in magnetic (electromagnetic protection devices, data storage device), optical (micro-lasers, micro-resonators, non-linear optics), and electrical devices (electro-optical, electrical, bio-sensors, bio-marker, magnetic device), in chemical industries (abrasives, filtration, sorbents, catalysis, fabrics, textiles, coatings, and paints) as well as in the fields of energy storage (batteries, fuel cells, solar cells, supercapacitors, storage of hydrogen) and distribution (gas purification, medication delivery, enzyme supporter, water decontamination, food analysis, distillation, anti-cancer treatment) [7–13]. On the other hand, most nanoporous materials are made through a long, expensive, and energy-intensive process that uses a lot of harmful ingredients and creates a lot of environmentally unfriendly (harmful) waste. Also, some porous materials can be harmful due to their intrinsic properties, especially in biomedical applications. Hence, the choice of these materials should be done very carefully by studying their side effects and compatibility with the human body. Because of its unique properties and best compatibility, as well as negligible side effects on the environment and humans, nanoporous silica is found to be a more suitable material. Porous silica materials are an important class of nano biomaterials due to their attractive properties such as facile synthesis, hassle-free handling, chemically and thermally stable structure, uniform pore size, high surface area, uniform pore distribution, high adsorption capacity, effortless tunability, non-toxicity, excellent bio-degradability, bio-compatibility [7, 14–17]. The aforementioned virtues make it most appropriate material for wide range of applications including biotechnology, biomedicine, chemical process industries, and in microelectronics. The most important features of silica, which make it versatile, are its shape, size, as well as structure of pores of the porous silica material. These properties can be precisely tuned by controlling synthesis parameters like amount of precursor of silica/silicates source (reactants), temperature, pressure, reaction time, synthesis route, varying the surfactant and catalyst concentration and changing the conditions of calcination, etc. [1, 17–20]. In recent decades, there are huge numbers of research articles published on technological and industrial applications of porous silica materials which clearly indicate their versatility. Kolbe *et al.*, synthesized first time highly dispersed non-porous silica particles in 1956 using the hydrolysis of silicon alkoxide by alcohol condensation method [21]. Stober *et al.* (1968), later reported the controlled production of monodispersed silica nano-spheres of micrometer size by hydrolyzing alkyl silicate in an alcoholic solution [22]. This is a very popular method for the synthesis of silica nano-particles and it is known as Stober's method in honor of Werner Stober, which is further modified and known as modified Stober method or sol-gel method. They were able to create spherical particles using ammonia as a base and a catalyst that controlled particle shape. The involvement of ammonia has been shown to be very important for the surfactant-free production of mesoporous silica nano-spheres with a controllable pore width approx. 100 to 230 nm. While, Cai *et al.* (2001), Mann *et al.* (2001), and Ostafin *et al.* (2002), were the first to effectively synthesize and report nano-sized mesoporous silica [23–25]. Over the last two decades, several independent groups throughout the world have rigorously synthesized and examined nanoporous silica materials with a wide range of surface and structural morphologies, pore sizes, diameters, and pore topologies. Multiple manipulations of synthesis circumstances, such as pH of the mixture of reactant during reaction, properties and types of surfactants or copolymers which are used as template, and concentrations and sources of silica, may lead to the synthesis of porous silica. Nanoporous silica materials are making waves in a lot of different fields, but they had made the biggest difference in health and energy fields. For example, they are used in precise drug delivery and cancer treatment, cleaning up of water and air, making highly sensitive biosensors and biomarkers, and making better devices for collecting energy from renewable sources and storing it. In this chapter, the types of nanoporous silica, its synthesis and use in various biomedical as well as in energy storage application will be discussed.

4.2 Methods Used for Synthesis of Nanoporous Silica

Unique physical, mechanical, and chemical qualities make silica materials so appealing. Currently there are large number of methods, including aqueous/non-aqueous, soft-templating/hard-templating, sol-gel, electrochemical etching, etc., available to synthesize the nonporous silica nano particles of different shapes, porosity, and sizes. Gerhard Kolbe discovered that tetraethyl orthosilicate (TEOS) may produce spherical particles when it reacts with ethanol as the solvent and ammonia as the catalyst. Werner Stöber then enhanced the precipitation process and produced silica spheres of few micrometers. This process is currently referred to as Stober synthesis. To produce monodisperse, organized, nanosized silica particles, and the Stober synthesis has evolved through various modifications over time. The nanoporous silica particles are capable of being produced in basic, acidic, and neutral environments, and by adjusting these parameters, one can design them into the desired shape, size, and porosity. The reaction settings were changed, producing particles of various sizes and shapes. Grun *et al.* (1997), were the first to modify Stober's way of synthesis of MCM-41 by using a cationic surfactant as a mould to make a spherical structure instead of a hexagonal one [26]. The idea of self-assembly, a fundamental approach for developing diverse nanostructures, was first introduced in 1991 by Dr. Whitesides of Harvard University [27]. The widely recognized ordered silica based mesoporous materials, namely MCM–50, MCM–48, and MCM–41 (collectively ascertained as MCM–41s), featuring consistent pore diameters ranging from 2 to 8 nm, were developed using a liquid-crystal template technique showcased by the Mobil corporation in the USA in 1992 [28]. Building upon this advancement, the University of California, Santa Barbara introduced a new category of well-ordered mesoporous materials in 1998. Named SBA-15, these materials featured larger pore widths spanning from ~ 7.5 to 32 nm and thicker walls measuring between ~ 3.1 and 6.4 nm [28]. The structure-directing agents used were block-copolymers with high molecular weights. The stability of mesoporous silica increases as the wall thickness rises, and the bigger holes enable the adsorption of large molecules while also significantly enhancing the mass transfer kinetics inside these pores. On the

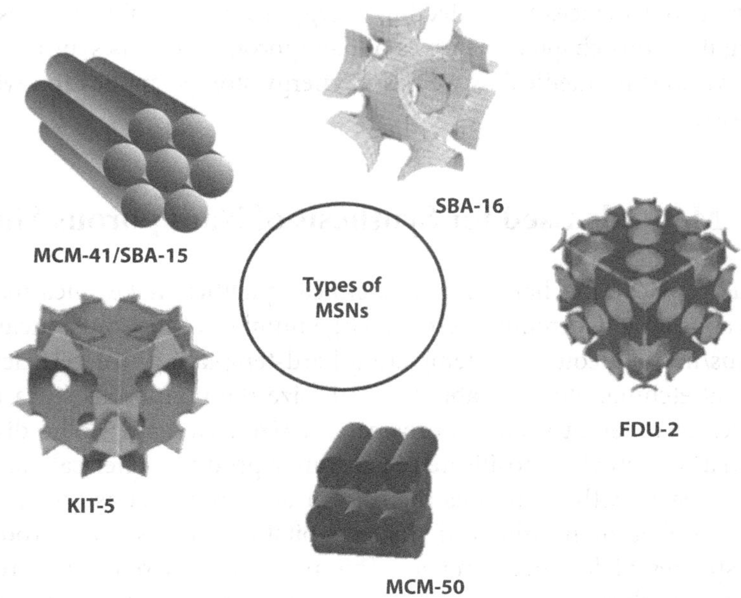

Figure 4.2 Various varieties of mesoporous silica.

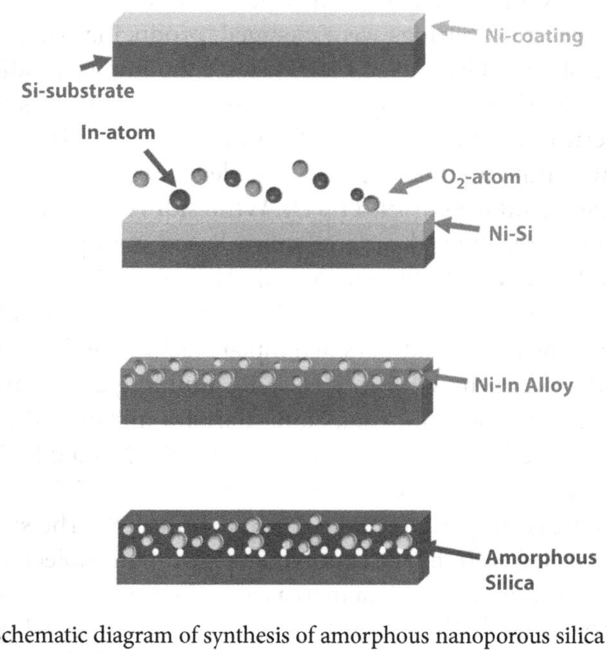

Figure 4.3 Schematic diagram of synthesis of amorphous nanoporous silica by metal droplet etching method.

basis of porous structure and pore size there are various type of silica nano particles; from them some famous mesoporous silica structures are shown in Figure 4.2. In their 2020 study, Luo *et al.* show how to synthesize large-scale, nanoporous, and amorphous silica structures using a straight forward and scalable metal droplet etching method (Figure 4.3) instead of the sol-gel method. The nanoporous amorphous silica was developed through the process of metal droplet etching, which led to the production of the vacancy. These synthesized nanostructures (nano-porous silica particles) provide a lot of ultraviolet light in their photoluminescence spectrum at room temperature.

4.3 Applications of Nanoporous Silica in Various Fields

4.3.1 Biomedical

Due to the potential of nanotechnology, nano-porous materials are being widely used in medicine and many other bio-medical applications [29]. The advantage of these materials is that they can be designed in a variety of shapes, sizes, and compositions. For the detection and measurement of a wide range of biological analytes, including triglycerides, glucose, cholesterol, different protein biomarkers, bacteria, viruses, and toxins found in food and water, there is a rising demand for extremely efficient and portable instruments [30]. In the medical industry, nano-porous materials are present and emerging to be a promising technology and are also expected to be of potential applications in the future. Also, the nano porous silica is one of most suitable porous materials for aforementioned applications due to its some attracting properties. Since its discovery in the early 1990s, there has been a lot of interest in biomedical applications of mesoporous silica. This is because it has well-defined mesoporous that are perfect for storing small molecules of drugs. The study by Gnanamoorthy, P. (2014), demonstrates that natural nanoporous silica materials derived from laboratory-cultured marine diatoms are a novel kind of drug delivery bio carrier for *in vitro* oral drug delivery biomedical treatments [10]. This study shows that drug molecules were adsorbed on the silica surface (foramen), within pores (cribrum), and also inside the interior hollow diatom structure (cribellum). A hierarchical two-layered nanotubular structure was developed by Zhe *et al.* using a vertically aligned mesoporous silica thin film (Figure 4.4) that was consistently produced against the walls of a titania nanotube array [31]. The produced thin-film displays more desirable performances as a result of the exposed mesoporous, including quick disintegration, effective loading,

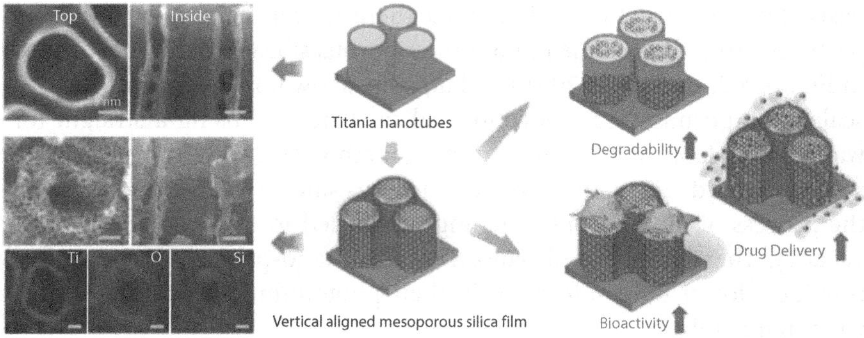

Figure 4.4 Schematic diagram of vertically aligned mesoporous silica film and drug delivery process.

and release of a small-molecule drug (dexamethasone), improved initial adhesion, and enhanced osteogenic differentiation of MC3T3-E1 cells. For the first time, a thin film of aligned mesoporous silica has been successfully deposited on a nanoporous substrate. Based on the results, this may be a good way to cover the surface of bone implants to give them bioactive properties. The intelligent nano-porous drug-loaded silica particles are capable of being transported to particular organs or cells, where they can release medications in a regulated manner in response to a variety of internal and

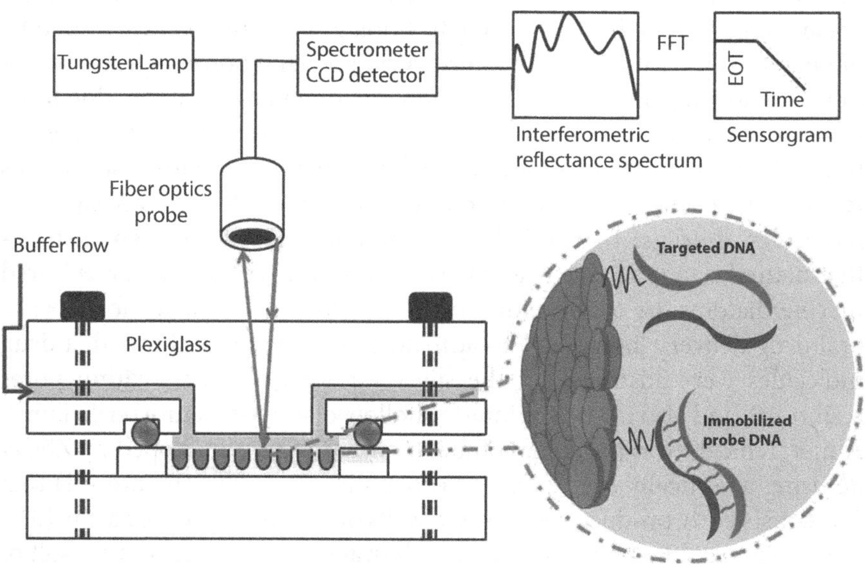

Figure 4.5 Working diagram of silica base bio-sensor.

external triggers, such as specific antibodies or ligands, at specific pH, or by external triggering such as using ultrasonic, electrical, magnetic, or light irradiation, etc. Along with safe and easy drug delivery, nano-porous silica plays an important role in the fabrication of cost-effective and efficient bio-sensors. The majority of porous silica-based biosensors employ an experimental setup (Figure 4.5) based on optical reflectivity. A bifurcated fiber optic cable emits white light, which illuminates the silica surface. A charge-coupled device (CCD) is used to detect reflected light collected by an optical fiber cable. The reflectance spectrum is subjected to a fourier transform (FFT) to obtain the effective optical thickness (EOT), which is then tracked over time *via* a sensorgram [29]. The close-up reveals an illustration of a bio-recognition process that takes place inside the porous silica layer, in this case, a hybridization of DNA takes place between probe DNA (immobilized) and target DNA of the solution. Changes in the reflectivity spectra show these layer-level events. These changes can be measured and are the basis for porous silica optical detection.

4.3.2 Water Decontamination

In the present scenario, water pollution remains a serious global issue. It is found that it is extremely difficult to remove the impurities of heavy metals and ions from water. Various ways are being used for the elimination of heavy metal ions from water and wastewater such as filtration by using membranes, reverse osmosis, precipitation or flotation, ion exchange, coagulation and adsorption technology, etc. In comparison to the other purification methods listed above, which has greater maintenance and operational expenses, adsorption technology is the most effective at removing harmful heavy metal ions and bacteria from water while being inexpensive and ecologically safe [17, 19, 28, 32–35]. In the adsorption technique for the removal of impurities, the number of adsorbents, such as activated carbon nanotubes, clays, resins, ash of rice husk, zeolites neem bark, grape stem waste, biomaterials, and silica (nanoporous), have been recently used. Due to their unique structure, silica nanoparticles have many advantages over the aforementioned other materials for adsorption applications. These include, a high specific surface area, a chemically, thermally and mechanically stable porous structure, a porous structure that can be modified or changed, silanol groups on the surface that make it easy to change, less damage to the environment, a low production cost, a simple way to synthesize them, etc. However, based on its excellent adsorbing properties, nanoporous silicon may remove both bacterial and heavy metal contaminants from drinking water in one step because it is a strong, adaptable material that can eliminate pollutants largely based on size. Bhogineni *et al.* (2016), created nanoporous SBA-15 and used it to remove heavy metal-ions Cu (II) and Ni (II) from environmental water samples. The main thing that affects the adsorption capability and selectivity of SBA-15 nanoporous sorbents is how the surface of the material is changed by functionalization of materials using organic functional groups [19]. This process adds nitrogen (N) oxygen (O), and sulfur (S) atoms that can serve as coordination sites for the metal ions. An *in situ* interfacial polymerization method was used by Jun *et al.* to synthesize a thin-film (membrane) of nanocomposite using MCM-41 (porous silica) nanoparticles [34]. This method utilizes solutions containing organic trimesoyl chloride and aqueous m-phenyl diamine. While, as fillers, non-porous spherical silica nanoparticles (concentrations from 0 to 0.1wt%) and porous MCM-41 (100 nm) were used to prepare the membrane, which was used in water filtration, it was found that this prepared thin film showed much better water permeability and high salt rejection than normal filter membranes without porous silica. An efficient and cost-effective method of separating organic and inorganic contaminants

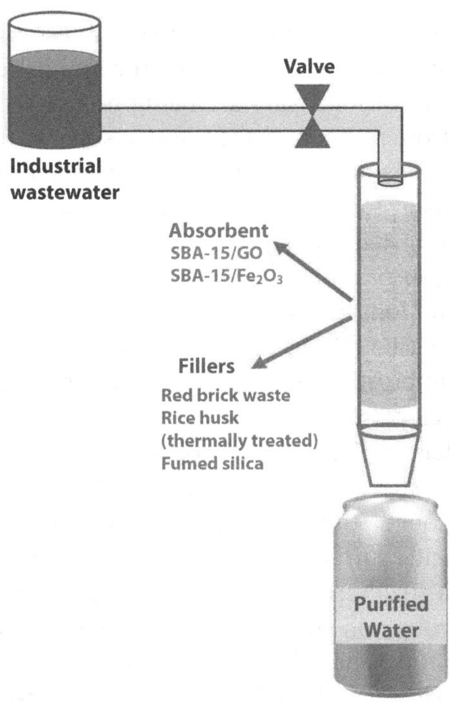

Figure 4.6 Schematic diagram of decontamination of industrial wastewater using nano silica (SBA-15) based composite material.

from industrial wastewater is provided by Eman F *et al.* in 2021 [36]. To improve the sorption characteristics, a novel fixed-bed column (Figure 4.6) was built using SBA-15 based nanocomposites as adsorbents, in which rice husk (thermally treated), fumed silica and waste red brick were used as fillers. It was found that a prepared nanoporous silica-based composite is capable of cleaning contaminants from wastewater effectively. Based on several reports, it was found that nanoporous silica functionalized with different organic compounds or trapped in any polymer hosts can be used as an adsorbent more and more often due to their high selectivity and excellent chemical stability for absorbing heavy metal ions as well as salt with better water permeability.

4.3.3 Energy

Conventional non-renewable energy sources (fossil fuels) are rapidly losing their reserves due to their finite resources, daily rising demand, and indiscriminate exploitation. Additionally, the enormous demand for electric and hybrid cars as well as portable electronic gadgets necessitates the use of several affordable, robust, environmentally friendly, effective, and renewable energy sources, such as geothermal, wind, solar, hydropower, etc. Rechargeable batteries and supercapacitors are two main ion-based devices that are capable of fully filling the gap between production (from renewable energy sources) and supply (to various devices). Herein, the versatility of nanoporous silica materials is also evident in energy harvesting and storage applications. Some of the major applications are shown in Figure 4.7. The scientific community is trying to develop a solid or semi-solid electrolyte to sort out the leakage, corrosion, and portability issues associated with liquid electrolyte used in energy storage devices. In this process, porous silica can play an important role not only as a host for liquid electrolytes but as a filler to improve the flexibility and conductivity of solid electrolytes [37–40]. Srivastava and Chandra presented the first work on the entrapment of an ionic solution (such as NH_4BF_4, KH_2PO_4, $CuSO_4$) in porous silica matrix (Srivastava & Chandra, 2001, 2002). Singh *et al.* (2010), found that when [BMIM] [PF_6] ionic liquid was incorporated into the porous SiO_2 matrix, the vibrations of aromatic ring of the ionic liquid ([BMIM] [PF_6]) changed more when it was confined to small pores of silica [11, 41]. They found that SiO_2 seems to prefer interacting with C-H groups of BMIM's cyclic ring over those found in the tail of its alkyl groups, according to theoretical estimates of interaction. Verma *et al.* (2014), demonstrated a typical process for synthesizing porous silica with high ionic liquid loading (85 to 92%) utilizing a non-hydrolytic

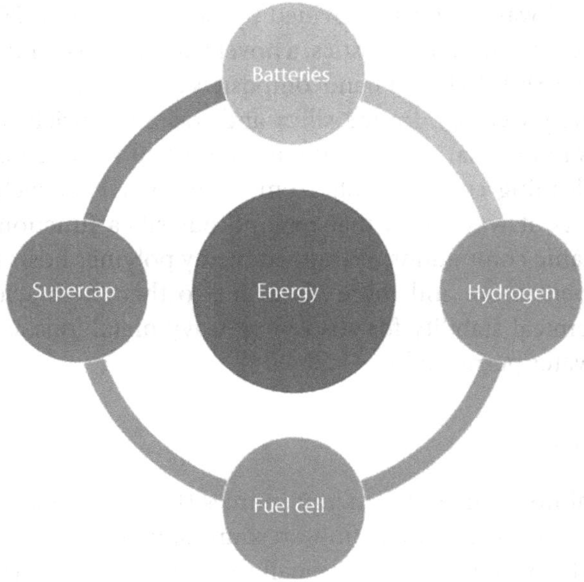

Figure 4.7 Schematic diagram of utilization of silica nanoparticles in energy storage device application.

sol-gel method [42]. The ionic liquid was enclosed inside its fold by the silica, which may then be pulled into a structure like a tubular spring. Interactions of ionic liquids with the pore wall surfaces of silica have been discovered to alter the characteristic properties of ionic liquids, including glass transition temperature, vibrational bands associated with the ring of ionic liquid, and fluorescence properties. A physical immobilization approach has been reported by Tripathi *et al.* (2017), in which they prepared quasi-solid-state electrolytes based on the ionic liquid ([EMIM][FSI]) trapped into an ordered mesoporous silicon matrix MCM-41 [43]. Activated carbon electrodes and quasi-solid-state electrolytes were used to fabricate a supercapacitor assembly that can function up to 2 V and has an excellent specific capacitance of 30 F g^{-1}. At a current density of 1 mA cm^{-2}, the supercapacitor has a specific energy density of 17 Wh kg^{-1} and a power density of 1000 W kg^{-1}. It has good cycle performance at room temperature. Also, Tripathi and Singh (2018) demonstrated that salt ionic liquid solutions (LiTFSI and EMIMTFSI) incorporated into MCM-41 mesoporous silica can be successfully used as an electrolyte in lithium-ion battery systems (Figure 4.8) [39]. In Figure 4.8, it is shown how a quasi-solid-liquid electrolyte is made and how lithium ions are moved through the MCM-41 silica matrix by the ionic liquid solution that has lithium ions

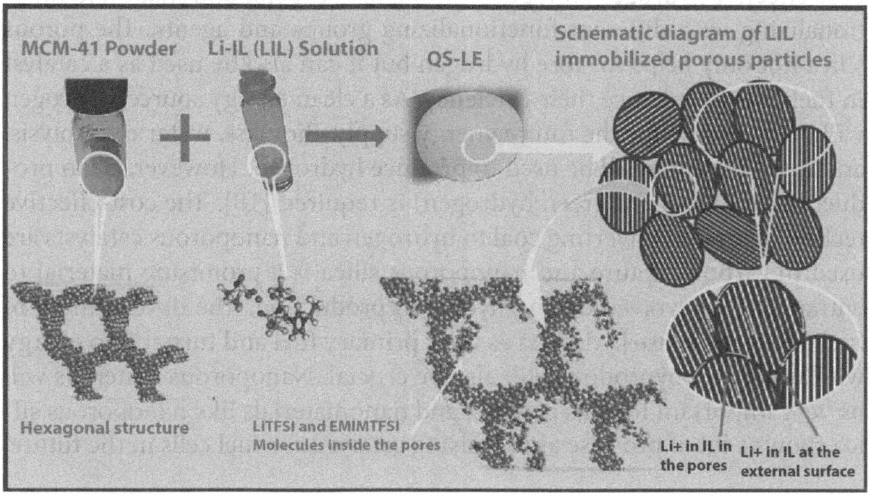

Figure 4.8 Schematic representation of the development of quasi solid-liquid electrolyte (QS-LE) and the pathway for lithium-ions transport performed by the lithium-ion containing ionic liquid (IL) solution in MCM-41 silica matrix.

in it. The dissolution of intermediated polysulfides during the redox reaction is a big problem in lithium-sulfur batteries. This dissolution is also the reason why lithium-sulfur batteries only last for a few cycles. To overcome this issue, nanoporous silica is used in various ways, such as a host for the encapsulation of sulfur as well as an interlayer to entrap polysulfides [44]. A dense, lightweight multi-functional interlayer was developed by Xiao *et al.* (2019), by using a nitrogen-functionalized 2D mesoporous silica nanoplate and a straightforward casting method. This technique successfully reduced polysulfide dissolution while promoting Li-ion diffusion [45]. The findings of much research demonstrate that porous silica-based batteries and supercapacitors are appropriate candidates for energy storage applications such as backup power, electric and hybrid cars, portable electronic gadgets, etc. The porous silica shows its versatility in energy storage device applications.

Not only does nanoporous silica aid in energy conservation, but it also aids in the production of green and environmentally friendly energy. By being used as fuel for fuel cells, hydrogen is the most promising fuel for the production of green energy. Alfonso, *et al.* (2020) have looked into how well different ratios of functionalized precursors affect the ability of ordered mesoporous silica to hold H_2 [46]. They discovered that porous silica is the most suitable material for safe, environmentally friendly, and cost-effective hydrogen storage by tuning the pore size and shape and functionalizing with different functionalizing groups and agents. The porous silica not only helps to store hydrogen but it can also be used as a catalyst in fuel cells to enhance their efficiency. As a clean energy source, hydrogen will be necessary for the future energy supply. Biomass, water electrolysis, and fossil fuels may all be used to produce hydrogen. However, clean production of hydrogen (green hydrogen) is required [18]. The cost-effective technologies for converting coal to hydrogen and nanoporous catalysts are used in carbon capture and nanoporous silica is a promising material to entrap carbon evolved during hydrogen production. The development of fuel cells, which use hydrogen as their primary fuel and turn it into energy with water as a byproduct, will also be crucial. Nanoporous materials will be very important for this process, and nanomaterials like nanoporous silica shows a lot of promise as promising catalysts for fuel cells in the future.

4.4 Conclusion

The area of nanomaterials and nanotechnology (nanoionics) has been growing literally explosively, particularly in the past two decades, ever since 1959, when Richard Feynman coined the word "nanotechnology" for the first time in his renowned lecture entitled "There's Plenty of Room at the Bottom". When compared to traditional bulk materials, nanomaterials have two major advantages: (1) their extremely small size (nano), large specific surface area, and large surface-to-volume ratio, which are useful for many interface-related applications; and (2) their uniquely tuneable physical, chemical, optical, mechanical, and electronic properties, which can be sensibly altered by governing their size (pore as well as particle), shape, surface morphology, and crystal orientation. Because of this, the idea of moving to the nanoscale has opened up a lot of new possibilities that would be unthinkable with traditional bulk materials. The porous silica is immerged as a most promising nano material with large number of applications. The benefits of the nanometric size allow for a variety of applications. They can freely circulate in the bloodstream and be phagocytosed by cells under localized treatments. Porous silicon particles are appropriate for less invasive cancer treatment. In bio-sensors and bio-marker applications, the use of silica not only enhances the accuracy of the sensor but also reduces its cost and enhances its availability for people from the poor economic sector. Also, in the area of energy conservation and green energy production nano-porous silica is immersed as multifunctional material. For electrode materials, it can be used as a coating material to prevent the degradation of electrodes during electrochemical reactions and help to enhance the life of the devices. It can be also used as a host matrix of sulfur in the cathode material of lithium-sulfur batteries. There are numerous reports stating that, nano-porous silica particles are one of the best filler materials which can be used in high-performance solid electrolyte preparation for batteries and supercapacitors. The versatility of porous silica in energy storage device applications leads us to conclude that porous silica-based batteries and supercapacitors are suitable candidates for energy storage applications such as backup power, electric and hybrid electric vehicles, portable electronic devices, etc.

Acknowledgement

Rajendra Kumar Singh, one of the authors, expresses gratitude to the Institute of Eminence (IoE, UGC India) for their financial support in conducting this research. R. K. Tiwari appreciates the IoE for awarding him the JRF (Junior Research Fellowship).

References

1. Jianrong, C., Yuqing, M., Nongyue, H., Xiaohua, W., Sijiao, L., Nanotechnology and biosensors. *Biotechnol. Adv.*, 22, 505–518, 2004.
2. Harraz, F.A., Porous silicon chemical sensors and biosensors: A review. *Sens. Actuators, B*, 202, 897–912, 2014.
3. Liang, J., Liang, Z., Zou, R., Zhao, Y., Heterogeneous Catalysis in Zeolites, Mesoporous Silica, and Metal–Organic Frameworks. *Adv. Mater.*, 29, 1–21, 2017.
4. Liu, Z., Fujita, N., Miyasaka, K., Han, L., Stevens, S.M., Suga, M., Asahina, S., Slater, B., Xiao, C., Sakamoto, Y., Anderson, M.W., Ryoo, R., Terasaki, O., A review of fine structures of nanoporous materials as evidenced by microscopic methods. *J. Electron Microsc. (Tokyo)*, 62, 109–146, 2013.
5. de Sousa dos Santos, E.L.V., Cruz, J.N., da Costa, G.V., de Sá, E.M.F., da Silva, A.K.P., Fernandes, C.P., de Faria Mota Oliveira, A.E.M., Duarte, J.L., Bezerra, R.M., Tavares, J.F., da Costa, T.S., dos Anjos Ferreira, R.M., dos Santos, C.B.R., Souto, R.N.P., Essential Oil of Ocimum basilicum against Aedes aegypti and Culex quinquefasciatus: Larvicidal Activity of a Nanoemulsion and *In Silico* Study. *Separations*, 11, 97, 2024.
6. Porrang, S., Davaran, S., Rahemi, N., Allahyari, S., Mostafavi, E., How Advancing are Mesoporous Silica Nanoparticles? A Comprehensive Review of the Literature. *Int. J. Nanomed.*, 17, 1803–1827, 2022.

7. Murugan, B., Sagadevan, S., Anita Lett, J., Fatimah, I., Fatema, K.N., Oh, W.C., Mohammad, F., Johan, M.R., Role of mesoporous silica nanoparticles for the drug delivery applications. *Mater. Res. Express*, 7, 102002, 2020.
8. Pathak, Y., Sutariya, V., Hirani, A.A., *Nano-Biomaterials For Ophthalmic Drug Delivery*, vol. 1, p. 627, Springer International Publishing Switzerland, Springer Cham, 2016.
9. Auerbach, S.M., Fan, W., Monson, P.A., Modelling the assembly of nanoporous silica materials. *Int. Rev. Phys. Chem.*, 34, 35–70, 2015.
10. Gnanamoorthy, P., Anandhan, S., Prabu, V.A., Natural nanoporous silica frustules from marine diatom as a biocarrier for drug delivery. *J. Porous Mater.*, 21, 789–796, 2014.
11. Singh, M.P., Singh, R.K., Chandra, S., Ionic liquids confined in porous matrices: Physicochemical properties and applications. *Prog. Mater Sci.*, 64, 73–120, 2014.
12. Wu, S.H. and Lin, H.P., Synthesis of mesoporous silica nanoparticles. *Chem. Soc. Rev.*, 42, 3862–3875, 2013.
13. Sarfraz, M.F.H., Zubair, M., Aslam, B., Ashraf, A., Siddique, M.H., Hayat, S., Cruz, J.N., Muzammil, S., Khurshid, M., Sarfraz, M.F.H., Hashem, A., Dawoud, T.M., Avila-Quezada, G.D., Abd_Allah, E.F., Comparative analysis of phyto-fabricated chitosan, copper oxide, and chitosan-based CuO nanoparticles: antibacterial potential against Acinetobacter baumannii isolates and anticancer activity against HepG2 cell lines. *Front. Microbiol.*, 14, 1188743, 2023.
14. Alvarez, S.D., Derfus, A.M., Schwartz, M.P., Bhatia, S.N., Sailor, M.J., The compatibility of hepatocytes with chemically modified porous silicon with reference to *in vitro* biosensors. *Biomaterials*, 30, 26–34, 2009.
15. Lee, J., Han, S., Hyeon, T., Synthesis of new nanoporous carbon materials using nanostructured silica materials as templates. *J. Mater. Chem.*, 14, 478–486, 2004.
16. Krishna Prasad, N.V., Babu, T.A., Sarma, S.R.K.N., Ramesh, S., Nirisha, K., Mathew, T., Madhavi, N., Role of Porous Nanomaterial'S in Water Purification, Electronics, Drug Delivery and Storage: a Comprehensive Review. *J. Optoelectron. Biomed. Mater.*, 13, 11–22, 2021.
17. Mohammed, R.H., Rashad, E., Huo, R., Su, M., Chow, L.C., Pore-size engineered nanoporous silica for efficient adsorption cooling and desalination cycle. *npj Clean Water*, 4, 1–10, 2021.
18. Jibowu, T., A Review on Nanoporous Metals. *Front. Nanosci. Nanotechnol.*, 2, 165–168, 2016.
19. Sreenu, B., Sharma, P., Seshaiah, K., Singh, A.P., Synthesis and characterization of nanoporous silica SBA-15 diaminocyclohexane and its application in removal of Cu(II) and Ni(II) from aqueous solution. *Desalin. Water Treat.*, 57, 15397–15409, 2016.
20. Gardner, D.S., Holzwarth, C.W., Liu, Y., Clendenning, S.B., Jin, W., Moon, B.K., Pint, C., Chen, Z., Hannah, E.C., Chen, C., Wang, C., Mäkilä, E., Chen, R.,

Aldridge, T., Gustafson, J.L., Integrated on-chip energy storage using passivated nanoporous-silicon electrochemical capacitors. *Nano Energy*, 25, 68–79, 2016.
21. Kolbe, G., Das Komplexchemische Verhalten der Kiesels aure, Friedrich-Schiller-Universität Jena, 1956.
22. Stöber, W., Fink, A., Bohn, E., Controlled growth of monodisperse silica spheres in the micron size range. *J. Colloid Interface Sci.*, 26, 62–96, 1968.
23. Fowler, C.E., Khushalani, D., Mann, S., Facile synthesis of hollow silica microspheres. *J. Mater. Chem.*, 11, 1968–1971, 2001.
24. Cai, Q., Luo, Z.S., Pang, W.Q., Fan, Y.W., Chen, X.H., Cui, F.Z., Dilute solution routes to various controllable morphologies of MCM-41 silica with a basic medium. *Chem. Mater.*, 13, 258–263, 2001.
25. Nooney, R.I., Thirunavukkarasu, D., Yimei, C., Josephs, R., Ostafin, A.E., Synthesis of nanoscale mesoporous silica spheres with controlled particle size. *Chem. Mater.*, 14, 4721–4728, 2002.
26. Grün, M., Lauer, I., Unger, K.K., The synthesis of micrometer- and submicrometer-size spheres of ordered mesoporous oxide MCM-41. *Adv. Mater.*, 9, 254–257, 1997.
27. Whitesides, G.M., Mathias, J.P., Seto, C.T., Molecular self-assembly and nanochemistry: A chemical strategy for the synthesis of nanostructures. *Science (80-.)*, 254, 1991.
28. Li, R., Zhang, L., Wang, P., Rational design of nanomaterials for water treatment. *Nanoscale*, 7, 17167–17194, 2015.
29. Jane, A., Dronov, R., Hodges, A., Voelcker, N.H., Porous silicon biosensors on the advance. *Trends Biotechnol.*, 27, 230–239, 2009.
30. Dhanekar, S. and Jain, S., Porous silicon biosensor: Current status. *Biosens. Bioelectron.*, 41, 54–64, 2013.
31. Li, Z., He, Y., Klausen, L.H., Yan, N., Liu, J., Chen, F., Song, W., Dong, M., Zhang, Y., Growing vertical aligned mesoporous silica thin film on nanoporous substrate for enhanced degradation, drug delivery and bioactivity. *Bioact. Mater.*, 6, 1452–1463, 2021.
32. Mălăeru, T., Codescu, M.M., Chiţanu, E., Georgescu, G., Banciu, C.A., Dascălu, R.C., Pătroi, D., Marinescu, V., Borbath, I., Synthesis and characterization of silica nanoparticles with modified surface for hydrophobic coatings. *J. Optoelectron. Adv. Mater.*, 24, 74–81, 2022.
33. Jackson, E.A. and Hillmyer, M.A., Nanoporous membranes derived from block copolymers: From drug delivery to water filtration. *ACS Nano*, 4, 3548–3553, 2010.
34. Yin, J., Kim, E.S., Yang, J., Deng, B., Fabrication of a novel thin-film nanocomposite (TFN) membrane containing MCM-41 silica nanoparticles (NPs) for water purification. *J. Membr. Sci.*, 423–424, 238–246, 2012.
35. Muzammil, S., Neves Cruz, J., Mumtaz, R., Rasul, I., Hayat, S., Khan, M.A., Khan, A.M., Ijaz, M.U., Lima, R.R., Zubair, M., Effects of Drying Temperature

and Solvents on *In Vitro* Diabetic Wound Healing Potential of Moringa oleifera Leaf Extracts. *Molecules*, 28, 710, 2023.
36. Aboelfetoh, E.F., Zain Elabedien, M.E., Ebeid, E.Z.M., Effective treatment of industrial wastewater applying SBA-15 mesoporous silica modified with graphene oxide and hematite nanoparticles. *J. Environ. Chem. Eng.*, 9, 104817, 2021.
37. Gupta, A.K., Verma, Y.L., Singh, R.K., Chandra, S., Studies on an ionic liquid confined in silica nanopores: Change in T g and evidence of organic-inorganic linkage at the pore wall surface. *J. Phys. Chem. C*, 118, 1530–1539, 2014.
38. Kruk, D., Wojciechowski, M., Florek-Wojciechowska, M., Singh, R.K., Dynamics of ionic liquids in confinement by means of nmr relaxometry—emim-fsi in a silica matrix as an example. *Materials (Basel)*, 13, 1–16, 2020.
39. Tripathi, A.K. and Singh, R.K., Development of ionic liquid and lithium salt immobilized MCM-41 quasi solid-liquid electrolytes for lithium batteries. *J. Energy Storage*, 15, 283–291, 2018.
40. Tripathi, A.K. and Singh, R.K., Immobilization induced molecular compression of ionic liquid in ordered mesoporous matrix. *J. Phys. D. Appl. Phys.*, 51, 075301, 2018.
41. Singh, M.P., Singh, R.K., Chandra, S., Properties of ionic liquid confined in porous silica matrix. *ChemPhysChem*, 11, 2036–2043, 2010.
42. Verma, Y.L., Singh, R.K., Oh, I.K., Chandra, S., Ionic liquid template assisted synthesis of porous nano-silica nails. *RSC Adv.*, 4, 39978–39983, 2014.
43. Tripathi, A.K., Verma, Y.L., Shalu, Singh, V.K., Balo, L., Gupta, H., Singh, S.K., Singh, R.K., Quasi solid-state electrolytes based on ionic liquid (IL) and ordered mesoporous matrix MCM-41 for supercapacitor application. *J. Solid State Electrochem.*, 21, 3365–3371, 2017.
44. Tiwari, R.K., Singh, S.K., Gupta, H., Srivastava, N., Meghnani, D., Mishra, R., Patel, A., Tiwari, A., Tiwari, V.K., Singh, R.K., Multifaceted ethylenediamine and hydrothermal assisted optimum reduced GO-nanosulfur composite as high capacity cathode for lithium-sulfur batteries. *Electrochem. Sci. Adv.*, 2, e202100025, 2022.
45. Xiang, X., Wu, J.Y., Shi, Q.X., Xia, Q., Xue, Z.G., Xie, X.L., Ye, Y.S., Mesoporous silica nanoplates facilitating fast Li+ diffusion as effective polysulfide-trapping materials for lithium-sulfur batteries. *J. Mater. Chem. A*, 7, 9110–9119, 2019.
46. Policicchio, A., Conte, G., Stelitano, S., Bonaventura, C.P., Putz, A.M., Ianăşi, C., Almásy, L., Horváth, Z.E., Agostino, R.G., Hydrogen storage performances for mesoporous silica synthesized with mixed tetraethoxysilane and methyltriethoxysilane precursors in acidic condition. *Colloids Surf., A*, 601, 125040, 2020.

5

Bioinspired Nanoionics for Biomedical and Bioelectronic Applications

Masooma Siddiqui[1], Maroof Ali[1], Uzma[1], Azfar Jamal[2,3] and Mohd Imran Ahamed[1]*

[1]*Department of Chemistry, Faculty of Science, Aligarh Muslim University, Aligarh, U.P., India*
[2]*Department of Biology, College of Science Al-Zulfi, Majmaah University, Al-Majmaah, Saudi Arabia*
[3]*Health and Basic Science Research Centre, Majmaah University, Al-Majmaah, Saudi Arabia*

Abstract
Bioinspired nanoionics has emerged as a dynamic and interdisciplinary field at the interface of materials science, nanotechnology, and biology. This article provides a comprehensive overview of recent developments and emerging trends in bioinspired nanoionics, highlighting its applications in the biomedical and bioelectronic fields. Over the past decade, researchers have made great strides in the design and synthesis of biomimetic materials and devices that mimic natural ion transport processes. These innovations promise major advances in medical diagnostics, drug delivery, and bioelectronic interfaces. This article explores the key concepts, recent breakthroughs, and future prospects of bioinspired nanoionics and provides valuable insights for researchers and practitioners in the field. Bioinspired nanoionics for biomedical and bioelectronic applications heralds a new era at the interface of biology, nanotechnology, and materials science. In this interdisciplinary field, novel materials and technologies are engineered using biomimetic ion transport systems, such as channels, pumps, and exchangers. Recent developments have accelerated tissue engineering, medication administration, and bioimaging, while bioelectronic advancements promise to transform neuroprosthetics, energy storage, and ion-selective sensing. Scalability, safety, and ethical issues are still problems, which highlight the need for strong regulatory frameworks. The incorporation of bioinspired nanoionics into healthcare and technology promises

**Corresponding author*: gc2587@myamu.ac.in

Inamuddin, Tariq Altalhi, Mohammad Luqman and Jorddy Neves Cruz (eds.) Nanoionics: Fundamentals and Applications, (97–132) © 2025 Scrivener Publishing LLC

to transform medicine, diagnostics, and human-machine interfaces as we stand on the verge of tremendous achievements. To fully achieve the revolutionary potential of the synthesis of biological principles with cutting-edge materials engineering, continuous collaboration across disciplines is required.

***Keywords*:** Bioinspired nanoionics, bioelectronic therapeutics ionic circuitry, biomedical applications, neuroprosthetics, ethical considerations, bioelectronic skin, innovative healthcare technologies

5.1 Introduction

Bioinspired nanoionics draws inspiration from biological systems to create functional materials and devices with unprecedented properties [1–3]. It uses the principles of ion transport in living organisms, such as ion channels, pumps, and exchangers, to develop synthetic systems that can precisely control ion movement. This interdisciplinary approach has gained momentum in recent years with the potential applications of bioinspired materials and devices in various biomedical and bioelectronic contexts [4]. The intriguing convergence of biology and nanotechnology has given rise to a burgeoning field known as "bioinspired nanoionics," which transcends the boundaries of traditional disciplines and pushes scientific exploration to new frontiers [5–7].

We explore the astounding advancements of the past ten years in this remarkable journey, showcasing the promise of upcoming discoveries and outlining the crucial course of action (Figure 5.1). Biomimetic ion transport systems, which are subject to the same orchestration of nature's ion channels, pumps, and exchangers, are now paving the way for remarkable materials and gadgets inside this tangle of scientific spheres. In this study, we take a broad look into bioinspired nanoionics and investigate the complexity of material design as well as the wide range of emerging biomedical and bioelectronic applications (Figure 5.2). This review conveys an image of a scientific transformation, from drug delivery systems that can mimic biological precision to bioimaging methods that take advantage of the delicate cycles of ion transport and on to the development of artificial organs. Embarking on the journey into the realm of bioelectronic applications, we encounter a future where electronic skin communicates like human touch, and circuits dance with ions to power neuromorphic computing. Yet, as we delve deeper, challenges loom large—scalability, biocompatibility, ethics, and regulations demand our unwavering attention. We conclude by envisioning the transformative potential of bioinspired

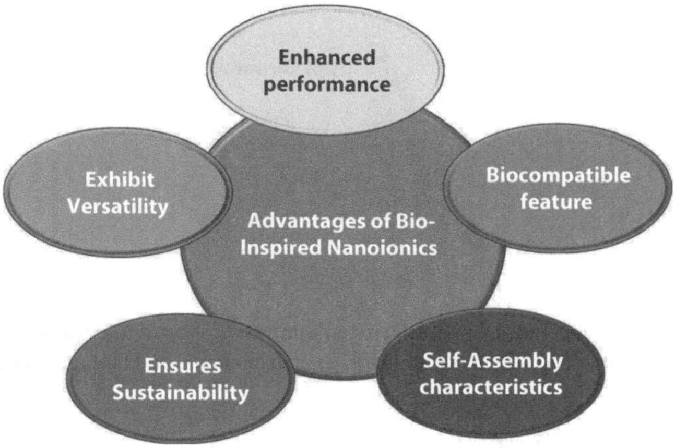

Figure 5.1 Advantages of bio-inspired nanoionics in biomedical and bioelectronic application.

nanoionics as it promises to redefine the landscape of medicine, diagnostics, and human-machine interfaces. The echoes of this symphony, where biology and nanotechnology harmonize, beckon us toward a new era of innovation and discovery, reminding us that the union of these disparate worlds can indeed yield boundless possibilities for a brighter, healthier, and more interconnected future [8–10].

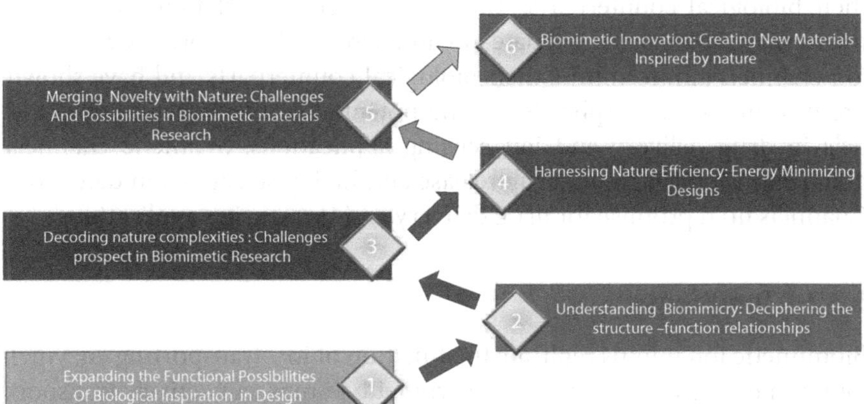

Figure 5.2 Implementation of bioinspired nanoionics for biomedical and bioelectronic application using intensive research composed of six steps.

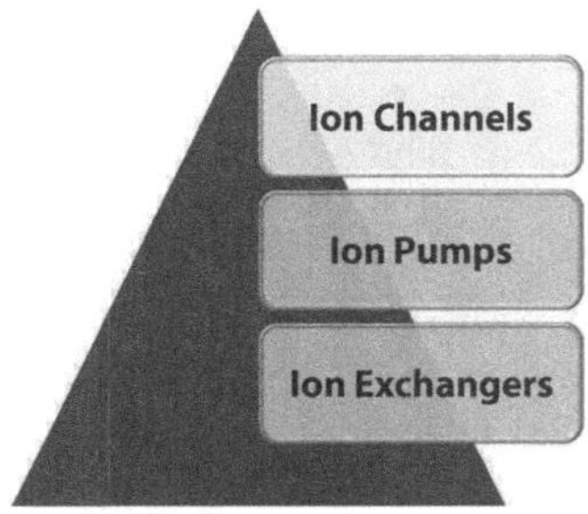

Figure 5.3 Biomimetic ion transport system: an innovative approach to mimicking natural ion channels for enhanced selectivity and efficiency in chemical and biological applications.

5.2 Biomimetic Ion Transport Systems

Focusing on Biomimetic Ion Transport Systems explanation description of ion exchangers, ion pumps, and ion channels are given. These systems draw inspiration from nature's elegant ion transport mechanisms, offering innovative solutions across various scientific and technological domains (Figure 5.3) [11].

5.2.1 Ion Channels

Ion channels are transmembrane proteins responsible for regulating ion flow across cell membranes. Researchers have successfully designed synthetic ion channels that mimic the selectivity and gating mechanisms of their biological counterparts. These biomimetic ion channels are transmembrane proteins that regulate ion flow across cell membranes. Synthetic ion channels aim to mimic their biological counterparts and have shown promise in various applications. The potential of biomimetic ion channels in drug delivery and ion sensing applications. Synthetic channels hold promise for targeted drug release and highly sensitive ion detection. Channels hold promise for drug delivery and ion sensing applications.

5.2.2 Ion Pumps

Biomimetic ion pumps replicate the function of ion-transporting enzymes found in biological systems. These synthetic pumps can actively transport ions against concentration gradients, enabling applications in drug delivery, energy harvesting, and controlled release systems. Biomimetic ion pumps replicate the function of ion-transporting enzymes found in biological systems. These synthetic pumps can actively transport ions against concentration gradients, and highlight recent breakthroughs in the development of biomimetic ion pumps, including the use of novel materials and advanced fabrication techniques. These innovations are poised to transform various industries [12].

5.2.3 Ion Exchangers

Ion exchangers inspired by natural systems are being developed for applications such as water purification and ion-specific sensing. These materials can selectively adsorb or release ions, offering solutions to various environmental and analytical challenges. Ion exchangers are synthetic materials designed to mimic the ion-selective properties found in biological systems. These materials can adsorb or release specific ions from solutions, offering applications in various fields. The academic paper discusses the following key aspects of biomimetic ion exchangers, highlighting the growing importance of interdisciplinary collaboration and the potential for further innovation in this exciting field. It underlines how these biomimetic systems bridge the gap between biology and technology, offering novel solutions to complex problems across various scientific disciplines. It provides a comprehensive exploration of these bioinspired materials and devices, shedding light on their principles, applications, recent advancements, and future potential. It serves as a valuable resource for researchers and practitioners, inspiring further exploration and innovation at the intersection of biology and materials science.

5.2.4 Biomimetic Ion Transport Systems in Drug Delivery

Mechanisms underlying the use of biomimetic ion transport systems in drug delivery. It explains how these synthetic systems can replicate the precision and selectivity observed in natural ion channels and pumps to administer therapeutic agents (Figure 5.4) [13, 14].

Figure 5.4 Process system of biomimetic ion transport.

• Targeted Drug Delivery
One of the central themes explored is the concept of targeted drug delivery. Biomimetic ion transport systems offer the ability to transport drugs to specific cells or tissues with remarkable accuracy, minimizing systemic side effects. The report discusses the potential for these systems to revolutionize cancer treatment, allowing for localized drug delivery to tumor cells while sparing healthy tissues.

• Controlled Release
The report delves into the controlled release capabilities of biomimetic ion transport systems. These systems can modulate the rate and timing of drug release, enabling personalized treatment regimens and prolonged therapeutic effects. This section highlights how such control can improve patient compliance and therapeutic outcomes.

• Advantages and Challenges
The academic paper provides an assessment of the advantages and challenges associated with using biomimetic ion transport systems in drug delivery. It underscores the potential for increased drug efficacy, reduced toxicity, and improved patient quality of life. However, it also acknowledges the complexities of designing and implementing these systems in clinical settings.

This demonstrates the promising potential of these systems in revolutionizing the field of medicine. By harnessing the principles of ion transport found in nature, researchers are paving the way for more precise, effective, and patient-friendly drug delivery methods. The report encourages further exploration of this exciting field, emphasizing the need for interdisciplinary collaboration between materials scientists, pharmacologists, and medical practitioners to translate these innovations into real-world medical applications.

5.2.5 Biomedical Applications

5.2.5.1 Drug Delivery

Bioinspired nanoionics have the potential to revolutionize drug delivery by providing precise control over the release of therapeutic ions. Smart drug delivery systems based on biomimetic ion channels and pumps can enhance the efficacy and reduce the side effects of medications (Figure 5.5) particularly in the domain of drug delivery. Biomimetic ion transport systems have emerged as transformative tools for precise and controlled drug

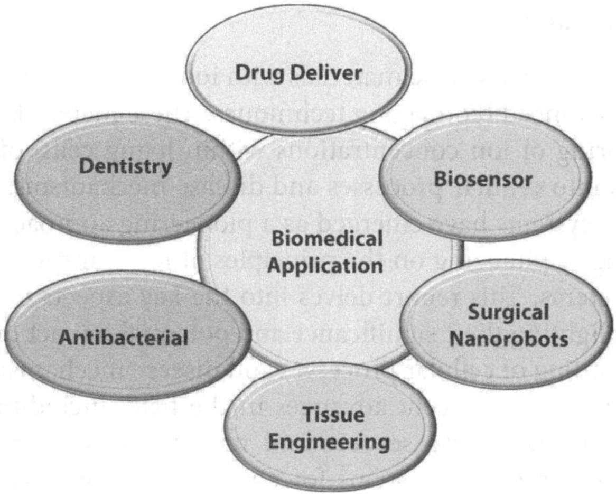

Figure 5.5 Biomedical applications in nanodevices: advancing precision medicine through targeted drug delivery, diagnostic imaging, and therapeutic interventions.

delivery, mimicking the ion transport mechanisms found in biological systems. This report delves into the key aspects of these applications, highlighting their significance and potential impact on the field of medicine. These systems can modulate the rate and timing of drug release, enabling personalized treatment regimens and prolonged therapeutic effects. This section highlights how such control can improve patient compliance and therapeutic outcomes [15–19].

Overall, it underscores the potential for increased drug efficacy, reduced toxicity, and improved patient quality of life. However, it also acknowledges the complexities of designing and implementing these systems in clinical settings. These have potential such as the development of novel biomimetic materials and techniques for drug encapsulation within the synthetic ion channels and pumps. It also explores ongoing research aimed at enhancing the targeting precision and therapeutic versatility of these systems. By harnessing the principles of ion transport found in nature, researchers are paving the way for more precise, effective, and patient-friendly drug delivery methods. The report encourages further exploration of this exciting field, emphasizing the need for interdisciplinary collaboration between materials scientists, pharmacologists, and medical practitioners to translate these innovations into real-world medical applications.

5.2.5.2 *Bioimaging*

Nanoparticles and nanoscale materials with ion-responsive properties can be used for advanced bioimaging techniques. These materials enable real-time monitoring of ion concentrations within living cells, offering valuable insights into cellular processes and disease mechanisms. Biomimetic ion transport systems have emerged as a pioneering approach in the field of bioimaging, capitalizing on the principles of ion transport observed in biological systems. This report delves into the key aspects of these applications, highlighting their significance and potential impact on advancing our understanding of cellular processes and disease mechanisms [20].

An overview of the recent advances in the field, including the development of advanced nano sensors and nanoprobes that integrate with biomimetic ion transport systems for enhanced bioimaging capabilities. It also explores ongoing research aimed at improving the spatial and temporal resolution of ion-specific bioimaging techniques. It demonstrates the significant potential of these systems in advancing our understanding of cellular processes and disease mechanisms. By harnessing the principles of ion transport found in nature, researchers are poised to revolutionize bioimaging techniques, offering real-time, high-resolution insights into cellular dynamics and disease progression [21]. The report encourages further exploration of this exciting field, emphasizing the need for collaboration between materials scientists, biologists, and medical researchers to translate these innovations into valuable tools for diagnostics and drug development [4, 6, 22].

5.2.5.3 *Tissue Engineering*

Bioinspired nanoionics plays a crucial role in tissue engineering by providing materials that can regulate ion transport across artificial tissues. These materials promote cell growth, differentiation, and overall tissue functionality, making them essential for regenerative medicine applications [23, 24].

Biomimetic ion transport systems have emerged as a pivotal tool in tissue engineering, capitalizing on the principles of ion transport observed in biological systems. This report delves into the key aspects of these applications, emphasizing their significance in promoting cell growth, differentiation, and overall tissue functionality, thereby driving advancements in regenerative medicine [24].

An overview of recent advances in the field, including the development of sophisticated biomaterials and 3D printing techniques integrates biomimetic ion transport systems for enhanced tissue engineering outcomes. It also explores ongoing research aimed at improving the integration of these systems into clinical applications. This segment on biomimetic ion transport systems for biomedical tissue engineering underscores the immense potential of these systems in advancing regenerative medicine. By harnessing the principles of ion transport found in nature, researchers are poised to revolutionize tissue engineering approaches, offering solutions for tissue repair, transplantation, and disease modeling. This report encourages further exploration of this exciting field, emphasizing the need for collaboration between materials scientists, biologists, and medical researchers to translate these innovations into clinical practice, thereby improving the quality of life for patients in need of regenerative therapies [25, 26].

5.2.6 Bioelectronic Applications

This chapter explores biomedical applications in the field of bioelectronics. Biomimetic ion transport systems, inspired by natural ion transport mechanisms, offer unprecedented opportunities in various bioelectronic applications (Figure 5.6). This report presents a comprehensive overview of three key areas: Ion-Selective Sensors, Neuroprosthetics, and Energy Storage, elucidating their principles, significance, and recent advancements [27–29].

5.2.6.1 Ion-Selective Sensors

Biomimetic ion-selective sensors find applications in healthcare, environmental monitoring, and food safety. These sensors can detect specific ions with high sensitivity and selectivity, offering valuable data for a range of industries. Ion-selective sensors play a pivotal role in healthcare, environmental monitoring, and food safety. These sensors are designed to detect

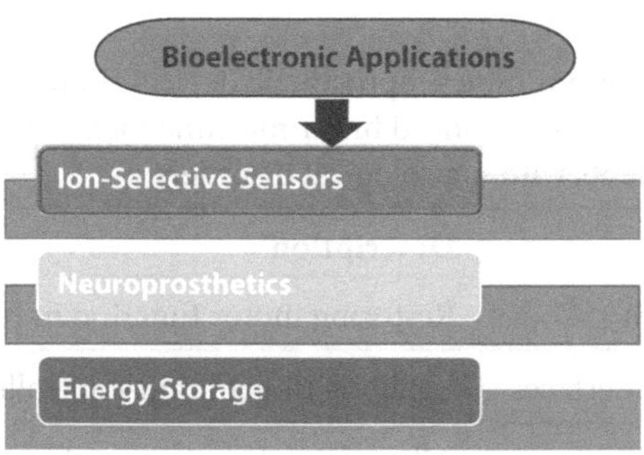

Figure 5.6 Bioelectronic applications in nanodevices.

Table 5.1 The diverse applications of ion-selective sensors.

Application	Description
Healthcare	Monitoring ion levels for diagnosis and treatment
Environmental	Tracking pollutant concentrations
Food Safety	Ensuring product safety and quality

specific ions with high sensitivity and selectivity, offering valuable data for a wide range of industries (Table 5.1). The academic paper explores the following aspects: [30–32].

➢ Recent Advances

There are recent breakthroughs in ion-selective sensor technology, such as the development of nanoscale sensors with improved sensitivity and stability. These innovations are poised to enhance our ability to monitor and control ion concentrations in various environments [33, 34].

5.2.6.2 Neuroprosthetics

Bioinspired nanoionics is driving innovation in neuroprosthetic devices. These devices interface with the nervous system, offering solutions for patients with paralysis or neurodegenerative conditions. Biomimetic ion channels and neural interfaces enable precise control and communication with neural circuits. Bioinspired nanoionics is driving innovation in neuroprosthetic devices (Table 5.2), which interface with the nervous system to offer solutions for patients with paralysis or neurodegenerative conditions. The paper explores the following aspects: [35, 36].

➢ Applications

Table 5.2 The applications of neuroprosthetic devices: Enhancing mobility and sensory functions through advanced brain-machine interfaces for improved quality of life in patients with neurological disorders.

Application	Description
Paralysis Treatment	Restoring motor function in paralyzed patients
Brain-Computer Interfaces	Enabling direct brain-controlled devices
Pain Management	Alleviating chronic pain through neural stimulation

➢ Recent Breakthroughs

Recent advancements in neuroprosthetic technology, including brain-computer interfaces, enable seamless communication between the human brain and external devices. These breakthroughs hold immense promise for enhancing the quality of life for individuals with neurological conditions.

5.2.6.3 *Energy Storage*

The development of bioinspired materials for energy storage devices, such as batteries and supercapacitors, is a growing area of research. Biomimetic ion transport systems can lead to high-performance energy storage solutions with improved efficiency and safety [9].

The development of bioinspired materials for energy storage devices, such as batteries and supercapacitors, is a growing area of research [37, 38]. Biomimetic ion transport systems can lead to high-performance energy storage solutions with improved efficiency and safety [39, 40].

➢ Recent Advances

Recent advancements in biomimetic materials for energy storage, such as flexible and environmentally friendly batteries are innovations paving the way for sustainable energy solutions. Biomedical and bioelectronic applications unveil a world of innovation and potential. By emulating nature's ion transport mechanisms, researchers are creating solutions that impact healthcare, neural interfaces, and energy storage. Collaborative efforts among materials scientists, biologists, and engineers are essential to harness the full potential of these systems for the benefit of society. As we embark on this interdisciplinary journey, the possibilities are boundless, promising a future where biomimetic ion transport systems drive advancements in healthcare, technology, and sustainability (Table 5.3).

Table 5.3 The applications of biomimetic ion transport in energy storage: Revolutionizing battery technology and capacitive systems for enhanced efficiency, longevity, and environmental sustainability.

Application	Description
Batteries	High-performance, longer-lasting batteries
Supercapacitors	Rapid energy storage and release
Energy Harvesting	Efficient conversion of environmental energy

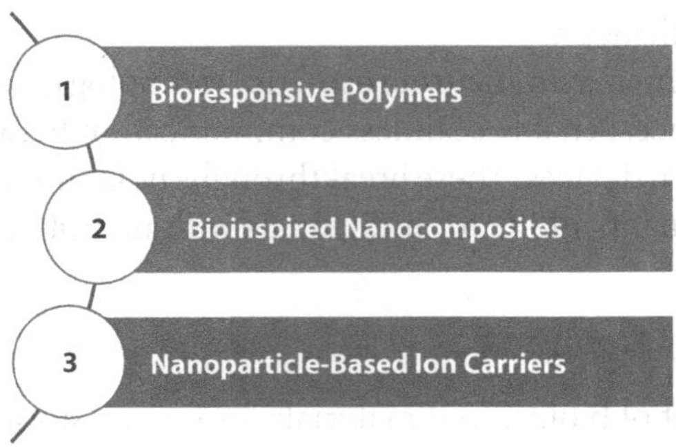

Figure 5.7 Types of biomimetic materials in bioinspired nanoionics.

5.3 Biomimetic Materials in Bioinspired Nanoionics

Biomimetic materials, the cornerstone of bioinspired nanoionics, have revolutionized the fields of biomedicine, bioelectronics, and materials science by drawing inspiration from nature's intricate design principles (Figure 5.7). In this comprehensive exploration, we delve into three remarkable categories of biomimetic materials: Bioresponsive Polymers, Bioinspired Nanocomposites, and Nanoparticle-Based Ion Carriers. We not only elucidate the underlying principles of each category but also provide real-world examples, tabular comparisons, and insights into their transformative potential [18, 41–43].

5.3.1 Bioresponsive Polymers

Discuss the development of polymers that change their structure or properties in response to ion concentration changes, offering applications in drug delivery and ion sensing [44–46].

- Principles: Bioresponsive polymers are smart materials designed to mimic biological processes by responding to specific environmental cues. These polymers change their properties, such as solubility or shape, in response to biological signals, making them ideal candidates for controlled drug delivery and tissue engineering.
    - Examples: A classic example is the use of temperature-responsive polymers in drug delivery systems. When injected into the body, these polymers undergo a reversible phase transition triggered by body temperature, releasing drugs precisely when and where they are needed.

5.3.2 Bioinspired Nanocomposites

Explore the fabrication of nanocomposite materials that incorporate bioinspired elements, enhancing their functionality for various biomedical and bioelectronic applications [47, 48].

- ➢ Principles: Bioinspired nanocomposites draw inspiration from the hierarchical structures found in nature, such as bone or nacre. These materials combine different components at the nanoscale to create robust, lightweight, and multifunctional materials. They are employed in applications ranging from structural materials to biomedical implants.
  - Examples: Composites reinforced with carbon nanotubes resemble the robust and lightweight structure of bones. These composites, which display outstanding mechanical capabilities due to the incorporation of carbon nanotubes into polymer matrices, are useful for biomedical and aerospace applications.

5.3.3 Nanoparticle-Based Ion Carriers

Examine the use of nanoparticles as carriers for ions, enabling controlled ion release and transport, with potential applications in targeted drug delivery and bioimaging [49].

- ➢ Principles: Nanoparticle-based ion carriers leverage the unique properties of nanoparticles to transport ions with precision. Inspired by biological ion channels, these carriers offer controlled and selective ion transport, with applications in drug delivery, sensing, and energy storage.
  - Examples: Lipid-based nanoparticles are used in drug delivery to encapsulate and transport therapeutic ions or molecules to specific targets in the body. These nanoparticles can be tailored to release their cargo in response to pH changes or other physiological cues.

- ➢ Comparison:
  - Principles: Bioresponsive polymers rely on environmental cues, bioinspired nanocomposites emulate natural structures, and nanoparticle-based ion carriers mimic biological ion channels (Table 5.4).

Table 5.4 Comparisons of biomimetic materials: Evaluating performance, durability, and functional advantages in diverse applications ranging from medical implants to environmental sensors.

Biomimetic material	Principles	Examples	Applications
Bioresponsive Polymers	Respond to environmental cues	Temperature-responsive polymers for drug delivery	Controlled drug release, tissue engineering
Bioinspired Nanocomposites	Emulate hierarchical natural structures	Carbon nanotube-reinforced composites for aerospace	Structural materials, biomedical implants
Nanoparticle-Based Ion Carriers	Mimic ion-selective biological channels	Lipid-based nanoparticles for drug delivery	Drug delivery, ion sensing, energy storage

- Examples: Temperature-responsive polymers, carbon nanotube-reinforced composites, and lipid-based nanoparticles exemplify the respective categories.
- Applications: Bioresponsive polymers find applications in controlled drug release and tissue engineering. Bioinspired nanocomposites are employed in structural materials and biomedical implants. Nanoparticle-based ion carriers serve in drug delivery, ion sensing, and energy storage.

Biomimetic materials in bioinspired nanoionics are propelling innovation by mimicking nature's design principles. These materials, each with its unique set of principles and applications, hold the promise of transforming industries ranging from healthcare to materials science. Their ability to respond to environmental cues, replicate hierarchical structures, and mimic biological ion transport opens doors to an array of groundbreaking technologies and solutions. As research in this field continues to evolve, we can anticipate even more exciting applications that harness the power of biomimicry to address complex challenges in science and engineering [50].

5.4 Biomedical Breakthroughs

5.4.1 Organelle-Targeted Drug Delivery

In the realm of biomedicine, one of the most transformative breakthroughs in recent years has been the development of organelle-targeted drug delivery systems. These cutting-edge technologies allow for the precise delivery of therapeutic agents to specific cellular organelles, unlocking new possibilities for the treatment of various diseases. This explanation will delve into the principles of organelle-targeted drug delivery, provide real-world examples, and culminate in a table summarizing these breakthroughs [40, 51].

➢ Examples of Organelle-Targeted Drug Delivery Breakthroughs (Table 5.5):

- Mitochondrial Targeting: Mitochondria are the powerhouses of the cell, and their dysfunction is implicated in numerous diseases, including cancer and neurodegenerative disorders. Breakthroughs in mitochondrial targeting have led to the development of mitochondria-specific drug delivery systems. For instance, researchers have designed liposomes coated with mitochondria-targeting peptides that can ferry antioxidants directly into the mitochondria, mitigating oxidative stress and protecting against neurodegeneration.
- Lysosomal Targeting: Lysosomes are cellular organelles responsible for breaking down cellular waste and foreign materials. Dysregulation of lysosomal function is linked to lysosomal storage disorders. Breakthroughs in lysosomal targeting have enabled the development of enzyme replacement therapies (ERTs). For example, ERTs for Pompe disease utilize engineered enzymes encapsulated within nanoparticles that specifically traffic to lysosomes, restoring enzymatic activity and ameliorating the disease.
- Nuclear Targeting: Targeting the cell nucleus is crucial for the treatment of diseases involving genetic material, such as cancer. Researchers have developed nanostructures functionalized with nuclear localization signals that facilitate nuclear uptake. This approach enables the delivery of gene-editing tools like CRISPR-Cas9 or therapeutic nucleic acids to target and modify specific genes associated with disease.

Table 5.5 Shows the organelle-targeted drug delivery breakthroughs: Advancements in precision medicine for enhanced therapeutic efficacy and reduced side effects in cellular treatments.

Organelle	Breakthrough	Example	Application
Mitochondria	Mitochondria-targeting liposomes	Delivery of antioxidants for neuroprotection	Neurodegenerative disease treatment
Lysosomes	Enzyme replacement therapies (ERTs)	Nanoparticles for Pompe disease	Treatment of lysosomal storage disorders
Nucleus	Nuclear localization signals	Gene-editing tools and therapeutic nucleic acids	Gene therapy and cancer treatment

Organelle-targeted drug delivery is a paradigm-shifting advancement in biomedicine, offering precise and effective treatments for a wide array of diseases. By harnessing the specificity of nanocarriers and surface modifications, researchers have unlocked the potential to treat diseases at their root, within specific cellular organelles. These breakthroughs have ushered in a new era of targeted therapies with enhanced efficacy and reduced side effects, promising improved patient outcomes and quality of life.

5.4.2 Theranostics: Simultaneous Therapy and Imaging

In the ever-evolving landscape of biomedicine, one of the most profound and game-changing innovations has been the emergence of theranostics. This revolutionary approach represents a fusion of therapy and diagnostics, enabling the simultaneous delivery of treatment and real-time monitoring of its effects. In this explanation, we'll delve into the principles of theranostics, provide real-world examples, and present a table summarizing these biomedical breakthroughs. Theranostics is rooted in the concept of personalized medicine, where treatment plans are tailored to individual patients based on their unique genetic, [52] and molecular action. It mainly depends on-

- Targeted Delivery: Theranostic agents are designed to specifically target diseased cells or tissues. They are equipped

with ligands or nanoparticles that recognize and bind to molecular markers unique to the disease.

- Real-Time Monitoring: These agents incorporate diagnostic components, such as imaging probes or sensors, which enable the continuous monitoring of treatment response. This ensures that therapy is both targeted and effective.

➢ Examples of Theranostics Breakthroughs:

- Cancer Theranostics: One of the most prominent applications of theranostics is in oncology. Researchers have developed nanoparticles that can simultaneously deliver chemotherapy drugs and imaging agents to cancer cells. For instance, gold nanoparticles loaded with chemotherapy drugs can be guided to tumor sites using surface-bound antibodies. These nanoparticles not only deliver the drug but also serve as contrast agents for imaging, allowing clinicians to monitor the tumors' response to treatment in real time.
- Cardiovascular Theranostics: Theranostics has also made significant strides in cardiovascular medicine. Nanoparticles loaded with drugs to reduce plaque buildup in arteries are being used in combination with imaging agents. These nanoparticles help to precisely deliver medications to atherosclerotic plaques while allowing clinicians to monitor changes in plaque size and composition.
- Neurological Disorders: In the realm of neurological disorders like Alzheimer's disease, theranostics holds promise. Researchers are developing nanoparticles that can target and remove toxic protein aggregates associated with neurodegeneration. Simultaneously, these nanoparticles carry imaging agents that enable tracking of protein clearance in real-time, providing valuable insights into disease progression and treatment efficacy.

Theranostics represents a groundbreaking shift in biomedical practice, where diagnosis and treatment are seamlessly intertwined (Table 5.6). These innovative approaches not only enhance treatment precision but also provide clinicians with the ability to adapt therapies in real time based on patient response. As theranostics continues to advance, it holds the potential to transform healthcare across a spectrum of diseases, leading to more effective, personalized, and efficient medical interventions.

Table 5.6 Shows the theranostics breakthroughs: Innovations in combined diagnostic and therapeutic technologies for personalized medicine and enhanced disease management.

Application	Breakthrough	Example	Significance
Cancer	Targeted drug delivery and imaging	Gold nanoparticles for cancer therapy	Precise treatment and real-time monitoring
Cardiovascular	Plaque-targeted drug delivery and imaging	Nanoparticles for atherosclerosis	Enhanced treatment and monitoring
Neurological Disorders	Protein aggregate clearance and imaging	Nanoparticles for Alzheimer's disease	Improved disease understanding and therapy

5.4.3 Artificial Biomimetic Organs

In the realm of biomedicine, the development of artificial biomimetic organs stands as one of the most remarkable and transformative breakthroughs of recent years. These engineered organs not only replicate the form and function of their natural counterparts but also offer groundbreaking solutions to address the chronic shortage of donor organs. This explanation will delve into the principles of artificial biomimetic organs, provide real-world examples, and culminate in a table summarizing these biomedical breakthroughs.

The core principles underlying these innovations include:

- Biocompatible Materials: The use of biocompatible materials that interact seamlessly with biological tissues to prevent immune rejection and promote integration.
- Functional Replication: Accurate replication of the functions of natural organs, including their ability to process fluids, generate electrical signals, or perform metabolic processes.
- Regulatory Control: Incorporation of regulatory mechanisms that allow artificial organs to respond to physiological cues, such as hormones or neurotransmitters, in real time.

➢ Examples of Artificial Biomimetic Organs Breakthroughs:

- Artificial Heart: A remarkable example is the development of artificial hearts that closely mimic the functions of the human heart. These devices utilize advanced materials, such as biocompatible polymers, to pump blood throughout the body effectively. The SynCardia temporary Total Artificial Heart is one such example, serving as a bridge to transplant for patients awaiting heart transplants.
- Artificial Kidney: Researchers have made significant strides in creating artificial kidneys that replicate the filtration and excretion functions of natural kidneys. Wearable artificial kidneys, such as the Wearable Artificial Kidney (WAK), are designed to provide continuous renal replacement therapy for patients with end-stage renal disease, improving their quality of life.
- Artificial Pancreas: In the realm of diabetes management, artificial pancreas systems combine continuous glucose monitoring with automated insulin delivery, mimicking the natural regulatory functions of the pancreas. The Medtronic MiniMed 670G is an example of such a system, offering improved glucose control and reducing the burden on individuals with diabetes.

In conclusion, artificial biomimetic organs represent a revolutionary leap forward in the field of biomedicine, offering new hope for patients in need of life-saving or life-enhancing interventions. These innovative creations, guided by the principles of biocompatibility, functional replication, and regulatory control, have the potential to significantly improve the quality of life for individuals facing organ failure or chronic diseases. As research and development in this field continue to advance, we can anticipate even more sophisticated artificial organs that closely mimic the intricacies of their natural counterparts.

5.5 Bioelectronic Innovations

In the captivating realm of "Bioinspired Nanoionics for Biomedical and Bioelectronic Applications," where science unfurls the secrets of nature's design, bioelectronic innovations emerge as a striking symphony between

Table 5.7 Shows the artificial biomimetic organs breakthrough: Revolutionary developments in synthetic organ design for improved organ replacement, regeneration, and functionality in transplant medicine.

Artificial organ	Breakthrough	Example	Significance
Artificial Heart	Biocompatible materials	SynCardia Total Artificial Heart	Bridge to heart transplant, improved survival
Artificial Kidney	Filtration and excretion	Wearable Artificial Kidney (WAK)	Continuous renal replacement therapy
Artificial Pancreas	Glucose regulation	Medtronic MiniMed 670G	Improved diabetes management

biology and technology (Table 5.7). This explanation embarks on a journey into the heart of bioinspired nanoionics, where the boundaries of human-machine interaction blur, and introduces the fascinating concept of bioelectronic skin—a transformative breakthrough that promises to reshape our world. Imagine a world where the boundary between humans and machines becomes indistinguishable, where electronic devices can feel, sense, and interact with the human body as naturally as our own organs do.

5.5.1 Bioelectronic Skin

The Symphony of Bioelectronic Skin:

Imagine a world where our skin, the body's largest sensory organ, not only perceives the environment but communicates directly with technology. This is the promise of bioelectronic skin—an innovation that bridges the organic and the synthetic in perfect harmony. At its heart, bioelectronic skin is a delicate and flexible electronic interface designed to mimic the functions of human skin. It senses temperature, pressure, touch, and even biochemical changes, just like our natural skin. It seamlessly integrates with our bodies, adhering like a second skin, and opens a gateway for us to interact with machines in ways we never thought possible.

The Transformative Potential:

Bioelectronic skin holds transformative potential across diverse domains:

Human-Machine Interaction: Think of bioelectronic skin as the bridge between humans and machines. It enables us to control devices with mere gestures, much like a conductor guiding an orchestra with a wave of the baton. Imagine turning on lights, adjusting thermostats, or even playing music by simply tapping your fingers on your own skin.
Healthcare and Prosthetics: For the medical field, bioelectronic skin offers a new dimension of patient care. It can monitor vital signs in real time, alerting healthcare providers to any anomalies. For amputees, bioelectronic skin can be integrated into prosthetics, providing sensory feedback and a more natural sense of touch.
Virtual and Augmented Reality: In the realm of entertainment and virtual reality, bioelectronic skin can take immersion to unprecedented levels. You can feel the warmth of a virtual sun on your skin or sense a gentle breeze while exploring a digital world.

The Ongoing Journey:
As we embark on this journey, it's crucial to acknowledge that bioelectronic skin is not just a futuristic dream; it's an ongoing reality. Researchers and innovators are continually pushing the boundaries of materials science, biocompatibility, and neurobiology to make bioelectronic skin more advanced, more accessible, and more integrated into our lives.

In this realm, where biology and electronics entwine, ethical considerations become vital. We must ponder questions of privacy, data security, and the impact of these innovations on society. As we move forward, bioelectronic skin calls upon us to strike a harmonious chord between technological advancement and ethical responsibility.

In conclusion, bioelectronic skin represents a symphony of innovation and possibility in the realm of bioinspired nanoionics. It invites us to reimagine our relationship with technology, our bodies, and the world around us. As this harmonious journey continues, we find ourselves on the precipice of a new era, where the lines between human and machine blur, and the enchanting melodies of bioelectronic skin lead us into a future brimming with wonder and transformation (Table 5.8).

5.5.2 Ionic Circuitry

➢ Ionic circuitry holds immense promise across various domains:

- Bioelectronic Implants: Consider a scenario where medical implants seamlessly integrate with our bodies, communicating and responding to physiological signals. Ionic circuitry

Table 5.8 Shows the bioinspired nanoionics for biomedical and bioelectronic applications: Nanoscale ionic devices with enhanced biocompatibility and functionality for medical diagnostics and therapeutic.

Field	Example	Name of application/ innovation	Function	Potential impact	Ethical considerations
Sensors and Diagnostics	Ion-Selective Electrodes	pH Sensors	Monitor pH levels in the body	Early detection of acidity-related disorders	Informed consent for surgery and data privacy
Neuroprosthetics	Myoelectric Prosthetics	Ionic-Controlled Prosthetic Limbs	Enable natural limb movement	Enhanced mobility and functionality	Data security and patient consent
Brain-Computer Interfaces	EEG Brain Interfaces	Neurofeedback Systems	Translate brain signals into actions	Assisting individuals with disabilities	Ethical use of neural data and consent
					Privacy of neural data and informed consent

paves the way for more efficient and biocompatible devices, promising a future where chronic conditions can be managed with unprecedented precision (Table 5.8).

- Sensors and Diagnostics: Imagine sensors that can monitor specific ions in real-time, providing valuable insights into our health. These sensors could detect changes in pH levels, ion concentrations, or even specific biomarkers, offering early warnings for various medical conditions.
- Neuroprosthetics and Brain-Computer Interfaces: Ionic circuitry has the potential to bridge the gap between man and machine in ways previously unimaginable. Think of neuroprosthetic limbs that respond as naturally as our own, or brain-computer interfaces that enable direct communication between our thoughts and external devices.

➢ Harmony with Nature: The Ethical Implications:
As we delve deeper into the world of ionic circuitry, we must also consider the ethical implications. How do we ensure that these innovations align with the principles of responsible research and application? How do we strike a balance between technological advancement and safeguarding the sanctity of the human body?

Questions of privacy, consent, and data security become paramount. It's not just about what we can achieve, but how we achieve it—in a way that respects the individual's autonomy and dignity. Ionic circuitry represents a symphony of innovation, where the language of ions converges with the artistry of electronics. It invites us to reimagine how we interact with and understand the human body, opening doors to a future where technology and biology coalesce in perfect harmony. As this symphony continues to unfold, let us remember that in this dance of innovation, it is not just about the steps we take, but also about the grace with which we take them.

5.5.3 Bioelectronic Therapeutics

Bioelectronic therapeutics is an emerging field at the intersection of biology, electronics, and medicine. It involves the development and application of electronic devices, often miniaturized and implantable, to modulate or interact with the body's electrical or biochemical systems for therapeutic purposes. These devices are designed to treat a wide range of medical conditions by targeting specific physiological pathways.

➢ Key aspects of bioelectronic therapeutics include:

- Electrical stimulation: This involves using electrical impulses to modulate neural or physiological pathways. For example, deep brain stimulation (DBS) is used to treat conditions like Parkinson's disease by implanting electrodes in the brain to regulate abnormal neural activity.
- Neurostimulation: Devices like spinal cord stimulators or peripheral nerve stimulators are used to alleviate chronic pain by modulating the transmission of pain signals along neural pathways.
- Biochemical modulation: Some bioelectronic therapies involve the controlled release of specific biochemical agents, such as drugs or neurotransmitters, to influence cellular or physiological processes.
- Sensory feedback: In prosthetics or neural interfaces, bioelectronic devices can provide sensory feedback to users, allowing them to feel or control artificial limbs or devices more naturally.
- Responsive therapies: These are systems that adapt and respond to real-time physiological data, such as glucose levels in diabetes management. They can adjust the delivery of therapy accordingly, enhancing its effectiveness.

Bioelectronic therapeutics is a rapidly evolving field with the potential to revolutionize the treatment of various medical conditions, including neurological disorders, chronic pain, metabolic diseases, and more. These therapies offer precise and targeted approaches to managing health conditions, often with fewer side effects compared to traditional pharmaceutical interventions. As research and technology in this field advance, it holds promise for improving the quality of life for many patients.

At its core, bioelectronic therapeutics leverages the principles of ion transport, mimicking nature's own ion channels and receptors. It's like composing a musical score where ions become the notes, creating a healing melody that resonates with the body's needs.

➢ The Transformative Potential:
Bioelectronic therapeutics holds the potential to revolutionize healthcare in astonishing ways:

- Neurological disorders: Imagine a world where bioelectronic devices could restore harmony in the brain's electrical signals, providing relief to individuals with epilepsy or depression. It's like conducting a symphony of neural impulses, bringing the brain's rhythm back to equilibrium.
- Pain management: Consider a therapy that modulates ion channels to alleviate chronic pain. It's like fine-tuning an instrument, creating a soothing resonance that drowns out the cacophony of discomfort.
- Metabolic diseases: Picture a treatment that restores the balance of ions in cells, offering hope to those with metabolic disorders. It's like a nutritional symphony, ensuring that cells receive the right nutrients in perfect harmony.

➢ Harmony in Healing: Ethical Considerations:
As we explore the potential of bioelectronic therapeutics, we must also consider the ethical dimensions. How do we ensure that these innovations are used responsibly and ethically? How do we strike a balance between technological advancement and the well-being of individuals?

Questions of informed consent, data privacy, and equitable access become paramount. It's not just about what we can achieve, but how we achieve it—in a way that respects autonomy, preserves dignity, and benefits all. Bioelectronic therapeutics represent a symphony of innovation and healing. It invites us to reimagine healthcare as a harmony between electronics and biology, where the body's own melodies are fine-tuned to restore health and well-being. As this therapeutic symphony continues to play, let us remember that in this dance of healing, it's not just about the notes we play, but the harmony we create for the betterment of humanity.

5.6 Biocompatibility and Safety

Biocompatibility, in essence, is the art of creating harmonious symphonies between the nanoworld and the biological realm. It's about ensuring that the marvels crafted at the nanoscale don't just dazzle us with their ingenuity but also coexist harmoniously within the complex ecosystems of our bodies. It's akin to inviting a delicate guest into your home, ensuring that they feel at ease, causing no disruptions. In the realm of bioinspired nanoionics, biocompatibility isn't a mere formality; it's a cornerstone. Materials and devices must resonate with the human body, evoke no immune responses, and navigate bodily fluids and tissues with finesse. Imagine nanoparticles, inspired by nature's ion transport mechanisms, navigating the bloodstream like stealthy travelers, undisturbed and undisturbing, reaching their destination with precision.

➢ Safety: The Guardian of Innovation:
Safety is the guardian standing vigil at the gates of innovation in bioinspired nanoionics. It's the watchful eye ensuring that our breathtaking creations while pushing boundaries, do not unwittingly tread into hazardous terrain. Safety here is not an afterthought but an integral part of the design process, akin to a safety net stretched beneath a highwire act.

In this world, where artificial biomimetic organs beat alongside natural ones, where ion-selective sensors rival human senses, and where neural interfaces blur the lines between biology and electronics, safety standards are paramount. Imagine the careful testing and scrutiny that neural interfaces must undergo to ensure they don't disrupt the delicate balance of the brain. Or consider the rigorous evaluations of ion carriers to guarantee they don't inadvertently trigger harmful responses within the body.

➢ The Future Calls, Safety:
As we progress through this assessment, we see a future in which bioinspired nanoionics transforms medicine, advances technology, and opens up new horizons of opportunity. Biocompatibility and safety are not barriers to this future; rather, they are enablers. They reassure us that these ideas are solid, dependable solutions that will find a place in the lives of patients and consumers, not just fun experiments. So, in the enchanting world of bioinspired nanoionics, where the biological and the nanoscale intertwine, biocompatibility and safety become our guides, offering assurance that the dazzling creations emerging from labs will not only astound us with their brilliance but also enhance our lives with their safety and efficacy. As we venture deeper into this nanoscale frontier, let us remember that with every innovation, biocompatibility, and safety remain steadfast companions, ensuring that our journey remains one of wonder, not of risk.

5.7 Ethical and Regulatory Consideration

In the enchanting world of "Bioinspired Nanoionics for Biomedical and Bioelectronic Applications," where scientific innovation explore at the intersection of nature and technology, ethical and regulatory considerations stand as our guiding stars. This review embarks on a journey through the ethereal landscape of nanoionics, where the boundaries between biology and nanotechnology blur. Here, we explore the delicate interplay of ethics and regulations, illuminating the path towards responsible and sustainable progress in healthcare and technology (Figure 5.8).

➢ Ethical Considerations: The Moral Compass of Nanoionics:

- Respect for autonomy: Nanoionics offers the potential for transformative medical interventions. Ethical considerations emphasize that individuals have the right to make informed decisions about their healthcare, including whether to embrace innovative treatments like neural interfaces or bioinspired organ implants. Informed consent is a paramount ethical principle.
- Beneficence and non-maleficence: The development of bio-inspired nanoionics should always prioritize the well-being of individuals. Innovations should aim to maximize benefits while minimizing harm. Ethical evaluations ensure that

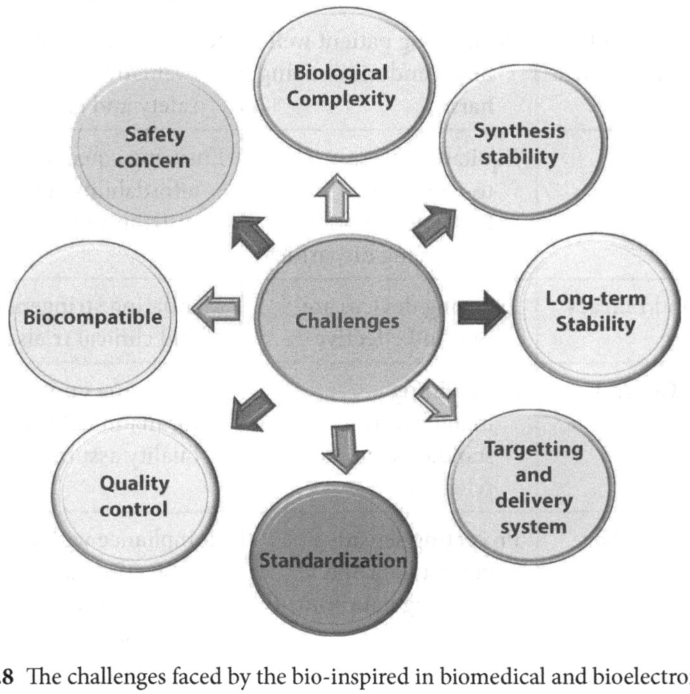

Figure 5.8 The challenges faced by the bio-inspired in biomedical and bioelectronic application.

devices are safe, effective, and do not lead to unintended consequences.

- Justice: Equal access to nanoionic technologies is essential. Ethical considerations include equitable distribution, affordability, and addressing health disparities to ensure that these innovations do not exacerbate societal inequities.

➢ Regulatory Considerations: The Guardians of Safety:

- Safety and efficacy testing: Stringent regulatory evaluations are essential to ensure the safety and efficacy of nanoionic applications. Rigorous testing and clinical trials are mandated to assess the performance, risks, and benefits of bioinspired technologies (Table 5.9).

Table 5.9 Shows the ethical and regulatory considerations: Comprehensive examination of moral, legal, and social implications surrounding emerging technologies to ensure responsible development and deployment in various industries and societal contexts.

Aspect	Ethical consideration	Regulatory consideration
Beneficence and Non-Maleficence	Prioritizing patient well-being and minimizing harm	Rigorous testing and evaluation to assess safety and efficacy
Justice	Equitable access to nanoionic technologies and addressing disparities	Ensuring equal access, affordability, and fair distribution
Safety and Efficacy Testing	Ensuring devices are safe and effective	Mandating stringent testing and clinical trials
Quality Control	Maintaining manufacturing standards and device reliability	Establishing guidelines for manufacturing and quality assurance
Privacy and Data Security	Protecting sensitive patient data and ensuring data security	Compliance with data protection laws and informed consent for data use

- Quality control: Manufacturing standards and quality control processes are crucial to maintain the reliability and consistency of nanoionic devices. Regulatory bodies establish guidelines to uphold these standards.
- Privacy and data security: Nanoionics often involve the collection and transmission of sensitive patient data. Ethical and regulatory frameworks must ensure strict data protection, informed consent for data use, and compliance with privacy laws.

5.8 Conclusion

Bioinspired nanoionics has opened up new possibilities for biomedical and bioelectronic applications. By leveraging the principles of ion transport found in biological systems, researchers are creating innovative materials and devices that have the potential to transform medicine, diagnostics, and technology [5–7]. As the field continues to evolve, collaboration between materials scientists, biologists, and engineers will be essential in harnessing the full potential of bioinspired nanoionics for the benefit of society. This paper serves as a comprehensive resource for those interested in exploring and contributing to this cutting-edge field [24, 53].

As we conclude our exploration of "Bioinspired Nanoionics for Biomedical and Bioelectronic Applications," we find ourselves at the intersection of remarkable achievements, ongoing research endeavors, and the vast landscape of future prospects. This paper has been a journey through the fascinating realm of biomimicry at the nanoscale, where nature's ion transport mechanisms inspire innovative solutions with the potential to transform healthcare, bioelectronics, and beyond. As we reflect on the present, acknowledge the ongoing work, and gaze into the future, the implications and possibilities are both awe-inspiring and promising. Achievements of the Present: We are currently living in a time of notable developments in the field of bioinspired nanoionics. Biomedical and bioelectronic applications are being revolutionized by materials and technologies that researchers have ingeniously developed using biological ion transport mechanisms. With previously unheard-of accuracy, ion-selective sensors have now become widely used in a variety of fields, from environmental monitoring to healthcare diagnostics. Robotics and human-machine interactions will be revolutionized by the development of bioelectronic skins that imitate human ion-sensing abilities. The field of bioinspired nanoionics has seen remarkable progress in drug delivery, with biomimetic nanocarriers precisely targeting specific cell organelles and diseases. Theranostic approaches combining therapy and real-time imaging are on the cusp of transforming personalized medicine. Artificial biomimetic organs are no longer confined to the realm of fantasy, as they have the potential to alleviate the organ transplant crisis. In the field of bioelectronics, ionic circuits are on the rise, promising advances in neuromorphic computing and signal processing and blurring the lines between biology and electronics.

Ongoing Research: Simultaneously, the field is characterized by ongoing research, where the promise of the present meets the potential of the future. The scalability and manufacturability of bioinspired nanoionics are subjects of intense study, with researchers striving to translate laboratory innovations into scalable, real-world solutions. The quest for enhanced biocompatibility and safety remains a paramount concern, driving efforts to ensure that these materials and devices are suitable for *in vivo* applications. As bioinspired nanoionics is progressively incorporated into healthcare and technology, ethical and regulatory concerns are looming large on the horizon, necessitating the need for guidelines and ethical research procedures. In order to reap the potential benefits without jeopardizing safety and privacy, ongoing research aims to find a balance between scientific innovation and ethical responsibility.

Future Prospects: As we peer into the future, the prospects of bioinspired nanoionics are radiant with promise. The integration of biology and nanotechnology holds transformative potential. In biomimetic materials, researchers are exploring self-healing and adaptive properties, mimicking nature's ability to repair and adapt. The development of bioinspired electronic skins that can sense not only ions but also complex biochemical signals could usher in a new era of human-machine interaction [54–56].

In bioelectronic applications, the concept of ionic circuitry is poised to advance, with ionic components enabling computation and information processing that emulate biological neural networks. The fusion of bioinspired nanoionics with artificial intelligence could give rise to advanced diagnostic tools, personalized therapies, and novel insights into complex biological processes. The pursuit of sustainable energy solutions through bioinspired materials for energy storage could revolutionize renewable energy utilization, contributing to a greener future [6, 57–59].

In conclusion, this chapter on "Bioinspired Nanoionics for Biomedical and Bioelectronic Applications" unveils a dynamic field at the confluence of nature and technology. The present showcases groundbreaking achievements, while ongoing research promises to address challenges and expand the horizons of possibility. The future holds the potential to reshape healthcare, bioelectronics, and beyond, unlocking novel solutions to age-old problems and pushing the boundaries of what is achievable. As we navigate this interdisciplinary realm, we are reminded that the synergy between biology and nanotechnology can yield boundless opportunities for innovation, discovery, and progress [1, 7, 16, 47, 60–64].

References

1. Zheng, Y., Omar, R., Hu, Z., Duong, T., Wang, J., Haick, H., Bioinspired Triboelectric Nanosensors for Self-Powered Wearable Applications. *ACS Biomater. Sci. Eng.*, 9, 2087, 2023.
2. Wu, G., Hui, X., Hu, L., Bai, Y., Rahaman, A., Yang, X.F., Chen, C., Recent advancement of bioinspired nanomaterials and their applications: A review. *Front. Bioeng. Biotechnol.*, 10, 952523, 2022.
3. Xu, S., He, L., Chen, F., Dang, W., Xie, S., Li, L., Cui, C., Yang, Y., Liu, Y., Tan, W., Bioinspired molecular engineering of bivalent aptamers by ligand-induced self-dimerization. *Chem. Commun.*, 59, 4790, 2023.
4. Löffler, S., Libberton, B., Richter-Dahlfors, A., Organic bioelectronic tools for biomedical applications. *Electron.*, 4, 879, 2015.
5. Diniz, M.S.F., Mourão, M.M., Xavier, L.P., Santos, A.V., Recent Biotechnological Applications of Polyhydroxyalkanoates (PHA) in the Biomedical Sector—A Review. *Polymers (Basel)*, 15, 4405, 2023.
6. Muzammil, S., Neves Cruz, J., Mumtaz, R., Rasul, I., Hayat, S., Khan, M.A., Khan, A.M., Ijaz, M.U., Lima, R.R., Zubair, M., Effects of Drying Temperature and Solvents on *In Vitro* Diabetic Wound Healing Potential of Moringa oleifera Leaf Extracts. *Molecules*, 28, 710, 2023.
7. Harun-Ur-Rashid, M., Jahan, I., Foyez, T., Imran, A., Bin Bio-Inspired Nanomaterials for Micro/Nanodevices: A New Era in Biomedical Applications. *Micromachines*, 14, 1786, 2023.
8. *Bio-Inspired Materials for Biomedical Applications*, MDPI Books, 2021.
9. Bhattacharya, P., Du, D., Lin, Y., Bioinspired nanoscale materials for biomedical and energy applications. *J. R. Soc. Interface*, 11, 20131067, 2014.
10. Burley, S.K., Berman, H.M., Christie, C., Duarte, J.M., Feng, Z., Westbrook, J., Young, J., Zardecki, C., RCSB Protein Data Bank: Sustaining a living digital data resource that enables breakthroughs in scientific research and biomedical education. *Protein Sci.*, 27, 316, 2018.
11. Ponomarenko, N.S., Kokhan, O., Pokkuluri, P.R., Mulfort, K.L., Tiede, D.M., Examination of abiotic cofactor assembly in photosynthetic biomimetics: site-specific stereoselectivity in the conjugation of a ruthenium(II) tris(bipyridine) photosensitizer to a multi-heme protein. *Photosynth. Res.*, 143, 99, 2020.

12. Pongampai, S., Charoonsuk, T., Pinpru, N., Pulphol, P., Vittayakorn, W., Pakawanit, P., Vittayakorn, N., Triboelectric-piezoelectric hybrid nanogenerator based on BaTiO3-Nanorods/Chitosan enhanced output performance with self-charge-pumping system. *Compos. Part B Eng.*, 208, 108602, 2021.
13. Hu, D., Ren, Q., Li, Z., Han, S., Ding, L., Lu, Z., Zhang, L., Unveiling the mechanism of an amelogenin-derived peptide in promoting enamel biomimetic remineralization. *Int. J. Biol. Macromol.*, 253, 127322, 2023.
14. Kim, Y.K., Hwang, S.H., Seo, H.J., Jeong, S.M., Lim, S.K., Effects of biomimetic cross-sectional morphology on the piezoelectric properties of BaTiO3 nanorods-contained PVDF fibers. *Nano Energy*, 97, 107216, 2022.
15. Le, A.T., Ahmadipour, M., Pung, S.Y., A review on ZnO-based piezoelectric nanogenerators: Synthesis, characterization techniques, performance enhancement and applications. *J. Alloys Compd.*, 844, 156172, 2020.
16. AlTowireb, S.M. and Goumri-Said, S., Core-Shell structures for the enhancement of energy harvesting in piezoelectric Nanogenerators: A review. *Sustain. Energy Technol. Assess.*, 55, 102982, 2023.
17. Carmo Bastos, M.L., Silva-Silva, J.V., Neves Cruz, J., Palheta da Silva, A.R., Bentaberry-Rosa, A.A., da Costa Ramos, G., de Sousa Siqueira, J.E., Coelho-Ferreira, M.R., Percário, S., Santana Barbosa Marinho, P., Marinho, A.M.R., de Oliveira Bahia, M., Dolabela, M.F., Alkaloid from Geissospermum sericeum Benth. & Hook.f. ex Miers (Apocynaceae) Induce Apoptosis by Caspase Pathway in Human Gastric Cancer Cells. *Pharmaceuticals*, 16, 765, 2023.
18. Rafique, S., Kasi, A.K., Aminullah, Kasi, J.K., Bokhari, M., Shakoor, Z., Fabrication of Br doped ZnO nanosheets piezoelectric nanogenerator for pressure and position sensing applications. *Curr. Appl. Phys.*, 21, 72, 2021.
19. Roy, K., Jana, S., Mallick, Z., Ghosh, S.K., Dutta, B., Sarkar, S., Sinha, C., Mandal, D., Two-Dimensional MOF Modulated Fiber Nanogenerator for Effective Acoustoelectric Conversion and Human Motion Detection. *Langmuir*, 37, 7107, 2021.
20. Deng, Z., Zhou, B., Xu, Y., Jin, C., Liu, W., Liu, B., Sun, J., Yang, J., Recent progresses of organic photonic synaptic transistors. *Flexible Printed Electron.*, 7, 24002, 2022.
21. Liu, Y., Zhou, Y., Liu, X., Dong, F., Wang, C., Wang, Z., Wasserstein GAN-Based Small-Sample Augmentation for New-Generation Artificial Intelligence: A Case Study of Cancer-Staging Data in Biology. *Engineering*, 5, 156, 2019.
22. Kriener, K., Lala, R., Homes, R.A.P., Finley, H., Sinclair, K., Williams, M.K., Midwinter, M.J., Mechanical Characterization of the Human Abdominal Wall Using Uniaxial Tensile Testing. *Bioengineering*, 10, 1213, 2023.
23. Acharya, R., Dutta, S.D., Patil, T.V., Ganguly, K., Randhawa, A., Lim, K.T., A Review on Electroactive Polymer–Metal Composites: Development and Applications for Tissue Regeneration. *J. Funct. Biomater.*, 14, 523, 2023.

24. Shahane, K., Kshirsagar, M., Tambe, S., Jain, D., Rout, S., Ferreira, M.K.M., Mali, S., Amin, P., Srivastav, P.P., Cruz, J., Lima, R.R., An Updated Review on the Multifaceted Therapeutic Potential of Calendula officinalis L. *Pharmaceuticals*, 16, 611, 2023.
25. Saretzki, G., Telomerase inhibition as cancer therapy. *Cancer Lett.*, 194, 209, 2003.
26. Hurley, L.H., DNA and its associated processes as targets for cancer therapy. *Nat. Rev. Cancer*, 2, 188, 2002.
27. Bathe, M., Chrisey, L.A., Herr, D.J.C., Lin, Q., Rasic, D., Woolley, A.T., Zadegan, R.M., Zhirnov, V.V., Roadmap on biological pathways for electronic nanofabrication and materials. *Nano Futures*, 3, 12001, 2019.
28. Shih, C.P., Tang, X., Kuo, C.W., Chueh, D.Y., Chen, P., Design principles of bioinspired interfaces for biomedical applications in therapeutics and imaging. *Front. Chem.*, 10, 990171, 2022.
29. Hu, J., Dun, G., Geng, X., Chen, J., Wu, X., Ren, T.L., Recent progress in flexible micro-pressure sensors for wearable health monitoring. *Nanoscale Adv.*, 5, 3131, 2023.
30. Singh, V.V., Nigam, A.K., Batra, A., Boopathi, M., Singh, B., Vijayaraghavan, R., Applications of Ionic Liquids in Electrochemical Sensors and Biosensors. *Int. J. Electrochem.*, 2012, 1, 2012.
31. Makhoul-Mansour, M.M., Maraj, J.J., Sarles, S.A., Bioderived materials for stimuli-responsive, adaptive, and neuromorphic systems: A perspective. *J. Compos. Mater.*, 57, 659–678, 2023.
32. Dong, Z., Xie, C., Li, X., Wu, P., Probing the Tl+-induced structural transformation of G-rich DNA to G-quadruplex with a label-free thioflavin T-based biosensor. *Sens. Actuators B Chem.*, 378, 133210, 2023.
33. Burley, S.K., Bhikadiya, C., Bi, C., Bittrich, S., Chen, L., Crichlow, G.V., Christie, C.H., Dalenberg, K., Di Costanzo, L., Duarte, J.M., Dutta, S., Feng, Z., Ganesan, S., Goodsell, D.S., Ghosh, S., Green, R.K., Guranovic, V., Guzenko, D., Hudson, B.P., Lawson, C.L., Liang, Y., Lowe, R., Namkoong, H., Peisach, E., Persikova, I., Randle, C., Rose, A., Rose, Y., Sali, A., Segura, J., Sekharan, M., Shao, C., Tao, Y.P., Voigt, M., Westbrook, J.D., Young, J.Y., Zardecki, C., Zhuravleva, M., RCSB Protein Data Bank: Powerful new tools for exploring 3D structures of biological macromolecules for basic and applied research and education in fundamental biology, biomedicine, biotechnology, bioengineering and energy sciences. *Nucleic Acids Res.*, 49, D437, 2021.
34. Niu, Q., Wei, H., Hsiao, B.S., Zhang, Y., Biodegradable silk fibroin-based bio-piezoelectric/triboelectric nanogenerators as self-powered electronic devices. *Nano Energy*, 96, 107101, 2022.
35. Lone, S.A., Lim, K.C., Kaswan, K., Chatterjee, S., Fan, K.P., Choi, D., Lee, S., Zhang, H., Cheng, J., Lin, Z.H., Recent advancements for improving the performance of triboelectric nanogenerator devices. *Nano Energy*, 99, 107318, 2022.

36. Richter, D., Sakharov, S., Forsén, E., Mayer, E., Reindl, L., Fritze, H., Thin film electrodes for high temperature surface acoustic wave devices. *Procedia Eng.*, 25, 168, 2011.
37. Qin, Y., Wang, X., Wang, Z.L., Microfibre-nanowire hybrid structure for energy scavenging. *Nature*, 451, 809, 2008.
38. Zhang, M., Gao, T., Wang, J., Liao, J., Qiu, Y., Yang, Q., Xue, H., Shi, Z., Zhao, Y., Xiong, Z., Chen, L., A hybrid fibers based wearable fabric piezoelectric nanogenerator for energy harvesting application. *Nano Energy*, 13, 298, 2015.
39. Qiu, Y., Yang, D.C., Yin, B., Lei, J.X., Zhang, H.Q., Zhang, Z., Chen, H., Li, Y.P., Bian, J.M., Liu, Y.H., Zhao, Y., Hu, L.Z., Branched ZnO nanotrees on flexible fiber-paper substrates for self-powered energy-harvesting systems. *RSC Adv.*, 5, 5941, 2015.
40. Lee, Y., Kim, S., Kim, D., Lee, C., Park, H., Lee, J.H., Direct-current flexible piezoelectric nanogenerators based on two-dimensional ZnO nanosheet. *Appl. Surf. Sci.*, 509, 145328, 2020.
41. Yang, D., Qiu, Y., Wang, T., Song, W., Wang, Z., Xu, J., Feng, Q., Zong, Y., Sun, X., Growth of 3D branched ZnO nanowire for DC-type piezoelectric nanogenerators. *J. Mater. Sci. - Mater. Electron.*, 27, 6708, 2016.
42. Sam, A., Natraj, V., Marappan, G., Sivalingam, Y., Surya, V.J., Design and development of novel piezoelectric nanogenerator based on pH dependent ZnO nanostructures. *Mater. Lett.*, 294, 129798, 2021.
43. Janaszewska, A., Klajnert-Maculewicz, B., Marcinkowska, M., Duchnowicz, P., Appelhans, D., Grasso, G., Deriu, M.A., Danani, A., Cangiotti, M., Ottaviani, M.F., Multivalent interacting glycodendrimer to prevent amyloid-peptide fibril formation induced by Cu(II): A multidisciplinary approach. *Nano Res.*, 11, 1204, 2018.
44. Sankar Ganesh, R., Sharma, S.K., Sankar, S., Divyapriya, B., Durgadevi, E., Raji, P., Ponnusamy, S., Muthamizhchelvan, C., Hayakawa, Y., Kim, D.Y., Microstructure, structural, optical and piezoelectric properties of BiFeO3 nanopowder synthesized from sol-gel. *Curr. Appl. Phys.*, 17, 409, 2017.
45. Saravanakumar, B. and Kim, S.J., Growth of 2D ZnO nanowall for energy harvesting application. *J. Phys. Chem. C*, 118, 8831, 2014.
46. Sankar ganesh, R., Navaneethan, M., Mani, G.K., Ponnusamy, S., Tsuchiya, K., Muthamizhchelvan, C., Kawasaki, S., Hayakawa, Y., Influence of Al doping on the structural, morphological, optical, and gas sensing properties of ZnO nanorods. *J. Alloys Compd.*, 698, 555, 2017.
47. Siddiqui, S., Kim, D.I., Duy, L.T., Nguyen, M.T., Muhammad, S., Yoon, W.S., Lee, N.E., High-performance flexible lead-free nanocomposite piezoelectric nanogenerator for biomechanical energy harvesting and storage. *Nano Energy*, 15, 177, 2015.
48. Zahid, M., Khalid, T., Rehan, Z.A., Javed, T., Akram, S., Rashid, A., Mustafa, S.K., Shabbir, R., Mora-Poblete, F., Asad, M.S., Liaquat, R., Hassan, M.M., Amin, M.A., Shakoor, H.A., Fabrication and characterization of sulfonated

graphene oxide (Sgo) doped pvdf nanocomposite membranes with improved anti-biofouling performance. *Membranes (Basel)*, 11, 749, 2021.

49. Tong, Z., Zhang, L., Liao, W., Wang, Y., Gao, Y., Extraction, identification and application of gliadin from gluten: Impact of pH on physicochemical properties of unloaded- and lutein-loaded gliadin nanoparticles. *Int. J. Biol. Macromol.*, 253, 126638, 2023.
50. Yang, Y., Yang, X., Tan, Y., Yuan, Q., Recent progress in flexible and wearable bio-electronics based on nanomaterials. *Nano Res.*, 10, 1560, 2017.
51. Aminullah, Kasi, A.K., Najma, B., Kasi, J.K., Rafique, S., Bokhari, M., Fabrication of piezoelectric nanogenerator using 3D-ZnO nanosheets and optimization of charge storage system. *Mater. Res. Bull.*, 123, 110711, 2020.
52. Sultana, A., Alam, M.M., Garain, S., Sinha, T.K., Middya, T.R., Mandal, D., An Effective Electrical Throughput from PANI Supplement ZnS Nanorods and PDMS-Based Flexible Piezoelectric Nanogenerator for Power up Portable Electronic Devices: An Alternative of MWCNT Filler. *ACS Appl. Mater. Interfaces*, 7, 19091, 2015.
53. Ravindran, P., Vidya, R., Kjekshus, A., Fjellvåg, H., Eriksson, O., Theoretical investigation of magnetoelectric behavior in BiFe O3. *Phys. Rev. B: Condens. Matter Mater. Phys.*, 74, 2006.
54. Cheng, K., Wallaert, S., Ardebili, H., Karim, A., Advanced triboelectric nanogenerators based on low-dimension carbon materials: A review. *Carbon N. Y.*, 194, 81, 2022.
55. Cafarelli, A., Marino, A., Vannozzi, L., Puigmartí-Luis, J., Pané, S., Ciofani, G., Ricotti, L., Piezoelectric Nanomaterials Activated by Ultrasound: The Pathway from Discovery to Future Clinical Adoption. *ACS Nano*, 15, 11066–11086, 2021.
56. Rovisco, A., Dos Santos, A., Cramer, T., Martins, J., Branquinho, R., Águas, H., Fraboni, B., Fortunato, E., Martins, R., Igreja, R., Barquinha, P., Piezoelectricity Enhancement of Nanogenerators Based on PDMS and ZnSnO3Nanowires through Microstructuration. *ACS Appl. Mater. Interfaces*, 12, 18421, 2020.
57. Sarkar, D., Das, N., Saikh, M.M., Biswas, P., Das, S., Das, S., Hoque, N.A., Basu, R., Development of a Sustainable and Biodegradable Sonchus asper Cotton Pappus Based Piezoelectric Nanogenerator for Instrument Vibration and Human Body Motion Sensing with Mechanical Energy Harvesting Applications. *ACS Omega*, 6, 28710, 2021.
58. Zhou, L., Liu, D., Ren, L., Xue, H., Li, B., Niu, S., Liu, Q., Han, Z., Ren, L., Reconfigurable Fiber Triboelectric Nanogenerator for Self-Powered Defect Detection. *ACS Nano*, 16, 7721, 2021.
59. Xi, Y., Song, J., Xu, S., Yang, R., Gao, Z., Hu, C., Wang, Z.L., Growth of ZnO nanotube arrays and nanotube based piezoelectric nanogenerators. *J. Mater. Chem.*, 19, 9260, 2009.
60. Wang, L., Fei, Z., Wu, Z., Ye, Y., Qi, Y., Wang, J., Zhao, L., Zhang, C., Zhang, Y., Qin, G., Jiang, Z., Maeda, R., Wearable bending wireless sensing with

autonomous wake-up by piezoelectric and triboelectric hybrid nanogenerator. *Nano Energy*, 112, 108504, 2023.

61. Yan, J., Liu, M., Jeong, Y.G., Kang, W., Li, L., Zhao, Y., Deng, N., Cheng, B., Yang, G., Performance enhancements in poly(vinylidene fluoride)-based piezoelectric nanogenerators for efficient energy harvesting. *Nano Energy*, 56, 662, 2019.
62. Badole, M., Vasavan, H.N., Saxena, S., Das, A.K., Srihari, V., Kumar, S., Piezoelectric properties and structural evolution in La- and Al-modified K0.5Bi0.5TiO3 ceramics. *J. Alloys Compd.*, 944, 169204, 2023.
63. Yue, R., Ramaraj, S.G., Liu, H., Elamaran, D., Elamaran, V., Gupta, V., Arya, S., Verma, S., Satapathi, S., Hayawaka, Y., Liu, X., A review of flexible lead-free piezoelectric energy harvester. *J. Alloys Compd.*, 918, 165653, 2022.
64. Rao, Z., Wang, X., Mao, S., Qin, J., Yang, Y., Liu, M., Ke, C., Zhao, Y., Sun, B., Flexible Memristor-Based Nanoelectronic Devices for Wearable Applications: A Review. *ACS Appl. Nano Mater.*, 6, 18645, 2023.

6

Nanoionics in Biomedical Applications: Diagnostic and Therapeutic Approaches

Tasnim Mahzabin Tanha[1], Md. Ahad Ali[1,2] and Md. Abu Bin Hasan Susan[1,3]*

[1]Department of Chemistry, University of Dhaka, Dhaka, Bangladesh
[2]Department of Chemistry, Jashore University of Science and Technology, Jashore, Bangladesh
[3]Dhaka University Nanotechnology Centre (DUNC), University of Dhaka, Dhaka, Bangladesh

Abstract

Nanoionics, the transport of ions within nanoscale dimensions, is catalyzing rapid transformation in biomedical science and medical technology. Exploration of the fascinating domain requires a thorough discussion of nanoionic materials, their contemporary applications, and integration into methods of medical diagnosis, treatments, and therapy of diseases. In the diagnostic sector, although integration of nanoionics is a novel approach and still in the rudimentary stage, it is showing promising potential for the detection and approximation of disease risk factors. Ability of detecting the levels of certain proteins, or lack thereof, enables nanoionics to be a useful diagnostic tool, implying profound opportunities for feasibility and efficiency. Therapeutic applications also await significant benefits from the incorporation of nanoionics. Tuneable multifunctionality, biocompatibility, and trans-membrane ion transport of biological nanoionics open avenues for treatment of various life-threatening diseases, as well as innovative applications in drug delivery and sensory devices. This chapter gives a summary of the various contributions of nanoionics, from its ongoing integration in modern medicines to its potential in enhancing various disease diagnoses. Advancing research of nanomedicine and nano ion transport promises a future where integration of the two will achieve revolutionary progress in the biomedical field.

Keywords: Nanoionics, fast-ion transport, diagnosis, bovine serum albumin, cancer therapy, trans-membrane ion transport, nanoscale

**Corresponding author*: susan@du.ac.bd

Inamuddin, Tariq Altalhi, Mohammad Luqman and Jorddy Neves Cruz (eds.) Nanoionics: Fundamentals and Applications, (133–160) © 2025 Scrivener Publishing LLC

6.1 Introduction to Nanoionics

Electrical and ionic characteristics of device components can be significantly improved by reducing the length to nanoscale [1, 2]. Significant fraction of grain boundaries in a substance differentiates nanostructured substances from normal polycrystalline materials. In nanostructured materials, surface properties begin to determine bulk properties, but ions further away from internal surfaces also encounter altered environments [3–6]. As a result, material characteristics that were previously unavailable to engineering have suddenly become amenable to human modification, allowing for customization of fundamental material specifications.

Nanoionics refers to the transfer of ions in nanoconfinement. The phrase or notion of nanoionics is not entirely new. In 1992, the concept and word "nanoionics" were first adopted [7]. At its inception, nanoionics was concerned with ion transport of only nanosolid systems. The notion of nanoionics must be way more generalized, due to its connection to ion transport and deposit in liquids. Current focus of materials choices is too narrowly focused on solid-state materials. A broader range of material choices is needed, including liquids, biological materials, and hybrids. Furthermore, nanoionics can be applied to many fields other than electrochemistry, from energy transfer to information transmission, both in artificial systems and biological bodies.

The strength of interfacial effects might cause anomalous instances in ion transport when compared to bulk systems in nanoscale space [8, 9]. Ion mobility may be restricted by spatial confinement, and/or their interactions with the nanoconfinement surface and each other may be altered. The interfaces are so tightly spaced in a solid-state nanoionic regime that they can become substantial, if not predominate, in their influence on ion transport [10]. Intermolecular interactions like hydrogen-bond, ion-dipole, and van der Waals interactions may take place if solvent molecules surround the ions [11]. Numerous processes, including supercapacitors, batteries, biochemistry, micro- and nano-fluidics, and membrane separation, depend heavily on these ion transport attributes.

6.2 Types of Nanoionics

Nanoionics is broadly termed as the study and application of all solid-state nanoscale systems containing fast ion transport processes. Nanoionic

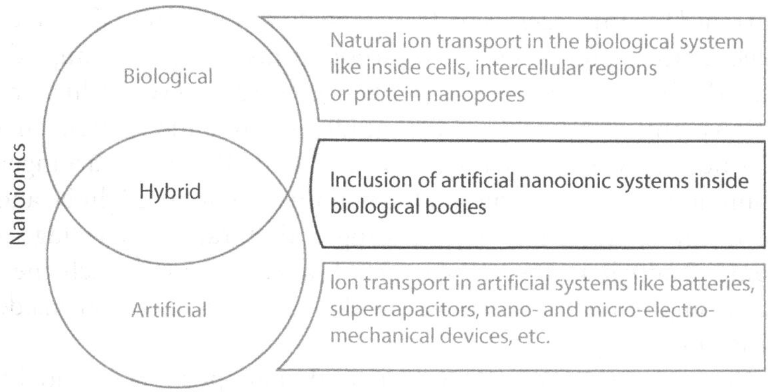

Figure 6.1 Types of nanoionics.

systems are broadly categorized into three types: biological nanoionics, artificial nanoionics, and biological-artificial hybrid nanoionics (Figure 6.1).

6.2.1 Biological Nanoionics

Biological nanoionics is an essential part for the living organism which occurs naturally inside biological bodies. Life depends on well-regulated ion transport throughout cells and organelles. Nanoionics are the fundamental biomolecular underpinning of all electrical events in initiating, processing, and transmitting information inside a biological brain system. Ion fluxes significantly control many processes in single cells, including the cell cycle, pH balance, activity potential, signaling, and volume adjustments. They also control the immune system, reproduction, muscular contraction, and production in higher-living species [12]. For instance, nanoionics normally regulates a cell volume precisely and tightly to maintain its average survival and performance. Volume control occurs in a single cell through a variety of mechanisms, such as active nanoionics that pump ions across a concentration gradient and membrane-bound ion channels that allow passive ion transfer between the internal cytosol and external fluid [13]. The electric eel, or *Electrophorus*, a freshwater fish native to the neotropics, is another such example. When assaulted by natural predators, they can produce as high as 600 V and 1 A. Specialized electrical organs in them are based on peristaltic channels. An individual charge-generating organ component may produce a tiny action voltage of around +65 mV. 1000 units coupled in parallel and series produce a concurrent elevated potential and current [14, 15]. Adenosine triphosphate (ATP) is transformed from solar energy through certain bacteriorhodopsins using ion pumps driven by light which causes protons (ions) to move from lower to higher concentrations. This electrical potential gradient drives the ATPase enzyme, which is necessary for the synthesis of ATP [16]. The spatial arrangement of calcium ions in the dendrites in the mammalian brain might be another crucial factor in information processing and storage. *Via* voltage-gated nanoionic channels in the cellular membrane, Ca^{2+} ions reach the dendrites, causing quick local changes in the Ca^{2+} concentration inside the dendritic branches [17].

Precise transport of ions through peptide nanopores is essential to all these vital information and energy functions. Two main categories of amino acid ion channels move ions across and out of cell membranes: ion pumps and channels. Ion channels stand for passive conduits allowing ion movements depending on electrical potential and concentration, while ion pumps actively move ions toward concentration gradients using energy sources like ATP [12].

Ion channels that often possess a specificity filter are pathways that allow the transport of a specific kind of ion. They may allow the passage of several ionic species under certain circumstances; for instance, specific ion channels gated by ligands enable the transport of Na^{+} and Ca^{2+} ions (Figure 6.2). Ion channel gates control the incontinent ion fluxes, allowing them to open only when necessary. Many biological processes depend on the consequent ionic current produced when charged ions pass through ion channels. Meanwhile, particular roles are played by various ion fluxes. For instance, the concentrations of Ca^{2+} ions themselves often convey information, while the K^{+} or Na^{+} ionic current fluxes function as physical signals. Cl^{-} ion flow current usually stabilizes membrane potentials in

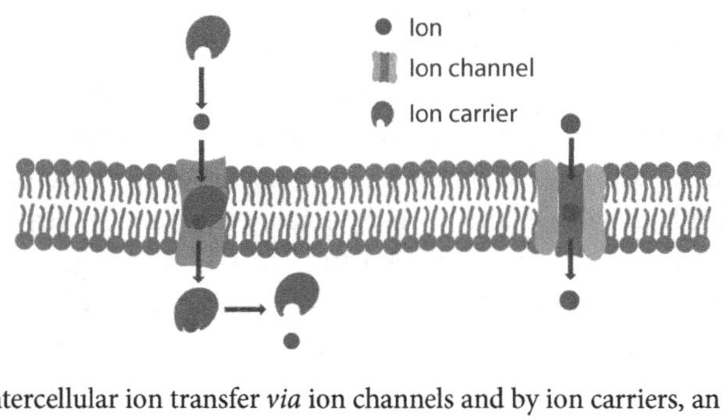

Figure 6.2 Intercellular ion transfer *via* ion channels and by ion carriers, an example of well-known biological nanoionic system.

skeletal muscle fibers and at inhibitory synapses. All protein transporters that can perform thermodynamic uphill transport—also known as "active transport"—are referred to as ion pumps. They are primary or secondary, depending on the energy source. Primary ion pump energy is derived from various processes, including light, redox reactions, and ATP hydrolysis. Ion pumps of the secondary type take advantage of the energy retained in other ions in the form of electrochemical gradients. They are sometimes referred to as co-transporters. Important insights into the activities of ion pumps and channels have been acquired from their contrasting behaviors, which include the regular integration of a reaction mechanism reminiscent to enzymes, movement of ions at low speed through pumps, and active thermodynamic ascending transport. On the other hand, particular surface characteristics and fine structure lead to carefully regulated functions of ion transport. The Streptomyces lividans' potassium filter contains two K^+ ions separated by approximately 7.5 Å and only one water molecule; the Na-K channel is nonselective and the filter allows passage of only one fully hydrated Na^+ ion; and in calmodulin, each calcium channel has the ability to bind two Ca^{2+} ions simultaneously [18].

In this context, ion transport-related activities of biological nanoionics have been addressed. This biological nanoionics, or energy systems related to biological ion transport and information processors, offer the perfect paradigm to address current energy and communication-related contemporary issues. It is infinitely possible to create artificial devices that mimic parts of these structures and functions, such as novel energy-efficient ion transport-based computer architectures instead of electron transport-based ones.

6.2.2 Artificial Nanoionics

As nanoscience and nanotechnology have advanced, artificial ion-transport devices have grown more integrated and smaller, going beyond biological systems. Many side effects accompany the shrinking sizes of the components of a device. Firstly, its contribution increases when the interface is connected to ion transport. "Nanosized" causes a notable rise of atoms at the phase boundary, which differ in coordination and electron state of the atoms at the bulk. As a consequence, rates of ion-diffusion quickly rise, and the diffusion flow volume grows [19]. Interactions like the van der Waals and hydrogen-bonding forces become important for a solid-liquid interface [20]. These variations also affect the spatial charge carrier distributions within the nanostructure, the ion-carrier phases, and the sorption behavior. Secondly, novel possibilities for the rearrangements of charge carriers under nanoconfinement are made possible by the modifying thermodynamics of the objects confined in the nano dimension. Thirdly, the interface (solid/liquid or solid/solid) energy levels will be changed. These phenomena produce the landmark of ion transport in nano systems, and ultimately dictates the applicability of the device.

6.2.3 Biological-Artificial Hybrid Nanoionics

As discussed for artificial and biological nanoionics, ions may be utilized as transport channels to create a variety of information and energy-processing systems, both *in vitro* and *in vivo*. The burgeoning question is whether a hybrid nanoionics system based on biology and artificial intelligence be fabricated or not. Since most human-machine interfaces (HMIs) rely on electron transport, HMIs constitute a typical system to exchange information [21, 22]. However, in systems of life, nerve cells interact with one another through the production and transfer of action potentials as well as ions and chemical substances known as neurotransmitters [23]. Consequently, the development of hybrid nanoionics based on biological and artificial principles and ion transport is imperative. In this scenario, direct neuromodulation can be achieved without the need for electrical to ionic transformation.

The majority of electronic platforms now in use function as automatic devices, meaning they can track cellular activity without offering any meaningful input [24, 25]. On the other hand, dynamic interactive management may be achieved by creating an active engagement with biological tissues, rather than only documenting biological processes. At this point, the artificial neural circuit must be able to close by producing outputs in addition to receiving biologically encoded inputs for processing.

Two distinct hybrid nanoionics systems are- signal input and signal output systems (Figure 6.3). First of all, an input signal comes from an outside stimulus that correlate to a biological sense system. Certain artificial nanoionics devices may be activated by these outside signals, releasing certain ions or molecules (neurotransmitters). When the voltage is higher than the threshold, an artificial ionic gateway that is voltage-gated, for instance, will unlock the gate for mass transit. An action potential will be produced post-synapse if the released neurotransmitter (such as dopamine) is able to activate biological nanopores, which will allow Na^+ or K^+ ions to pass across the transmembrane. This creates a hybrid nanoionics system that is biologically inspired and functions similarly to its biological equivalent. In a similar vein, biological–artificial hybrid nanoionics can likewise convert biological signals into artificial signals. The "read" and "write" of biological

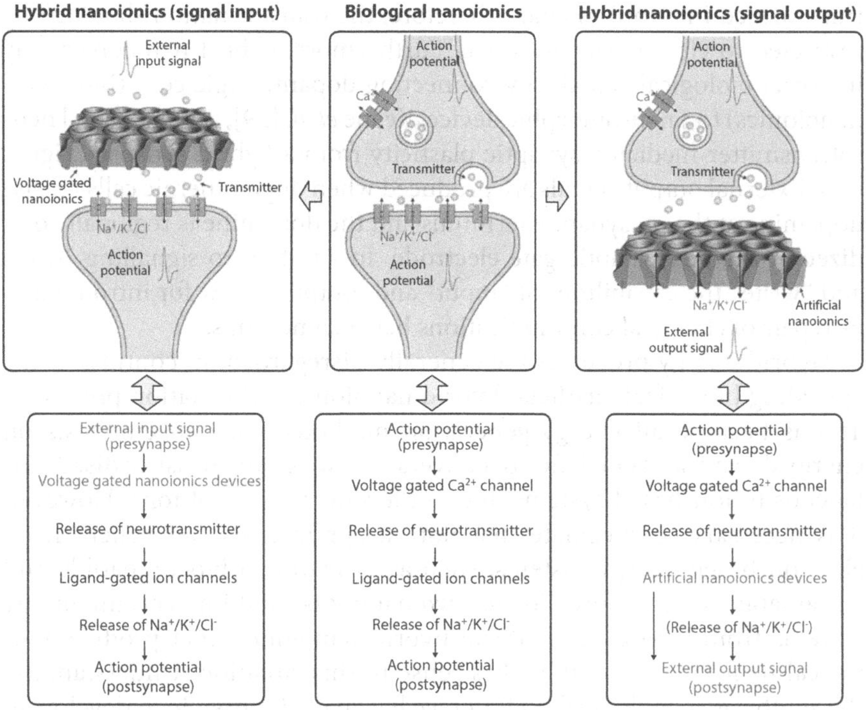

Figure 6.3 Illustration of the production of potentials for biological action in a network of neuron and biological-artificial hybrid nanoionics [130].

signals are represented by these two scenarios. We may presume that true "two-way" human-machine communication can be achieved by an ionic device that combines these two features.

Ion-selective membranes can convert electrical signals to ionic signals, activating and inhibiting nerves with ion concentrations modulated electrically *in situ* along a nerve. This is an example of how dynamic research results demonstrate that signal transduction between biological and artificial systems can be realized by biological–artificial hybrid nanoionics [26]. An everyday use of organic electronic ion pumps is the ion or molecule flow regulation for plant or cell stimulation, an additional artificial nanoionic device fabricated from an intrinsically conductive polymer [27, 28]. The ability of these hybrid nanoionics to convert electrical addressing impulses into brain-stem responses *via* ionic or molecular signaling represent a significant advancement in biology-technology interface.

By incorporating artificial nanoionics devices into biological systems, other environmental stimuli, including light, stress, and heat, may also be transformed into ionic signals; therefore, the transformation shouldn't be restricted to electric signals alone. Furthermore, hybrid nanoionics may interpret biological signals. By connecting dopaminergic cells (biological nanoionics) to a neuromorphic device, Keene *et al.* [29], accomplished neurotransmitter-mediated synaptic plasticity from a hybrid synapse (Figure 6.4). External output signals are produced when dopaminergic cells release dopamine at the presynaptic terminus and the dopamine is regionally oxidized at the postsynaptic gate electrode. In addition to signalling "read" and "write", the capabilities of "input" and "output" allow for introduction or repair of chemical communications between neurons.

Hybrid energy processes are generally disregarded, in contrast to the appealing biological–artificial hybrid nanoionics information processes. This is because all energy-generation methods place an emphasis on energy density and efficiency of conversion, two characteristics disadvantageous to ion-based systems due to the slow motion of ions. However, it is necessary to speculate that not all applications can benefit from electron-based energy systems, such as providing a biocompatible and sustainable energy source for *in-vivo* nanobots. Building certain energy systems from biological–artificial hybrid nanoionics that produce biological nanoionics energy makes sense in this situation. Understanding the synthesis of biological ATP *in vivo* has made it possible to develop an energy source based on hybrid nanoionics that may cause ion transport to go axially, producing the concentration gradient needed for a molecular motor to produce ATP [30].

The slow movement of ions, particularly in the sub-nanoscale region, has recently been questioned [31]. Superfast ion movement is conceivable in such a small space, for example, using Newton's ionic cradle model [32]. As a result, the energy density and conversion efficiency of the ionic energy system might be raised. However, these ideas need to be validated before their application can be applied to problems with nanomaterials and nanostructure.

6.3 General Applications of Nanoionics

Nanoionics is being used in a diverse field of applications, from energy storage to building electronic devices. The ultra-small size and ion transport properties enable nanoionics with a unique set of characteristics, making them useful in electronic, medical, and energy sectors alike. Utilizing ion transfer, nanoionics devices enable restructuring at the atomic level, unlike the traditional semiconductors in nanoelectronics; this is termed

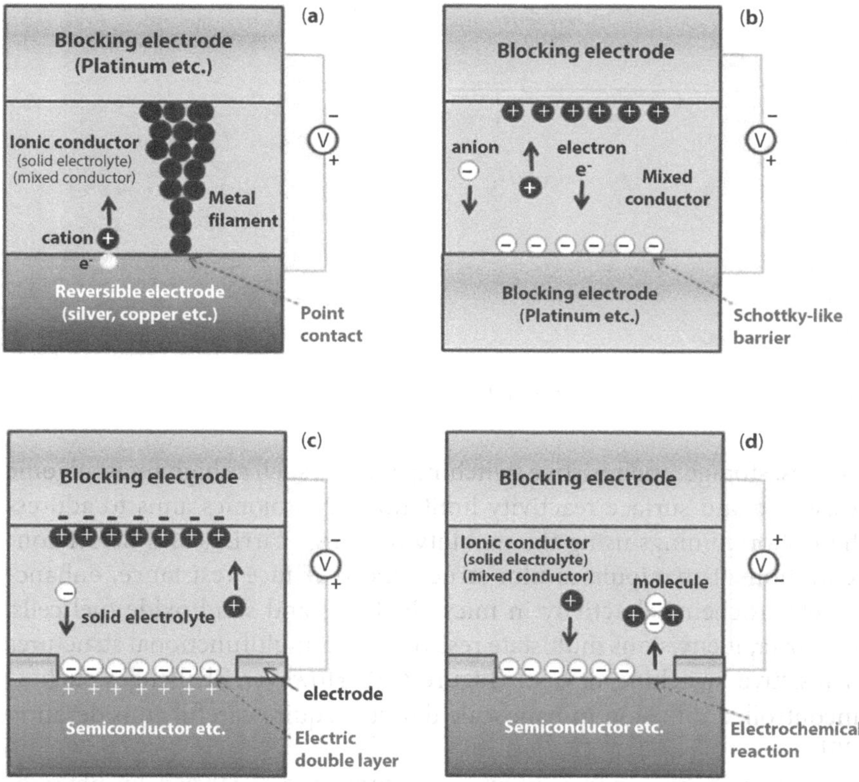

Figure 6.4 Local ion transfer control and electrochemical reactions in a nanostructure enabling multifunctionality, by different methods [33].

as "nano architectonics." This ability allows them adaptable structure and functionality (Figure 6.4), with possible applications in self-learning, artificial intelligence, and neural functions. These devices are promising for brain-inspired computing, such as artificial synapses and decision-making systems [33].

Yamaguchi *et al.* highlighted that gas adsorption at the interface is affected by increased surface charge density, explored chemical functions at interfaces, and proposed applications of nanoionics. This work discusses potential applications of nanoionics in devices like energy conversion, electrochemical, and sensing devices, suggesting prospects for advancements in nanoscale fabrication technologies, solid-state ionics, and defect chemistry [34].

Functional materials with nanoionics effects were also integrated into real devices with metal-oxide thin films highlighting challenges in energy

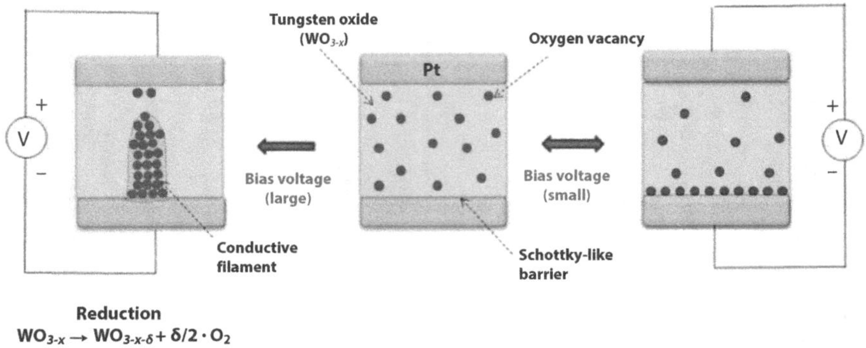

Figure 6.5 Multifunctional ion transport device [33].

sources, storage, and resistive switching devices, such as high internal ionic resistance and surface reactivity limitations. Nanoionics aims to address these shortcomings using the mobility of charge carriers and defect concentration. It manipulates film structures to reduce resistance, enhancing electrochemical activity in micro batteries and solid oxide fuel cells. Moreover, it envisions multistate resistivity and multifunctional structures in resistive switching devices (Figure 6.5). However, challenges such as uncontrolled diffusion in nanoscale devices require careful consideration [35].

Fabrication of nanoionics supercapacitors was catalyzed by physicochemical nanosystems, Ag(Cu)I-M, exhibiting transition and rare-earth metals in high concentrations. This fueled the growth of high-capacity sub-voltage devices, suggesting promising advancements in energy storage, battery technology, and computing devices [36].

Impacts of size effects on materials have also been studied. These studies explored the limitations of downsizing interfacially dominated mesoscopic systems and the transition to atomistic-scale heterogeneities, holding potential for groundbreaking insights and bridging nanoscience with structural chemistry in understanding nano heterogeneity and structural complexity [37].

Nanoionics also has potentials in developing room-temperature devices and self-sustaining structures. Nanoionics has been compared to nanoelectronics and the reliance of ion transport of electrochemical devices has been reported. The influence of size effects and interfaces on storage and transport implied novel perspectives of nanoionics in room-temperature devices and self-sustaining structures (Figure 6.6) [38].

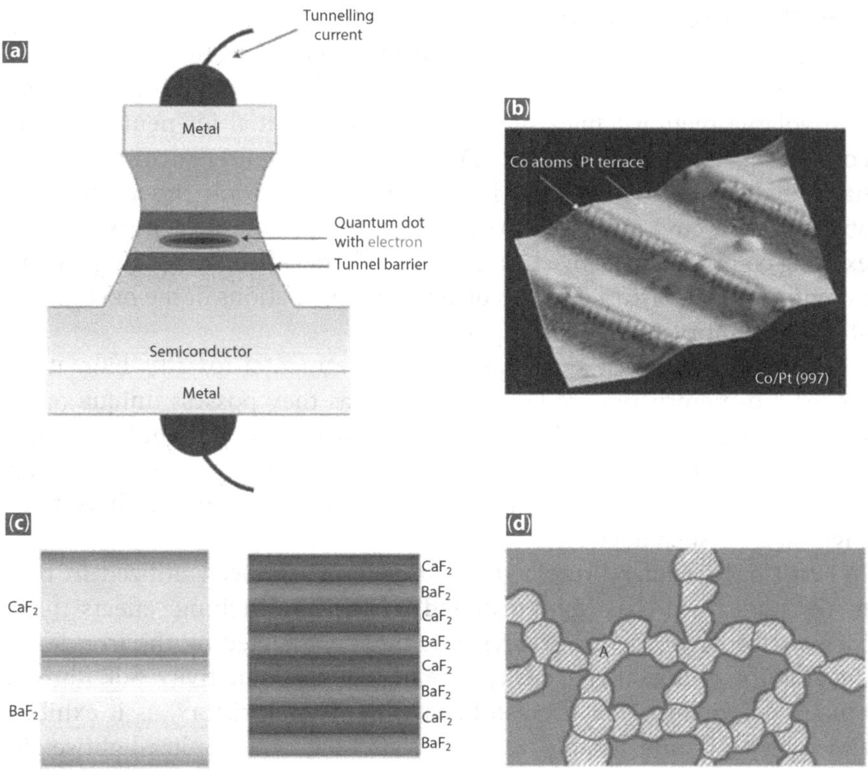

Figure 6.6 Interfacially controlled systems: (a) nanoelectronics, (b) nanomagnet, (c) artificial ion conductor, and (d) ion conductor comprising insulator A in an ion-conducting matrix in nanoionics [38].

6.4 Applications of Nanoionics in Diagnosis

While nanotechnology is a widely appreciated branch in the medical field, nanoionics is still a developing sector. The unique ion transport phenomena in nanoionics can be of much importance in biomedical applications. As of now, nanoionics-based memristors are being explored as a detecting tool for important proteins.

Memristors are non-volatile memory devices containing two-terminals. These memory resistors differ from traditional ones because they can "remember" the amount of charge having flowed through them and retain this memory even when the power is turned off, making memristors an attractive option for emerging memory and computing technologies [39]. Since the first physical device was developed in the Hewlett-Packard (HP)

Labs, they have gained attention for their diverse areas, such as memory storage, neuromorphic computing, and advanced electronic circuits. Some important characteristics of memristors include: resistance variation, non-volatile memory, high switching speed, potential for neuromorphic computing, and simple design. Despite great promise, memristors still have challenges to overcome, like complex fabrication processes, variability issues, environmental toxicity, and integration of memristors into existing computing designs. Current research focuses on unlocking the full potential of memristors and exploring their applications in the next generation of memory and computing devices.

To address complex synthesis and environmental toxicity, biomaterials for bio-memristors are being explored, as they possess unique natural structures, mechanical superiority, environmental friendliness, and biocompatibility. Owing to their natural abundance, simple fabrication processes and low cost, ferritin, egg albumen, and silk fibroin have been suggested for such devices.

Ferritin, a cellular protein that stores iron, has been utilized in bio-memristors, and has demonstrated resistive switching effects [40]. Memristive performance of ferritin is directly affected by the iron loading amount, and its high-density core improves conductivity. Silk fibroin, a natural protein fiber, has also been used in memristors, as it exhibits reversible and non-volatile resistive switching. These ultra-lightweight memristor devices showcase flexibility and biocompatibility [41, 42].

Egg albumen, a protein obtained directly from fresh eggs, has also been contemplated for memristor devices. The proposed device offers a simple fabrication process and reliable switching properties. Protein composition of albumen provides resistive switching behaviors, and its proton conductivity has been used in synaptic devices [43].

Detection of bovine serum albumin (BSA) utilizing hybrid devices based on TiO_2 in combination with graphene oxide is one significant diagnostic application of nanoionics [44]. Accurate detection and quantification of bio-molecules are vital in medical diagnostics. In such regard, memristors offer a promising method for identifying biomolecules by exploiting their varying resistance state. Devices such as resistive random-access memory and switching [45, 46] are electronic in nature opposite to memory [47], logic [48], and sensing [49]. Functionality of a memristive device is accredited to adapting states. Especially for sensory applications, nano-sensing has shown great interest since it offers the potential for real-time monitoring, personalized therapy, and minimal invasion.

Migration of oxygen ions in resistive switching (RS) memristors modifies the restive properties non-volatilely. In this context, TiO_2 and TiO_2 + graphene oxide-based memristors are explored for detecting BSA. Besides targeted delivery of drugs [50, 51], BSA binds with a good number of drugs, inorganic and organic species, radicals, and metal ions. It also finds applications for immunosorbent assay, immunohistochemistry, and immunoblots. Various diseases in cattle families may be caused by low levels of BSA, making quantification of BSA vital for understanding the probability and severity of certain diseases. A couple of methods of BSA detection are- triangular silver nanoplates using spectrophotometric method [52], and Mach-Zehnder interferometer based optical fiber. But these technologies are cumbersome and labor intensive. Thus, TiO_2 based RS memristor with high sensitivity and low voltage of operation has been explored [53].

When BSA is added to a TiO_2 RS memristor, its primary surface groups i.e. $Ti_2 = O-$, Ti-OH, and $Ti_2 = OH$ interact with different binding sites of BSA including $R\text{-}NH_3^+$ and R-COO− to cause significant modifications in the transport and optical effects of the memristor. Some electrostatic interactions are-

$$R\text{-}NH_3^+ + Ti\text{-}OH \rightarrow Ti\text{-}OH_2^+ : NH_2\text{-}R$$

$$R\text{-}COO^- + Ti\text{-}OH \rightarrow Ti\text{-}OH \ldots {}^-OOC\text{-}R$$

$$R\text{-}NH_3^+ + Ti_2 = O^- \rightarrow Ti_2 = O^- \, {}^+NH_3\text{-}R$$

This process is sensitive up to 4 mg/mL. These devices exhibit great bipolar resistive switching with convincing high precisions. Increased performance in the graphene oxide device is accredited to the prevention of multi-dimensional and randomly grown conductive paths. Despite limitations such as the use of toxic solvents in processing and limitations in forming micro/nanostructures, these are promising for development of emerging resistive random-access memory devices to detect biomolecules. These mechanisms can be further probed for the diagnosis of other proteins, enzymes, hormones, etc. as well as for diagnosis of certain abnormalities and diseases in biological bodies.

6.5 Applications of Nanoionics in Therapeutics

In drug discovery, artificial transmembrane nanoionics are appealing because, theoretically, they could be employed as novel therapeutics by initiating an antibiotic response by alleviating intracellular pH gradients, disrupting cellular ion homeostasis, or promoting apoptosis (programmed cell death). Artificial nanoionics involving selective ion sensing and transport has been developed. Nevertheless, there has yet to be much success in converting basic transport-process studies into *in vitro* functional systems. Several artificial transmembrane nanoionics have been produced to operate *via* channel or mobile carrier processes. The selective identification and transmembrane transport of critical ions have been made more feasible with structural modification and either distinct chemical substances or self-assembled nanomaterials.

Large proteins that create pores through lipid membranes make up natural ion channels, which can be categorized according to the mechanism of gating they use (e.g., activation mediated by voltage [54], ligand [55], mechanical stimuli [56, 57], light [58], as well as the ion selectivity (e.g., Na^+, Cl^-) [59]. Their involvement in several physiological and pathological processes has led to a growing interest in therapeutic intervention [60]. Surprisingly, 19% of medications target gate-activated and ligand-activated ion channel proteins [61]. Researches on ion channel modification have gained popularity and now includes both synthetic compounds [62] and biomolecules, including lipids [63–65], toxins [66], antibodies and nanobodies [67], and peptides derived from venom [68, 69]. Ion channel malfunction is associated with a number of pathologies that are collectively referred to as channelopathies (Figure 6.7). These include a wide range of diverse disorders that affect the immune response, the skeletal muscle, the kidneys, and the heart, among many other tissues [70], the central nervous system, and the kidneys, among many others. There are other causes of channel failure, including post-translational abnormalities [71, 72] and mutations [73–75].

Ion homeostasis is partially regulated by a family of molecules known as nanoionics transporters. They can pass across cell membranes and are mobile. They momentarily bind ions to facilitate ion passage across the lipid barrier (Figure 6.7). These include synthetic and natural biomolecules (such as polyethers and crown ethers) that have been suggested as stimulants and anticancer agents [76, 77]. Examples of the former are peptides and antibiotics. They have the further benefit of immunomodulation action [78] and have been suggested as antimicrobial agents [78–81]. Unfortunately, unresolved issues with formulation and targeting have made their successful use in therapy difficult, leading to less-than-ideal outcomes and adverse consequences. Their targeted activation at the diseased location may be a viable remedy [82].

Channel blockers serve in treatment to address a wide range of illnesses. One well-known example of an anti-hypertensive medication is a specific

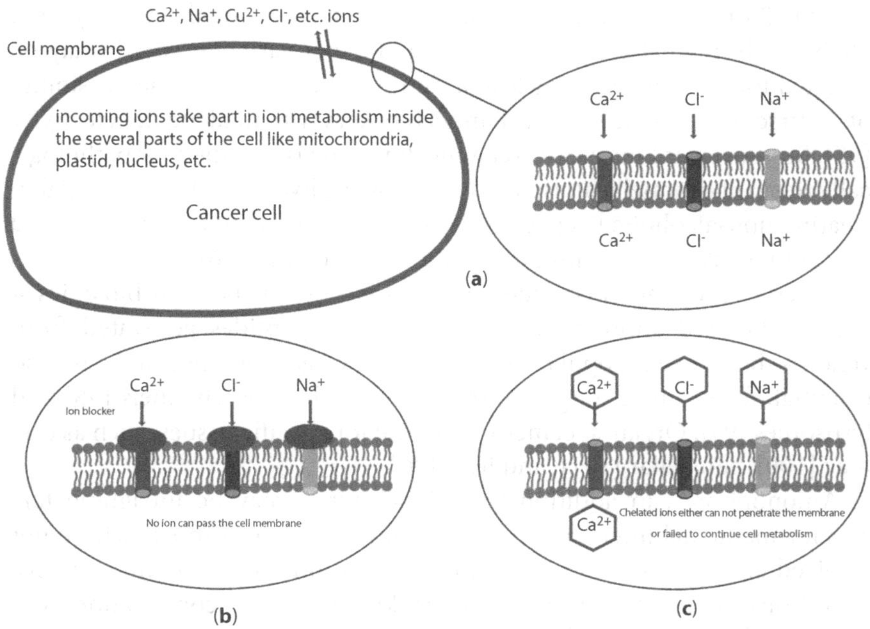

Figure 6.7 (a) Nanoionic transport of ions inside cancer cell; Mode of action of (b) ion channel blocker and (c) chelation of ions for cancer therapy.

Ca^{2+}-channel blocker [83]. Several ion transporter types are being investigated to treat ischemic stroke [84]. Transient receptor potential channels (TRP channels), for instance, serve as an inhibitory target for treatment of neurodegenerative diseases, depression, and epilepsy in addition to stroke [85]. Indeed, Ca^{2+}-channel blockers, used to treat hypertension, have a neuroprotective impact against depression and Alzheimer's disease [86]. Targeting dysregulation of Ca^{2+} homeostasis might be a potential treatment for neurodegeneration [87, 88]. There is also the use of Na^{+} and Ca^{2+} channel blockers to treat bipolar disorders, besides ion-related therapy involving administration of lithium salts which is effective for some patients [89]. Cellular hyperactivity is attenuated by Ca^{2+} channel blockers, which finds their use in psychiatry for the treatment of a variety of disorders [90]. Na-K-Cl co-transporters have the potential as therapeutic targets for a variety of ailments, such as hypertension, pain, epileptic attacks, and brain edema [91]. Moreover, chronic and neurogenic pain are treated with Ca^{2+} and Na^{+} channel blockers [92]. Ion channels based on acid-sensing may also be a useful therapeutic target for migraine [93].

Type-2 diabetes may have an ion channel dysfunctional aetiology [94]. Diabetes dysregulates a number of ion channel varieties, which may be targeted for therapy; nevertheless, a significant drawback is the possibility of adverse consequences from channel blockers with inadequate selectivity [95]. Moreover, as long as Ca^{2+}-channel targeting is specified for pathologically relevant channel isoforms produced by the liver, could be beneficial in treating non-alcoholic liver steatosis, another significant pathology linked to insulin resistance and influenced Ca^{2+} homeostasis [96].

When everything is considered, selectivity is crucial, and biologics—such as tailored antibodies, nanobodies, and peptides generated from venom—have been popular recently [97]. Organelle specificity is also essential, for example, targeting the mitochondria ion channels [98] and lysosomes [99]. Organ-specificity is critical for specific tissues, such as cardiac [100], cerebellar [101], and liver [96].

Anionophores, in addition to cationophores, may be intriguing targets for medicinal intervention. Specifically, Cl^- ion transporters have not received enough attention in treating several illnesses, including dry eye conditions, hypertension, osteoporosis, kidney stones, constipation, and secretory diarrhoeas [102].

6.5.1 Nanoionics in Cancer Therapy

Significant advancement has been achieved regarding the scientific knowledge of cancer during the last few decades. These days, the preferred techniques are often systemic treatments, such as anticancer medications. Alternate strategy might be the use of artificial transmembrane nanoionics transporters, that can induce apoptosis by interfering with intracellular ion homeostasis, otherwise altering the intracellular pH [103–105]. Furthermore, it is feasible to successfully combat medication resistance through the integration of nanoionics with other treatment techniques [106].

Nanoionics facilitates selective and controlled exchange of transmembrane ions across bilayers of lipid, which is essential for many biological functions. Transmembrane ion channels are essential in the genesis of cancer and for cell proliferation. This was first demonstrated in the instance of potassium channels, even though it is recommended for other channels including cations and chloride [107]. Development of artificial nanoionics that can imitate and better comprehend the roles of natural ion channels has thus received a surge of interest. In practice, numerous artificial nanoionics have been constructed to facilitate transit of certain ions, including K^+, Na^+, and Cl^- through the membrane [108–111].

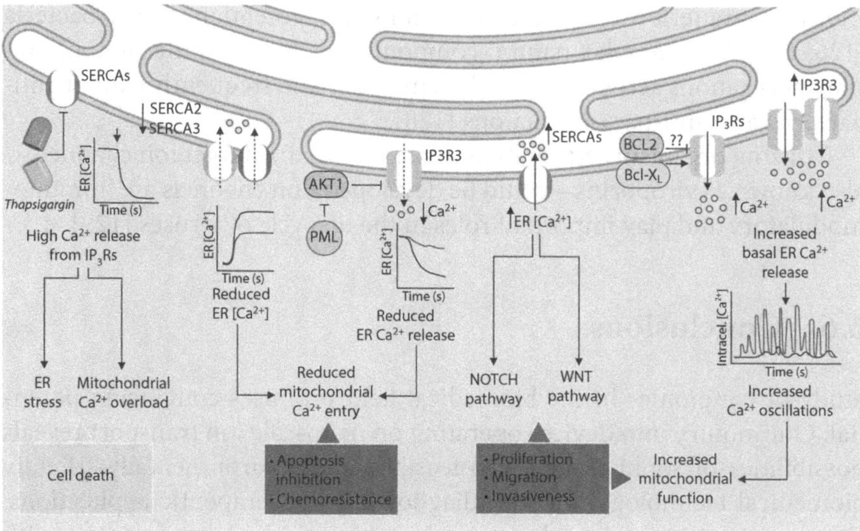

Figure 6.8 Redox homeostasis and Ca^{2+} signalling inside endoplasmic reticulum [123].

Nanoionics have also been made exploiting supramolecular interactions [112–114]. Synthetic ion transporters have often included traditional ion-binding motifs [115–118]. Squaramides, pillararenes, calixarenes, cucurbiturils, and calixpyrroles are only a few of the acyclic and macrocyclic hosts on which unimolecular transmembrane channels along with carriers have been explored [119–122].

Several diseases, including cancer, have been related to changes in the proteins and ion channels found in mitochondrial-associated endoplasmic reticulum membranes, which regulate important physiological functions including redox homeostasis and Ca^{2+} signalling (Figure 6.8) [123]. Similarly, lysosomes are underappreciated Ca^{2+} level regulators that may provide intriguing targets for anticancer drugs [123]. A number of ion channel blockers are under consideration for therapeutic management of cancer metastases [124]. It is crucial to have a thorough understanding of how each kind of ion channel expresses itself in order to design therapies that are both successful and low in adverse effects [125].

6.5.2 Nanoionics as Antibiotics

Ion homeostasis disruption has therapeutic advantages outside of cancer treatment. Additionally, bacteria can be good targets. Because of their ability to disrupt bacterial membranes and induce an ion imbalance, artificial transmembrane ion transporters demonstrate potential against bacteria [126, 127]. The complex natural compounds called polyether ionophores transport cations across biological membranes and frequently exhibit anti-bacterial and anti-parasitic actions [128].

Utilizing organism selectivity, antiviral drugs that target ion channels—also known as viroporins—could be developed. Ion channels are immunomodulatory and play important roles in the lifecycle of viruses [129].

6.6 Conclusions

Study of nanoionics in the biomedical field promises convincing potential. Our inquiry into devices operating on nanoscale ion transport reveals possibilities of sophisticated, eco-feasible, and environmentally friendly biomedical technology for both diagnostic and therapeutic applications. Converging solid and liquid-state ionics within the biological systems with artificial nanoionics creates scope for pushing the boundaries in both these sectors.

The rudimentary comprehension of ion dynamics within the nanostructures is facilitating the evolution of novel electronic components and sensory technologies. Nanoionics has transpired as a key player in the pursuit of convenient, compatible, and friendly devices. The potential for developing complex synaptic systems using nanoionics has great promise of imitating the human brain, opening up possibilities of treating many diseases. Especially, therapeutic applications wait to benefit significantly from the incorporation of nanoionics. Transmembrane ion transfer of nanoionics could bring about revolutionary progress in cancer treatment. If drug resistant bacteria respond to the incorporation of artificial nanoionics, the issue of superbugs could be addressed. Creating non-toxic, biocompatible, and active devices have the potential to be applied in other biomedical fields. Incorporating nanoionics in traditional diagnostic and therapeutic methods can make them faster, more precise, and targeted, besides addressing multiple issues in the contemporary techniques. Despite the optimism surrounding nanoionics, multiple challenges still remain. As this field is relatively new, the long-term impacts of nano-scale transport devices in biological systems are still unknown. Extensive research is needed to ensure safety, longevity, and effectiveness of these technologies. The dynamic journey of applying nanoionics in medical diagnosis and therapy presents ample opportunities and challenges, paving the way for a future where these technologies advance further into nanoionics frontier.

References

1. Szostak, J.W., Bartel, D.P., Luisi, P.L., Synthesizing life. *Nature*, 409, 387–390, 2001.
2. Dubyak, G.R., Ion homeostasis, channels, and transporters: An update on cellular mechanisms. *Am. J. Physiol. Adv. Physiol. Educ.*, 28, 143–154, 2004.
3. Bröer, S., Amino acid transport across mammalian intestinal and renal epithelia. *Physiol. Rev.*, 88, 249–286, 2008.
4. Gamper, N., Shapiro, M.S., Regulation of ion transport proteins by membrane phosphoinositides. *Nat. Rev. Neurosci.*, 8, 921–934, 2007.
5. Wang, X., Zhang, X., Dong, X.P., Samie, M., Li, X., Cheng, X., Goschka, A., Shen, D., Zhou, Y., Harlow, J., Zhu, M.X., Clapham, D.E., Ren, D., Xu, H., TPC proteins are phosphoinositide- Activated sodium-selective ion channels in endosomes and lysosomes. *Cell*, 151, 372–383, 2012.
6. Guo, J., Zeng, W., Chen, Q., Lee, C., Chen, L., Yang, Y., Cang, C., Ren, D., Jiang, Y., Structure of the voltage-gated two-pore channel TPC1 from Arabidopsis thaliana. *Nature*, 531, 196–201, 2016.
7. Despotuli, A.L., Nikolaichik, V.I., A step towards nanoionics. *Solid State Ion.*, 60, 275–278, 1993.
8. Zhan, H., Xiong, Z., Cheng, C., Liang, Q., Liu, J.Z., Li, D., Solvation-Involved Nanoionics: New Opportunities from 2D Nanomaterial Laminar Membranes. *Adv. Mater.*, 32, 1904562, 2020.
9. Xiao, K., Jiang, L., Antonietti, M., Ion Transport in Nanofluidic Devices for Energy Harvesting. *Joule*, 3, 2364–2380, 2019.
10. Maier, J., Ionic transport in nano-sized systems. *Solid State Ion.*, 175, 7–12, 2004.
11. Sparreboom, W., Van Den Berg, A., Eijkel, J.C.T., Principles and applications of nanofluidic transport. *Nat. Nanotechnol.*, 4, 713–720, 2009.
12. Gadsby, D.C., Ion channels versus ion pumps: The principal difference, in principle. *Nat. Rev. Mol. Cell Biol.*, 10, 344–352, 2009.
13. McEvoy, E., Han, Y.L., Guo, M., Shenoy, V.B., Gap junctions amplify spatial variations in cell volume in proliferating tumor spheroids. *Nat. Commun.*, 11, 6148, 2020.
14. Xu, J., Lavan, D.A., Designing artificial cells to harness the biological ion concentration gradient. *Nat. Nanotechnol.*, 3, 666–670, 2008.
15. Ramos, I.N.F., da Silva, M.F., Lopes, J.M.S., Cruz, J.N., Alves, F.S., do Rego, J.A.R., da Costa, M.L., de Assumpção, P.P., Barros Brasil, D.S., Khayat, A.S., Extraction, Characterization, and Evaluation of the Cytotoxic Activity of Piperine in Its Isolated form and in Combination with Chemotherapeutics against Gastric Cancer. *Molecules*, 28, 5587, 2023.
16. Inoue, K., Ono, H., Abe-Yoshizumi, R., Yoshizawa, S., Ito, H., Kogure, K., Kandori, H., A light-driven sodium ion pump in marine bacteria. *Nat. Commun.*, 4, 1678, 2013.

17. Catterall, W.A., Voltage-gated calcium channels. *Cold Spring Harbor Perspect. Biol.*, 3, a003947, 2011.
18. Doyle, D.A., Cabral, J.M., Pfuetzner, R.A., Kuo, A., Gulbis, J.M., Cohen, S.L., Chait, B.T., MacKinnon, R., The structure of the potassium channel: Molecular basis of K+ conduction and selectivity. *Science (80-.)*, 280, 69–77, 1998.
19. Wagemaker, M., Mulder, F.M., Van Der Ven, A., The role of surface and interface energy on phase stability of nanosized insertion compounds. *Adv. Mater.*, 21, 2703–2709, 2009.
20. Nouri, R., Guan W., Nanofluidic Charged-Coupled Devices for Controlled DNA Transport and Separation. *Nanotechnology*, 32, 345501, 2021.
21. Song, J.K., Do, K., Koo, J.H., Son, D., Kim, D.H., Nanomaterials-based flexible and stretchable bioelectronics. *MRS Bull.*, 44, 643–664, 2019.
22. Jo, Y.J., Ok, J., Kim, S.Y., Kim, T.I., Stretchable and Soft Organic–Ionic Devices for Body-Integrated Electronic Systems. *Adv. Mater. Technol.*, 7, 2001273, 2022.
23. Toyota, M., Spencer, D., Sawai-Toyota, S., Jiaqi, W., Zhang, T., Koo, A.J., Howe, G.A., Gilroy, S., Glutamate triggers long-distance, calcium-based plant defense signaling. *Science (80-.)*, 361, 1112–1115, 2018.
24. Yang, X., Zhou, T., Zwang, T.J., Hong, G., Zhao, Y., Viveros, R.D., Fu, T.M., Gao, T., Lieber, C.M., Bioinspired neuron-like electronics. *Nat. Mater.*, 18, 510–517, 2019.
25. Feiner, R., Engel, L., Fleischer, S., Malki, M., Gal, I., Shapira, A., Shacham-Diamand, Y., Dvir, T., Engineered hybrid cardiac patches with multifunctional electronics for online monitoring and regulation of tissue function. *Nat. Mater.*, 15, 679–685, 2016.
26. Song, Y.A., Melik, R., Rabie, A.N., Ibrahim, A.M.S., Moses, D., Tan, A., Han, J., Lin, S.J., Electrochemical activation and inhibition of neuromuscular systems through modulation of ion concentrations with ion-selective membranes. *Nat. Mater.*, 10, 980–986, 2011.
27. Isaksson, J., Kjäll, P., Nilsson, D., Robinson, N., Berggren, M., Richter-Dahlfors, A., Electronic control of Ca2+ signalling in neuronal cells using an organic electronic ion pump. *Nat. Mater.*, 6, 673–679, 2007.
28. Williamson, A., Rivnay, J., Kergoat, L., Jonsson, A., Inal, S., Uguz, I., Ferro, M., Ivanov, A., Sjöström, T.A., Simon, D.T., Berggren, M., Malliaras, G.G., Bernard, C., Controlling epileptiform activity with organic electronic ion pumps. *Adv. Mater.*, 27, 3138–3144, 2015.
29. Keene, S.T., Lubrano, C., Kazemzadeh, S., Melianas, A., Tuchman, Y., Polino, G., Scognamiglio, P., Cinà, L., Salleo, A., van de Burgt, Y., Santoro, F., A biohybrid synapse with neurotransmitter-mediated plasticity. *Nat. Mater.*, 19, 969–973, 2020.
30. Dong, H., Nie, R., Hou, X., Wang, P., Yue, J., Jiang, L., Assembly of F0F1-ATPase into solid state nanoporous membrane. *Chem. Commun.*, 47, 3102–3104, 2011.

31. Chen, L., Tu, B., Lu, X., Li, F., Jiang, L., Antonietti, M., Xiao, K., Unidirectional ion transport in nanoporous carbon membranes with a hierarchical pore architecture. *Nat. Commun.*, 12, 4650, 2021.
32. Hinds, B.J., Engineering small-ion transporter channels. *Science (80-.)*, 372, 459–460, 2021.
33. Terabe, K., Tsuchiya, T., Yang, R., Aono, M., Nanoionic devices enabling a multitude of new features. *Nanoscale*, 8, 13873–13879, 2016.
34. Yamaguchi, S., Nanoionics-Present and future prospects. *Sci. Technol. Adv. Mater.*, 8, 503, 2007.
35. Chiabrera, F., Garbayo, I., Tarancón, A., Nanoionics and interfaces for energy and information technologies, in: *Met. Oxide-Based Thin Film Structures*, Elsevier, 409–439, 2018.
36. Despotuli, A.L., Andreeva, A.V., Nanoionics: New materials and supercapacitors. *Nanotechnol. Russ.*, 5, 506–520, 2010.
37. Maier, J., Pushing nanoionics to the limits: Charge carrier chemistry in extremely small systems. *Chem. Mater.*, 26, 348–360, 2014.
38. Maier, J., Nanoionics: Ion transport and electrochemical storage in confined systems. *Nat. Mater.*, 4, 805–815, 2005.
39. Chua, L.O., Memristor—The Missing Circuit Element. *IEEE Trans. Circuit Theory*, 18, 507–519, 1971.
40. Zhang, C., Shang, J., Xue, W., Tan, H., Pan, L., Yang, X., Guo, S., Hao, J., Liu, G., Li, R.W., Convertible resistive switching characteristics between memory switching and threshold switching in a single ferritin-based memristor. *Chem. Commun.*, 52, 4828–4831, 2016.
41. Hota, M.K., Bera, M.K., Kundu, B., Kundu, S.C., Maiti, C.K., A natural silk fibroin protein-based transparent bio-memristor. *Adv. Funct. Mater.*, 22, 4493–4499, 2012.
42. Wang, H., Meng, F., Zhu, B., Leow, W.R., Liu, Y., Chen, X., Resistive Switching Memory Devices Based on Proteins. *Adv. Mater.*, 27, 7670–7676, 2015.
43. Chen, Y.C., Yu, H.C., Huang, C.Y., Chung, W.L., Wu, S.L., Su, Y.K., Nonvolatile bio-memristor fabricated with egg albumen film. *Sci. Rep.*, 5, 10022, 2015.
44. Sahu, D.P. and Jammalamadaka, S.N., Detection of bovine serum albumin using hybrid TiO^2 + graphene oxide based Bio - resistive random access memory device. *Sci. Rep.*, 9, 16141, 2019.
45. Carrara, S., Sacchetto, D., Doucey, M.A., Baj-Rossi, C., De Micheli, G., Leblebici, Y., Memristive-biosensors: A new detection method by using nanofabricated memristors. *Sens. Actuators, B*, 171–172, 449–457, 2012.
46. Tzouvadaki, I., Parrozzani, C., Gallotta, A., De Micheli, G., Carrara, S., Memristive Biosensors for PSA-IgM Detection. *Bionanoscience*, 5, 189–195, 2015.
47. Pavan, P., Bez, R., Olivo, P., Zanoni, E., Flash memory cells-an overview. *Proc. IEEE*, 85, 1248–1271, 1997.
48. Duan, X., Huang, Y., Lieber, C.M., Nonvolatile Memory and Programmable Logic from Molecule-Gated Nanowires. *Nano Lett.*, 2, 487–490, 2002.

49. Hu, J., Liu, S., Responsive polymers for detection and sensing applications: Current status and future developments. *Macromolecules*, 43, 8315–8330, 2010.
50. An, F.F., Zhang, X.H., Strategies for preparing albumin-based nanoparticles for multifunctional bioimaging and drug delivery. *Theranostics*, 7, 3667–3689, 2017.
51. Xie, L., Tong, W., Yu, D., Xu, J., Li, J., Gao, C., Bovine serum albumin nanoparticles modified with multilayers and aptamers for pH-responsive and targeted anti-cancer drug delivery. *J. Mater. Chem.*, 22, 6053–6060, 2012.
52. Zhang, L.L., Ma, F.F., Kuang, Y.F., Cheng, S., Long, Y.F., Xiao, Q.G., Highly sensitive detection of bovine serum albumin based on the aggregation of triangular silver nanoplates. *Spectrochim. Acta, Part A*, 154, 98–102, 2016.
53. Li, Z., Liao, C., Chen, D., Song, J., Jin, W., Peng, G.-D., Zhu, F., Wang, Y., He, J., Wang, Y., Label-free detection of bovine serum albumin based on an in-fiber Mach-Zehnder interferometric biosensor. *Opt. Express*, 25, 17105, 2017.
54. Zhang, J., Chen, X., Xue, Y., Gamper, N., Zhang, X., Beyond voltage-gated ion channels: Voltage-operated membrane proteins and cellular processes. *J. Cell. Physiol.*, 233, 6377–6385, 2018.
55. Phillips, M.B., Nigam, A., Johnson, J.W., Interplay between gating and block of ligand-gated ion channels. *Brain Sci.*, 10, 12, 928, 2020.
56. Murthy, S.E., Dubin, A.E., Patapoutian, A., Piezos thrive under pressure: Mechanically activated ion channels in health and disease. *Nat. Rev. Mol. Cell Biol.*, 18, 771–783, 2017.
57. Kefauver, J.M., Ward, A.B., Patapoutian, A., Discoveries in structure and physiology of mechanically activated ion channels. *Nature*, 587, 567–576, 2020.
58. Paoletti, P., Ellis-Davies, G.C.R., Mourot, A., Optical control of neuronal ion channels and receptors. *Nat. Rev. Neurosci.*, 20, 514–532, 2019.
59. Roux, B., Ion channels and ion selectivity. *Essays Biochem.*, 61, 201–209, 2017.
60. Liu, Y., Wang, K.W., Exploiting the diversity of ion channels: Modulation of ion channels for therapeutic indications, in: *Handb. Exp. Pharmacol.*, vol. 260, 187–205, 2019.
61. Santos, R., Ursu, O., Gaulton, A., Bento, A.P., Donadi, R.S., Bologa, C.G., Karlsson, A., Al-Lazikani, B., Hersey, A., Oprea, T.I., Overington, J.P., A comprehensive map of molecular drug targets. *Nat. Rev. Drug Discov.*, 16, 19–34, 2016.
62. Zhang, Y., Wang, K., Yu, Z., Drug Development in Channelopathies: Allosteric Modulation of Ligand-Gated and Voltage-Gated Ion Channels. *J. Med. Chem.*, 63, 15258–15278, 2020.
63. Poveda, J.A., Marcela Giudici, A., Lourdes Renart, M., Morales, A., González-Ros, J.M., Towards understanding the molecular basis of ion channel modulation by lipids: Mechanistic models and current paradigms. *Biochim. Biophys. Acta, Biomembr.*, 1859, 9, 1507–1516, 2017.

64. Thompson, M.J., Baenziger, J.E., Ion channels as lipid sensors: from structures to mechanisms. *Nat. Chem. Biol.*, 16, 1331–1342, 2020.
65. Harraz, O.F., Hill-Eubanks, D., Nelson, M.T., PIP2: A critical regulator of vascular ion channels hiding in plain sight. *Proc. Natl. Acad. Sci. U.S.A.*, 117, 20378–20389, 2020.
66. Kozlov, S., Animal toxins for channelopathy treatment. *Neuropharmacology*, 132, 83–97, 2018.
67. Stortelers, C., Pinto-Espinoza, C., Van Hoorick, D., Koch-Nolte, F., Modulating ion channel function with antibodies and nanobodies. *Curr. Opin. Immunol.*, 52, 18–26, 2018.
68. Norton, R.S., Chandy, K.G., Venom-derived peptide inhibitors of voltage-gated potassium channels. *Neuropharmacology*, 127, 124–138, 2017.
69. Chow, C.Y., Absalom, N., Biggs, K., King, G.F., Ma, L., Venom-derived modulators of epilepsy-related ion channels. *Biochem. Pharmacol.*, 181, 114043, 2020.
70. Sawarkar, R., Bhandarkar, S., Mendhi, S., More, S., Channelopathies an Approach to Elevate Level of Cure- A Review. *Int. J. Pharm. Sci. Rev. Res.*, 70, 65–74, 2021.
71. Terragni, B., Scalmani, P., Franceschetti, S., Cestèle, S., Mantegazza, M., Post–translational dysfunctions in channelopathies of the nervous system. *Neuropharmacology*, 132, 31–42, 2018.
72. Curran, J., Mohler, P.J., Alternative paradigms for ion channelopathies: Disorders of ion channel membrane trafficking and posttranslational modification. *Annu. Rev. Physiol.*, 77, 505–524, 2015.
73. Meisler, M.H., Hill, S.F., Yu, W., Sodium channelopathies in neurodevelopmental disorders. *Nat. Rev. Neurosci.*, 22, 152–166, 2021.
74. Fonseca, D.J., Vaz da Silva, M.J., Cardiac channelopathies: The role of sodium channel mutations. *Rev. Port. Cardiol. (Engl. Ed.)*, 37, 179–199, 2018.
75. Albury, C.L., Stuart, S., Haupt, L.M., Griffiths, L.R., Ion channelopathies and migraine pathogenesis. *Mol. Genet. Genomics*, 292, 729–739, 2017.
76. Kaushik, V., Yakisich, J.S., Kumar, A., Azad, N., Iyer, A.K.V., Ionophores: Potential use as anticancer drugs and chemosensitizers. *Cancers (Basel)*, 10, 360, 2018.
77. Steinbrueck, A., Sedgwick, A.C., Brewster, J.T., Yan, K.C., Shang, Y., Knoll, D.M., Vargas-Zúñiga, G.I., He, X.P., Tian, H., Sessler, J.L., Transition metal chelators, pro-chelators, and ionophores as small molecule cancer chemotherapeutic agents. *Chem. Soc. Rev.*, 49, 3726–3747, 2020.
78. Li, G., De Oliveira, D.M.P., Walker, M.J., The antimicrobial and immunomodulatory effects of Ionophores for the treatment of human infection. *J. Inorg. Biochem.*, 227, 111661, 2022.
79. Bharti, H., Singal, A., Raza, M., Ghosh, P.C., Nag, A., Ionophores as Potent Anti-malarials: A Miracle in the Making. *Curr. Top. Med. Chem.*, 18, 2029–2041, 2018.

80. Sarfraz, M.H., Zubair, M., Aslam, B., Ashraf, A., Siddique, M.H., Hayat, S., Cruz, J.N., Muzammil, S., Khurshid, M., Sarfraz, M.F., Hashem, A., Dawoud, T.M., Avila-Quezada, G.D., Abd_Allah, E.F., Comparative analysis of phyto-fabricated chitosan, copper oxide, and chitosan-based CuO nanoparticles: antibacterial potential against Acinetobacter baumannii isolates and anticancer activity against HepG2 cell lines. *Front. Microbiol.*, 14, 1188743, 2023.
81. Prabhakar, P.K., Bacterial Siderophores and Their Potential Applications: A Review. *Curr. Mol. Pharmacol.*, 13, 295–305, 2020.
82. Noh, D., Lee, H., Lee, S., Sun, I.C.,Yoon, H.Y., Copper-Based Nanomedicines for Cuproptosis-Mediated Effective Cancer Treatment. *Biomater. Res.*, 28, 0094, 2024.
83. Bhullar, S.K., Shah, A.K., Dhalla, N.S., Store-operated calcium channels: Potential target for the therapy of hypertension. *Rev. Cardiovasc. Med.*, 20, 139–151, 2019.
84. Hu, H.J., Song, M., Disrupted Ionic Homeostasis in Ischemic Stroke and New Therapeutic Targets. *J. Stroke Cerebrovasc. Dis.*, 26, 2706–2719, 2017.
85. Thapak, P., Vaidya, B., Joshi, H.C., Singh, J.N., Sharma, S.S., Therapeutic potential of pharmacological agents targeting TRP channels in CNS disorders. *Pharmacol. Res.*, 159, 105026, 2020.
86. Bergantin, L.B., The Interactions Between Alzheimer's Disease and Major Depression: Role of Ca^{2+} Channel Blockers and Ca^{2+}/cAMP Signalling. *Curr. Drug Res. Rev.*, 12, 97–102, 2020.
87. Tong, B.C.K., Wu, A.J., Li, M., Cheung, K.H., Calcium signaling in Alzheimer's disease & therapies. *Biochim. Biophys. Acta, Mol. Cell. Res.*, 1865, 1745–1760, 2018.
88. Popugaeva, E., Pchitskaya, E., Bezprozvanny, I., Dysregulation of neuronal calcium homeostasis in Alzheimer's disease – A therapeutic opportunity? *Biochem. Biophys. Res. Commun.*, 483, 998–1004, 2017.
89. Kato, T., Current understanding of bipolar disorder: Toward integration of biological basis and treatment strategies. *Psychiatry Clin. Neurosci.*, 73, 526–540, 2019.
90. Dubovsky, S.L., Applications of calcium channel blockers in psychiatry: pharmacokinetic and pharmacodynamic aspects of treatment of bipolar disorder. *Expert Opin. Drug Metab. Toxicol.*, 15, 35–47, 2019.
91. Garneau, A.P., Slimani, S., Fiola, M.J., Tremblay, L.E., Isenring, P., Multiple facets and roles of Na+-K+-Cl- cotransport: Mechanisms and therapeutic implications. *Physiology*, 35, 415–429, 2020.
92. Viswanath, O., Urits, I., Jones, M.R., Peck, J.M., Kochanski, J., Hasegawa, M., Anyama, B., Kaye, A.D., Membrane Stabilizer Medications in the Treatment of Chronic Neuropathic Pain: a Comprehensive Review. *Curr. Pain Headache Rep.*, 23, 1–9, 2019.
93. Karsan, N., Gonzales, E.B., Dussor, G., Targeted Acid-Sensing Ion Channel Therapies for Migraine. *Neurotherapeutics*, 15, 402–414, 2018.

94. Jacobson, D.A., Shyng, S.L., Ion Channels of the Islets in Type 2 Diabetes. *J. Mol. Biol.*, 432, 1326–1346, 2020.
95. Selvaraj, C., Selvaraj, G., Kaliamurthi, S., Cho, W.C., Wei, D.-Q., Singh, S.K., Ion Channels as Therapeutic Targets for Type 1 Diabetes Mellitus. *Curr. Drug Targets*, 21, 132–147, 2019.
96. Ali, E.S., Petrovsky, N., Calcium Signaling As a Therapeutic Target for Liver Steatosis. *Trends Endocrinol. Metab.*, 30, 270–281, 2019.
97. Wulff, H., Christophersen, P., Colussi, P., Chandy, K.G., Yarov-Yarovoy, V., Antibodies and venom peptides: new modalities for ion channels. *Nat. Rev. Drug Discov.*, 18, 339–357, 2019.
98. Szabo, I., Zoratti, M., Biasutto, L., Targeting mitochondrial ion channels for cancer therapy. *Redox Biol.*, 42, 101846, 2021.
99. Sterea, A.M., Almasi, S., El Hiani, Y., The hidden potential of lysosomal ion channels: A new era of oncogenes. *Cell Calcium*, 72, 91–103, 2018.
100. Jeevaratnam, K., Chadda, K.R., Huang, C.L.H., Camm, A.J., Cardiac Potassium Channels: Physiological Insights for Targeted Therapy. *J. Cardiovasc. Pharmacol. Ther.*, 23, 119–129, 2018.
101. Bushart, D.D., Shakkottai, V.G., Ion channel dysfunction in cerebellar ataxia. *Neurosci. Lett.*, 688, 41–48, 2019.
102. Verkman, A.S., Galietta, L.J.V., Chloride transport modulators as drug candidates. *Am. J. Physiol. Cell Physiol.*, 321, C932–C946, 2021.
103. Alfonso, I., Quesada, R., Biological activity of synthetic ionophores: Ion transporters as prospective drugs? *Chem. Sci.*, 4, 3009–3019, 2013.
104. da Silva, D.F., de Souza, J.L., da Costa, D.M., Costa, D.B., Moreira, P.O.L., da Fonseca, A.L., Varotti, F.P., Cruz, J.N., dos Santos, C.B.R., Alves, C.Q., Leite, F.H.A., Brandão, H.N., Antiplasmodial activity of coumarins isolated from Polygala boliviensis: *in vitro* and *in silico* studies. *J. Biomol. Struct. Dyn.*, 41, 22, 13383–13403, 2023.
105. Park, S.H., Park, S.H., Howe, E.N.W., Hyun, J.Y., Chen, L.J., Hwang, I., Vargas-Zuñiga, G., Busschaert, N., Gale, P.A., Sessler, J.L., Shin, I., Determinants of Ion-Transporter Cancer Cell Death. *Chem*, 5, 2079–2098, 2019.
106. Li, H., Valkenier, H., Thorne, A.G., Dias, C.M., Cooper, J.A., Kieffer, M., Busschaert, N., Gale, P.A., Sheppard, D.N., Davis, A.P., Anion carriers as potential treatments for cystic fibrosis: Transport in cystic fibrosis cells, and additivity to channel-targeting drugs. *Chem. Sci.*, 10, 9663–9672, 2019.
107. Kunzelmann, K., Ion channels and cancer. *J. Membr. Biol.*, 205, 159–173, 2005.
108. Benke, B.P., Aich, P., Kim, Y., Kim, K.L., Rohman, M.R., Hong, S., Hwang, I.C., Lee, E.H., Roh, J.H., Kim, K., Iodide-Selective Synthetic Ion Channels Based on Shape-Persistent Organic Cages. *J. Am. Chem. Soc.*, 139, 7432–7435, 2017.
109. Lang, C., Li, W., Dong, Z., Zhang, X., Yang, F., Yang, B., Deng, X., Zhang, C., Xu, J., Liu, J., Biomimetic Transmembrane Channels with High Stability

and Transporting Efficiency from Helically Folded Macromolecules. *Angew. Chem. Int. Ed.*, 55, 9723–9727, 2016.

110. Burns, J.R., Seifert, A., Fertig, N., Howorka, S., A biomimetic DNA-based channel for the ligand-controlled transport of charged molecular cargo across a biological membrane. *Nat. Nanotechnol.*, 11, 152–156, 2016.
111. Haynes, C.J.E., Zhu, J., Chimerel, C., Hernández-Ainsa, S., Riddell, I.A., Ronson, T.K., Keyser, U.F., Nitschke, J.R., Blockable Zn10L15 Ion Channels through Subcomponent Self-Assembly. *Angew. Chem. Int. Ed.*, 56, 15388–15392, 2017.
112. Talukdar, P., Bollot, G., Mareda, J., Sakai, N., Matile, S., Synthetic ion channels with rigid-rod π-stack architecture that open in response to charge-transfer complex formation. *J. Am. Chem. Soc.*, 127, 6528–6529, 2005.
113. Danial, M., Tran, C.M.N., Jolliffe, K.A., Perrier, S., Thermal gating in lipid membranes using thermoresponsive cyclic peptide-polymer conjugates. *J. Am. Chem. Soc.*, 136, 8018–8026, 2014.
114. Jung, M., Kim, H., Baek, K., Kim, K., Synthetic Ion Channel Based on Metal–Organic Polyhedra. *Angew. Chem.*, 120, 5839–5841, 2008.
115. Pfeifer, J.R., Reiß, P., Koert, U., Crown ether-gramicidin hybrid ion channels: Dehydration-assisted ion selectivity. *Angew. Chem. Int. Ed.*, 45, 501–504, 2006.
116. Vargas Jentzsch, A., Hennig, A., Mareda, J., Matile, S., Synthetic ion transporters that work with anion-π interactions, halogen bonds, and anion-macrodipole interactions. *Acc. Chem. Res.*, 46, 2791–2800, 2013.
117. Behera, H., Madhavan, N., Anion-Selective Cholesterol Decorated Macrocyclic Transmembrane Ion Carriers. *J. Am. Chem. Soc.*, 139, 12919–12922, 2017.
118. Yuan, L., Shen, J., Ye, R., Chen, F., Zeng, H., Structurally simple trimesic amides as highly selective anion channels. *Chem. Commun.*, 55, 4797–4800, 2019.
119. Busschaert, N., Kirby, I.L., Young, S., Coles, S.J., Horton, P.N., Light, M.E., Gale, P.A., Squaramides as Potent Transmembrane Anion Transporters. *Angew. Chem.*, 124, 4502–4506, 2012.
120. Si, W., Xin, P., Li, Z.T., Hou, J.L., Tubular Unimolecular Transmembrane Channels: Construction Strategy and Transport Activities. *Acc. Chem. Res.*, 48, 1612–1619, 2015.
121. Kim, D.S., Sessler, J.L., Calix[4]pyrroles: Versatile molecular containers with ion transport, recognition, and molecular switching functions. *Chem. Soc. Rev.*, 44, 532–546, 2015.
122. Kaucher, M.S., Harrell, W.A., Davis, J.T., A unimolecular G-quadruplex that functions as a synthetic transmembrane Na+ transporter. *J. Am. Chem. Soc.*, 128, 38–39, 2006.
123. Marchi, S., Giorgi, C., Galluzzi, L., Pinton, P., Ca^{2+} Fluxes and Cancer. *Mol. Cell*, 78, 1055–1069, 2020.

124. Fnu, G., Weber, G.F., Alterations of Ion Homeostasis in Cancer Metastasis: Implications for Treatment. *Front. Oncol.*, 11, 765329, 2021.
125. Seitter, H., Koschak, A., Relevance of tissue specific subunit expression in channelopathies. *Neuropharmacology*, 132, 58–70, 2018.
126. Leevy, W.M., Donato, G.M., Ferdani, R., Goldman, W.E., Schlesinger, P.H., Gokel, G.W., Synthetic hydraphile channels of appropriate length kill Escherichia coli. *J. Am. Chem. Soc.*, 124, 9022–9023, 2002.
127. Brea, R.J., Reiriz, C., Granja, J.R., Towards functional bionanomaterials based on self-assembling cyclic peptide nanotubes. *Chem. Soc. Rev.*, 39, 1448–1456, 2010.
128. Kevin, D.A., Meujo, D.A.F., Hamann, M.T., Polyether ionophores: Broad-spectrum and promising biologically active molecules for the control of drug-resistant bacteria and parasites. *Expert Opin. Drug Discov.*, 4, 109–146, 2009.
129. Gargan, S., Stevenson, N.J., Unravelling the immunomodulatory effects of viral ion channels, towards the treatment of disease. *Viruses*, 13, 2165, 2021.
130. Zhang, J., Liu, W., Dai, J., Xiao, K., Nanoionics from Biological to Artificial Systems: An Alternative Beyond Nanoelectronics. *Adv. Sci.*, 9, 2200534, 2022.

7

Nanoionics in Electronics and Optoelectronics: Advances and Applications

Most. Israt Jahan[1], Md. Enamul Kabir[1], Md. Abu Bin Hasan Susan[1,2] and Muhammed Shah Miran[1]*

[1]*Department of Chemistry, University of Dhaka, Dhaka, Bangladesh*
[2]*Dhaka University Nanotechnology Center, University of Dhaka, Dhaka, Bangladesh*

Abstract

This chapter explores the burgeoning field of nanoionics and its diverse applications in the realms of electronics and optoelectronics. Nanoionics, the manipulation of ion dynamics within nanostructures, has emerged as a pivotal catalyst for transformative advancements in technology. The investigation encompasses the comprehensive study of nanoionic materials, devices, and their integration into electronic and optoelectronic systems. In the domain of electronics, nanoionics has led to groundbreaking developments in electronic components and memory technologies. The ability to control ion dynamics within nanostructures has resulted in the creation of advanced electronic components, with a particular emphasis on high-density, ultra-fast memory systems. The implications of these advancements in data storage solutions are profound, offering unprecedented opportunities for scalability and efficiency. Optoelectronic applications stand to benefit significantly from the integration of nanoionics. The synergy between electronic and optical functionalities opens avenues for improved device performance and innovative applications in light-based technologies. This chapter encapsulates the multifaceted contributions of nanoionics, from its transformative impact on electronic components to its potential to reshape the landscape of optoelectronic applications. It also explores the exciting world of nanoionics, envisioning a future where electronic and optoelectronic technologies reach new heights, driven by revolutionary capabilities, and discussing potential challenges.

**Corresponding author*: shahmiran@du.ac.bd

Inamuddin, Tariq Altalhi, Mohammad Luqman and Jorddy Neves Cruz (eds.) Nanoionics: Fundamentals and Applications, (161–194) © 2025 Scrivener Publishing LLC

***Keywords*:** Nanoionics, electronics, optoelectronics, resistive switches, memristive devices, optical sensors, photo-assisted switches, solar cells

7.1 Introduction

"Nanoionics" is the study of phenomena, mechanisms, effects, properties, methods, and applications of procedures connected with ion transport in nano-confinement. It is a branch of nanotechnology and materials science that concentrates on the study and manipulation of ion mobility at the nanoscale dimension [1]. The advancement of nanotechnology and the increased likelihood of producing materials with artificially engineered nanoscale dimensions have given rise to novel approaches for discovering materials with elevated ionic conductivity. Surface effects, finite size effects, and interfaces can manifest and lead to improved ionic transport when the size of the solid is reduced, particularly in one dimension, to below 100 nanometers. Nanoionics is applied in different fields (Figure 7.1), including electronics, fuel cells, memristic device, solid-state batteries, sensors, energy storage, and many more, by understanding and controlling the movement of ions within nanoscale materials. Nature has chosen ions to play a crucial role in supporting essential physiological functions. At the molecular level, various types of ion channels contribute to the electrical activity of neurons, facilitating the transfer and processing of information at the foundational level of biological intelligence. Despite their slower

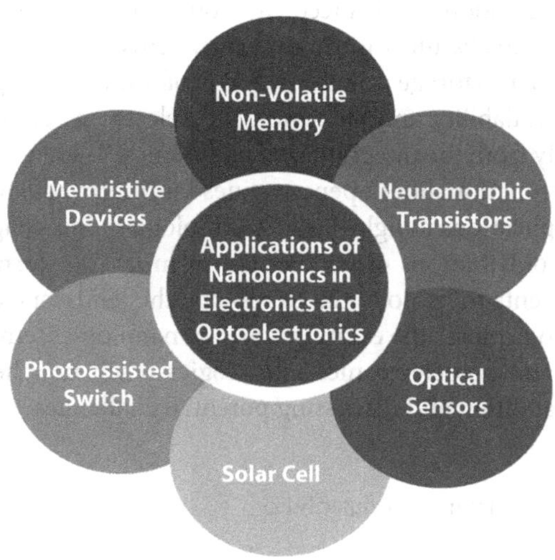

Figure 7.1 Some applications of nanoionics in electronics and optoelectronics.

speed compared to electrons, ions offer remarkable advantages, such as their weight, which enables them to block magnetic or electric disturbances and makes them largely resistant to electric field noise. In addition, although lighter mass allows for faster movement and lower energy consumption, it is counterintuitive to propose that smaller devices operating at scales below 2 nanometers should utilize heavier-mass information carriers. Another noteworthy benefit of heavier carriers is their reduced susceptibility to tunneling, in contrast to smaller devices with lighter mass, which face limitations in information storage duration due to quantum mechanical tunneling. Nanoionics primarily relies on the confinement effects stemming from quantum mechanics. The actual size effect leads to a situation in which the energy of electrons that are not bound to specific locations increases as the size decreases. Small size effects are important in electronic devices: consider varistors (where protecting boundaries of grains change to a conducive state when a bias is applied), Taguchi sensors (where adsorbed gases can alter the density of charge carriers on a semiconductor surface), or calcium doped lead titanate ceramics (where the resistance shows a positive temperature coefficient), where the relationship between interfacial chemistry and resistance to electricity is crucial [2].

Today, it is feasible to produce ultrathin films with thicknesses as small as a few nanometers while maintaining high crystalline quality and substantial lateral dimensions spanning several microns. Furthermore, it is possible to diminish the size of crystallites to each of the three extents to acquire polycrystalline materials and nano-crystalline materials with a moderate grain dimension falling within the range of 5 to 100 nanometers. Within nano-crystalline materials, the abundance of grain boundaries or interfacial regions can lead to significant alterations in ionic conductivity. The volume fraction of interfacial areas can reach as high as 50% by decreasing the crystallite size to just a few nanometers. A decreased coordination number compared to the bulk materials within these interfaces or grain boundaries of nanomaterials leads to amplified ionic or electronic diffusion [3–7]. H. L. Tuller identified a phenomenal boost in electronic conductivity in nano-crystalline CeO_2 materials that conduct both ions and electrons. This amelioration was attributed to the accumulation of oxygen vacancies at the grain boundary centers. When it comes to the electrolyte, nanoscale impacts are also crucial. Composites infused with LiI in nano-porous Al_2O_3 have secure conductivity values up to 10^{-3} Scm^{-1} at 60°C [8]. This makes them seem like a promising option for battery electrolytes. The application of nanostructured materials to optoelectronic devices is a topic of great attention these days, like lasers, solar cells, photodetectors, and light-emitting diodes (LEDs) because of their distinctive geometry. Nanostructures in compact dimensions can seamlessly integrate into diverse technological platforms, providing new chemical and physical characteristics for superior-accomplishment optoelectronic appliances. Utilizing novel nanostructures and leveraging their electrical and optical characteristics are essential for their forthcoming practical instrument uses. The latest rapid development in research related to two-dimensional layered nano-plasmonics and nanomaterials holds the potential to open the door to the creation of advanced photonic and optoelectronic devices in the future.

7.2 Development of Nanoionic Materials

The development of nanoionic materials involves the design, synthesis, and characterization of materials that exhibit unique ionic properties at the nanoscale. These materials are engineered at the nanoscale to exploit unique properties that emerge at this level. Introducing ionic dopants and defects is a common strategy to enhance the ionic conductivity of the material. This involves the intentional incorporation of ions, vacancies, or other defects to facilitate ion transport within the material. Nanoionic materials are incorporated into devices relevant to their intended applications. For example, in electronics, they might be used in transistors or memory devices, while in optoelectronics, they could be part of photodetectors or light-emitting devices.

7.2.1 Electronics

Nanoionic-based electronic devices leverage the unique properties of nanoscale ionic materials to achieve improved performance and novel functionalities. Ionic materials are substances that contain charged particles called ions, and nanoionic materials exhibit these properties at the nanoscale. This emerging field has the potential to revolutionize electronic devices in various applications. Here are some key aspects of nanoionic-based electronic devices. This section explores the intricate relationship between nanoionics, covering fundamental principles, materials, device architectures, and promising applications. Transition metal oxides, particularly those like NiO [9], TiO_2 [10, 11], Ta_2O_5 [12, 13], iron oxide [14, 15], and Nb_2O_5 [16], are extensively studied for resistive switching materials, particularly in resistive random-access memory (RRAM) devices. The components of the switching mechanism are the flow of oxygen ions and the generation of oxygen vacancies. A study focused on the resistive switching phenomena of polycrystalline Nb_2O_5 films deposited through e-beam reveals promising results for the application of non-volatile memory (NVM). Specifically, the Pt/Nb_2O_5/Pt and bare conductive paint (BCP)/Nb_2O_5/Pt device configurations demonstrate reversible bipolar resistive switching with less than ±1 V. Because of their exceptionally low operating voltage, improved endurance, and superior retention capabilities, Nb_2O_5 thin films have been projected to be highly advantageous for nonvolatile memories in electronic applications in the future. The deposition process for the top Pt electrode mirrors that of the Pt bottom electrode, creating a cross-layer structure as illustrated in Figure 7.2. The study identifies Nb_2O_5 as a promising candidate for NVM applications, emphasizing its low voltage requirements, improved endurance, and retention capabilities, all essential factors for the development of efficient electronic memories [17].

A theoretical model has been developed based on current-voltage (I-V) measurements shown in Figure 7.3, to explain the underlying physics of the resistive switching mechanism. The application of bias voltage/current triggered anion migration, influenced by the electric field and thermal effects, resulting in the creation and disruption of conducting nanofilaments. Positively charged oxygen vacancies or oxygen anions were recognized as the mobile species. The metal oxide layer, containing structural defects like dislocations and grain boundaries, facilitated ion diffusion due to their low diffusion energy [18]. Self-rectifying features of RRAM devices based on vacancy-modulated conductive oxide (VMCO), utilizing Ti/NiO_x/Al_2O_3/Pt structures have been explored. In this configuration, the NiO_x layer demonstrated characteristics of resistive switching modulated by vacancies, while the thin Al_2O_3 layers displayed typical tunnel barrier properties. Ti/NiO_x/Al_2O_3/Pt device structure has been fabricated

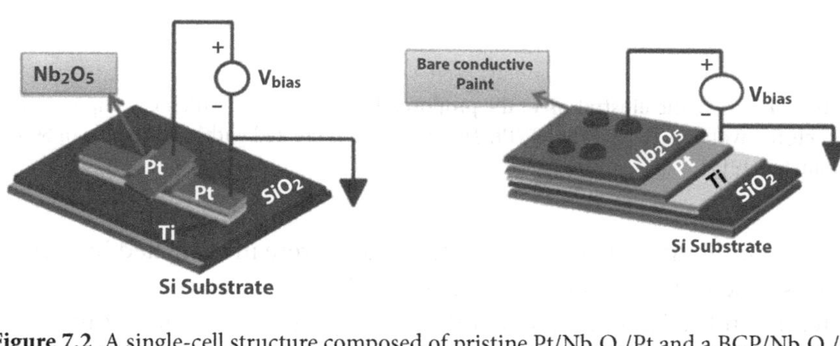

Figure 7.2 A single-cell structure composed of pristine Pt/Nb_2O_5/Pt and a BCP/Nb_2O_5/Pt stack was created using the e-beam evaporation technique [17].

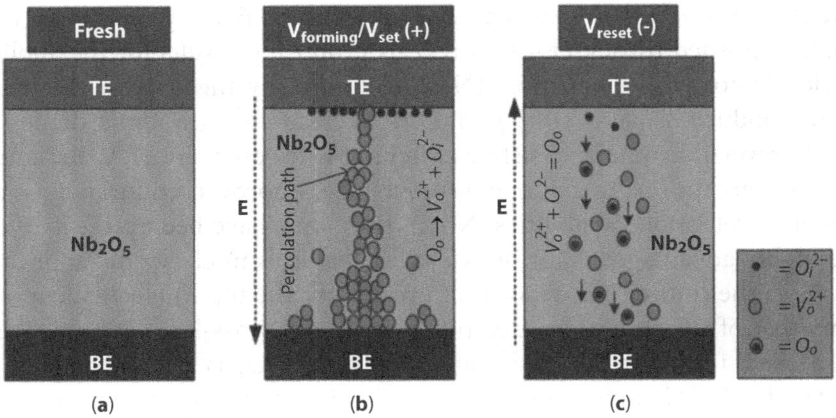

Figure 7.3 An essential model for comprehending resistive switching mechanism as inferred from I-V measurements. (a) The unaltered device, that is, before a bias voltage is applied, (b) While the process of electroforming or setting, (c) Upon reversing the bias potential, a reversal of the motion of oxygen ions occurred [18].

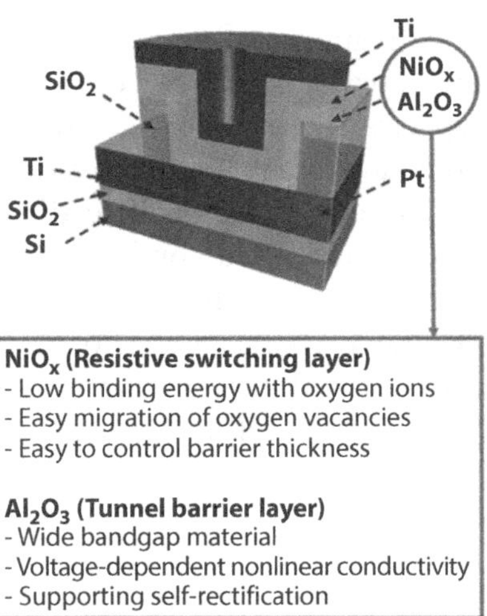

Figure 7.4 Schematic illustration of the proposed ReRAM device, which is composed of a structure involving Ti/NiO_x/Al_2O_3/Pt. This structure is created within a *via*-hole of 500 nanometers [9].

as shown in Figure 7.4. Self-rectifying features were investigated in RRAM devices based on VMCO with Ti/NiO_x/Al_2O_3/Pt structures. The resistive switching in the NiO_x layer was influenced by vacancies, while thin Al_2O_3 layers show typical tunnel barrier properties. The characteristics of I-V were impacted by the effective tunnel barrier thickness. Self-rectification was observed for 6 and 10 nm tunnel barrier thicknesses. In the positive-bias region, conduction involved direct tunneling, Schottky emission, PeF emissions, and FeN tunneling. Conversely, Schottky emission is governed in the negative-bias region. Ti/NiO_x/Al_2O_3/Pt-based VMCO RRAMs demonstrated a high on/off ratio, low operation current, self-compliance effects, and long retention with forming free and high rectification ratio [9]. In 3D vertical structures, self-rectification combined with superior NVM features can mitigate sneaking-path-current issues. Proton and lithium (Li) ions are commonly employed in ionic transistors due to their small size, leading to low energy consumption and enhanced endurance [19, 20]. Synaptic transistors, specifically those based on Li^+ ion, have been extensively studied, leveraging insights from battery applications [21]. WO_x is a key material for the channels in synaptic transistors, even with a minimal Li^+ ion injection exhibiting a noteworthy conductance increase [22]. Li/WO_x structures have been explored by various synaptic transistors. Within the realm of Li^+ ion-based synaptic transistors (LST), research has focused on achieving optimal synaptic characteristics, including injecting a consistent number of ions per pulse [23]. Strategies employed for this purpose encompass reducing the value of the open circuit potential [24], incorporating a barrier of ion tunneling [25], and suggesting novel electrolyte materials employing current pulses to avert channel conductance saturation and with heightened ion conductivity [26, 27]. The relationship between WO_x channel defects and synaptic characteristics is pivotal in determining Li-ion diffusivity. For non-volatile conduction states, the rate-controlling step is the Li-ion transport in the WO_x conduit. It is possible to achieve near-ideal synaptic properties by ensuring high ion diffusivity through defects.

A diagram of the Li-ion-based 3-T synaptic transistor is shown in Figure 7.5. Movement of Li^+ ions occurs between the cathode and the anode *via* an electrolyte in most battery configurations. Li_3PO_4 is the material of choice for the electrolyte layer in this configuration, which serves as a medium for Li-ion transport in addition to being a source of Li-ions [28]. Li_3PO_4 is frequently utilized due to its strong ion conductivity and its radio frequency sputtering deposition simplicity [29]. An overview of the synaptic properties seen in the three different WO_x stoichiometry channel types is given in Figure 7.6. A Li^+ ion-based ionic transistor with a significant conductance on/off ratio and a linear weight update was developed. The ideal $WO_{2.7}$ stoichiometry was attained by carefully controlling the oxygen deficit. A two-step voltage pulse approach was used for reducing the transitory conductance changes brought on by Li^+ ion diffusion in the electrolyte.

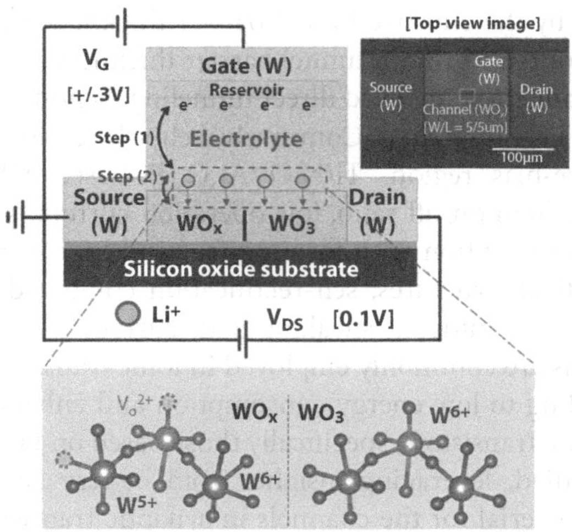

Figure 7.5 The transport of Li^+ ion in the electrolyte is followed by ion injection in the mechanism for the operation of the lithium-ion 3-T synaptic device. Channels with varying WO_x compositions were used to investigate the effects of channel imperfections on synaptic characteristics [28].

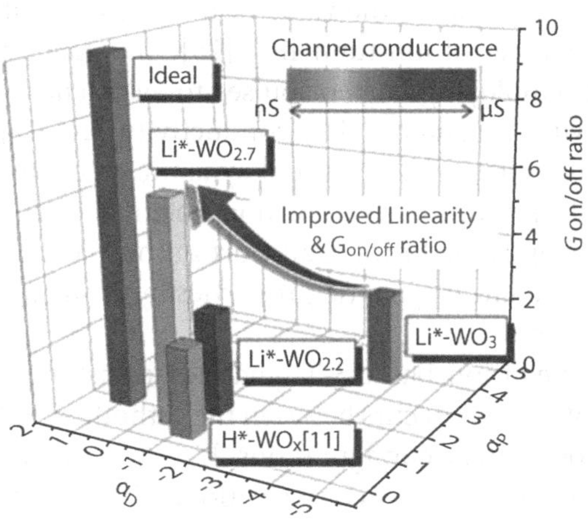

Figure 7.6 Synopsis of synaptic traits under varying conditions of channel stoichiometry [28].

Using a synaptic device with a tuned $WO_{2.7}$ channel showed remarkable pattern recognition accuracy. This means that enhanced ion diffusion behavior obtained by defect control approaches may open the way for the advancement of very successful on-chip retraining synapse electronics in the future [28].

7.2.2 Optoelectronics

The field of nanoionics has the potential to significantly impact the development of optoelectronics by enabling new materials and technologies. Nanoionics, which involves the manipulation of ions at the nanoscale, offers several advantages and opportunities for optoelectronic devices. Nanoionics facilitates the advancement of incentive materials with modified characteristics. These materials can elevate the proficiencies of optoelectronic devices, precisely quantum dot light-emitting diodes (QLEDs) [30]. Quantum dots (QDs) are semiconductor crystals with nanoscale proportions. The quantum captivity effect in QDs, ensuing from their small size analogous to the Bohr radius, causes distinct energy levels and atom-like spectra with confined emission, which leads to highly saturated colors, convenient for wide color extent displays. In addition, minimizing the size of QDs expands their band gap, shaping size-dependent optical characteristics such as fluorescence emission wavelength and absorption inception. The unique optical traits of QDs, including their confined emission and size-conditional properties, have incited noteworthy interest in solid-state lighting and flat-panel displays [31]. Over the past three decades, widespread research has scrutinized their implementation in electroluminescence and photoluminescence (PL). Similar to the flourishing trajectory of organics light-emitting diodes, the central challenge obstructing the merchandising of QLEDs lies in their device firmness. Despite the use of high-quality QDs with supreme photoluminescent solidity as substitutes for conventional phosphors in liquid crystal display backlighting, the functional lifetimes of most trendsetting QLEDs fall short of meeting pragmatic implementation prerequisites. However, there is encouraging news as recent years have seen remarkable advancements in materials and structures, with nanoionics playing a key role in developing materials that enhance energy efficiency. In the realm of optoelectronic devices, improved energy efficiency holds the promise of reducing power consumption and extending operational lifetimes. Significant progress was reported through the creation of an all-inorganic QLED, showcasing exceptional efficiency. This accomplishment entails applying an ultrathin Al_2O_3 coating layer to the surface of solution-processed NiO (s-NiO). NiOOH is well-secured on the s-NiO surface by the Al_2O_3 layer, which reduces exciton extinguishing. The schematic representation of the device structure can be observed in Figure 7.7(a), while Figure 7.7(b) illustrates the passivating mechanism. The Al_2O_3 layer effectively secures NiOOH on the s-NiO surface, thereby alleviating exciton extinguishing. Consequently, this makes it the top-performing all-inorganic QLED to date. Without the Al_2O_3 layer, this

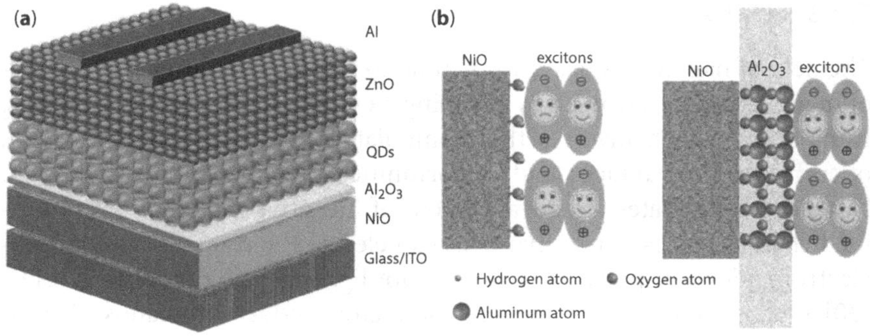

Figure 7.7 Diagram (a) illustrates the structure of the device, while diagram (b) depicts the mechanism involved in passivation [32].

enrichment causes a significant rise in the maximum efficiency of QLED. As a result, it is currently the best-performing all-inorganic QLED available [32].

Yang *et al.*, introduced a highly effective configuration for an inorganic QLEDs with the ITO/MoO_3/NiO/LiF/QDs/Al_2O_3/ZnO/Al structure. Through the insertion of dual inorganic securing sheets, they alleviated quenching caused by metal oxides and enhanced charge balance. This refinement led to a maximum external quantum efficiency and an extended operational lifespan [33]. Over the past decade, ZnO has accumulated significant zeal due to its newly revealed features and its potential for manipulation in emerging nanoscale applications, such as nanoionics devices, nano-lasers, and UV photodetectors [34]. Various QLEDs have demonstrated the use of colloidal ZnO nanocrystal films as electron injection layers. The electronic levels of energy of the nanocrystalline ZnO films improve the interfacial charge transfer processes, which in turn improves the emission stability of QD emitters [35]. The work showed that graphene oxide (GO) may be used to reliably and consistently manage PL. During this process, the sp^2/sp^3 domain portion of GO was altered, and GO and the mesoporous SiO_2 thin films displayed a notably reduced PL background. By adjusting the applied DC voltage, this device which is depicted in Figure 7.8, tunes the PL peak wavelength fast and effectively throughout a wide spectrum. Applications for the tuning process could be found in display technologies, nano biosensors, and lighting. The study deployed a PL source designed entirely of solid-state materials, as illustrated in Figure 7.9 [36]. In order to create this device, GO and a mesoporous SiO_2 thin film were used, along with a variety of assembly processes like spin coating,

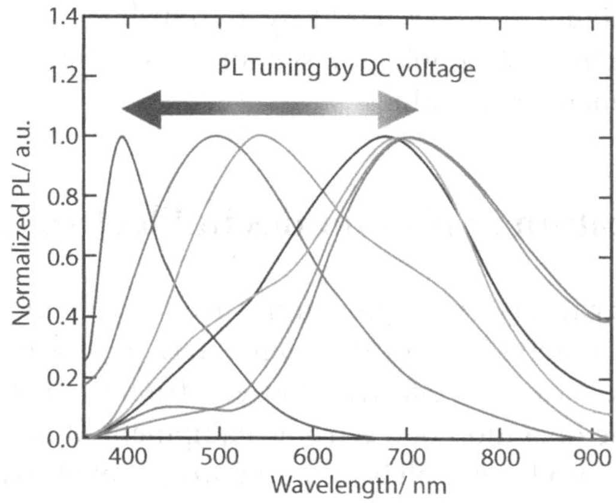

Figure 7.8 By changing the supplied DC voltage, the PL peak wavelengths were tuned over a large range (393 to 712 nm) in a matter of a few hundred seconds [36].

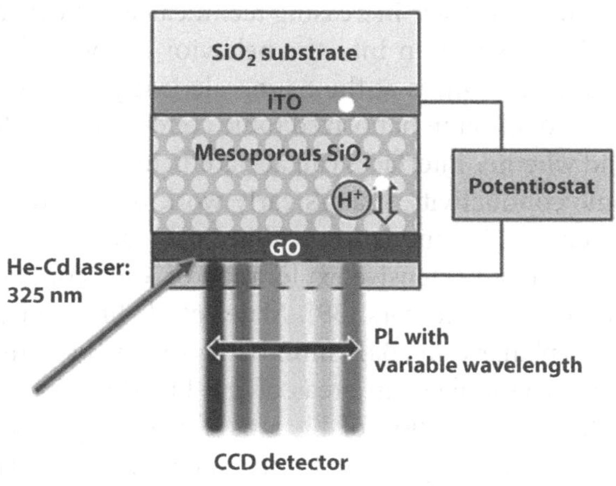

Figure 7.9 Diagrammatic illustration of a solid-state-technology PL source based on GO and a mesoporous silicon dioxide film that is thin [36].

electron beam deposition, the sol-gel method using template organics, and more [37]. Materials developed through nanoionics have applications in flexible and transparent optoelectronic devices, opening possibilities for use in flexible displays, wearable technology, and transparent solar cells. Nanoionics engineering holds the potential to revolutionize the field of optoelectronics by providing materials and tools for the development of more efficient, versatile, and high-performance optoelectronic devices.

As the field continues to evolve, it is likely to contribute to advancements in various optoelectronic technologies, spanning from lighting and displays to sensors and energy harvesting.

7.3 Application of Nanoionics in Electronics

Since electronic gadgets have a great influence on daily life and electronic materials are abundant, nanoscale science has mostly concentrated on nanoelectronics. However, effective and eco-friendly power sources are necessary for the operation of electronic equipment. It is anticipated that future electricity producers will be more environmentally friendly and efficient than those of the present. These changing needs are well suited for electrochemical devices, such as batteries and fuel cells, which involve a transformation of chemical energy into electrical energy [38]. Ionic transport has a critical function in the constituent materials of electrochemical devices, as evidenced by their increasing technical significance. Therefore, a more in-depth investigation into the behavior of ionic charge carriers in ion-conducting substances at the nanoscale is urgently needed. In the context of ion conduction in nanosized ionic conductors, the key question revolves around whether interfaces can act as efficient conduits, enhancing the overall ionic conductivity. Liang's early work demonstrated a significant enhancement in Li^+ conduction within LiI by dispersing small Al_2O_3 particles. This prompted extensive exploration of ionic conduction in various composite ionic conductors [39]. Maier *et al.*, highlighted increased F^- conductivity in planar CaF_2/BaF_2 films with growing periodicity of alternating layers. Li^+ accumulates in areas of the LiI:Al_2O_3 composite where there is a greater concentration of Li^+ ions, forming space-charge zones that improve Li^+ conductivity. When F^- ions in BaF_2 migrate to nearby CaF_2 in heterostructure multilayer films, vacancies are left in BaF_2, and the concentration of charge carriers at the interface increases, resulting in highly conductive pathways [40]. These examples show how the characteristics of solid-state ionic conductors can be effectively tailored using grain boundary engineering, which is accomplished through intentional control of interfacial defect chemistry.

7.3.1 Resistive Switching

Numerous nanoionics devices have been documented, highlighting enhanced functionalities resulting from the regulation of electrochemical reactions and ion transfers within nanostructures of ionic conductors.

To explore novel functionalities arising from the control of electrochemical reactions and ion transfers within nanostructures, attention was directed toward manipulating four distinct interfacial characteristics. Figure 7.10 schematically illustrates the use of a solid-state electric double layer (EDL) at the substrate/solid electrolyte interface, a Schottky-like barrier at the electrode/ionic conductor interface, a point contact among an electrode and a metal filamentous matter, and an electrochemical process within ions in the substrate and an ion conducting material [41]. The filament forms and eventually annihilates due to the redox action of the metallic ion in the electrolytic conductor. Point contact regulation facilitates switching operations by enabling regulated filament generation and annihilation at the level of the nanoscale and even to the atomic ranges.

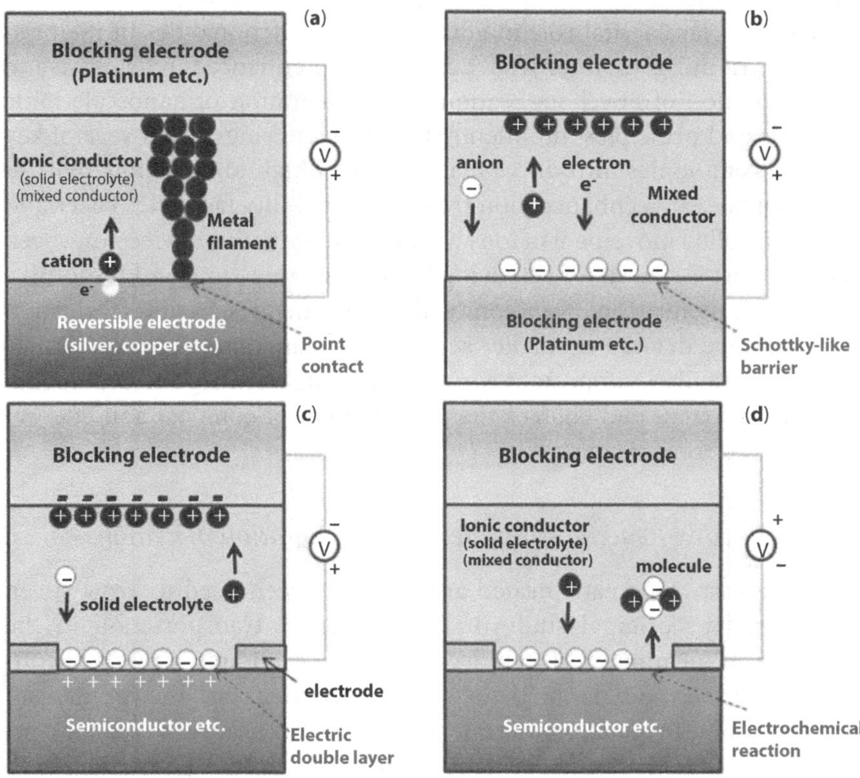

Figure 7.10 There are four methods for controlling localized ion transfers and electrochemical processes within nanostructures to achieve multifunctionality: (a) utilizing a point connection, (b) utilizing a Schottky-like obstruction, (c) incorporating a solid-state EDL, and (d) initiating an electrochemical process [41].

7.3.2 Memristive Devices

A memristive device, is a two-terminal NVM device that exhibits a specific type of electrical resistance change in response to applied voltage. Unlike traditional passive circuit elements such as resistors, capacitors, and inductors, memristors have the unique property of "memory" or "resistance memory." The resistance of a memristive device depends on the history of the current that has previously flowed through it, making it a crucial component in the development of novel computing and memory technologies. Even after the power is switched off, memristive devices maintain their resistance state. This non-volatile characteristic is crucial for applications in NVM, where data persistence is essential. Memristors exhibit resistive switching behavior, meaning that their resistance can be changed and controlled by applying an electric field. This property is discovered in the design of RRAM devices, enabling efficient data storage and retrieval. Nanoionics plays a vital role in advancing the functionalities of memristors, contributing to improved performance, enhanced reliability, and the exploration of novel applications. The integration of nanoscale ionic materials and principles into memristive devices brings about several key benefits. Nanoionics introduces materials with high ionic conductivity at the nanoscale. This enhanced ion transport capability facilitates faster and more controlled movement of ions within the memristive device. Improved ion transport is crucial for achieving high-speed switching and responsive behavior in memristors. Nanoionics allows for the design and fabrication of memristive devices at smaller scales. This reduction in device dimensions is particularly important for achieving higher integration densities in memory arrays and enables the development of compact and energy-efficient electronic systems.

7.3.2.1 Redox Reactions Initiated by the Migration of Cations

For more than 400 years, heated argentite has been used to grow silver. After Faraday's original study of Ag_2S electrical transportation in the beginning of the nineteenth century, an extensive amount of research constituted the thermodynamic theory of ionic conduction. The mechanism of silver stubble production was uncovered by using a galvanic cell for controlling the stoichiometric aberration of Ag^+ ions into Ag_2S crystals [42]. The use of solid electrolytes in the metal-insulator-metal (MIM) system, where conduction is dependent on metal cations, allows control over the creation and elimination of a metallic thread. The MIM system forms an ion-conducting 'I' surface with an electrolyte that is solid, an inert metallic counterpart electrode, and an electrochemically responsive metallic electrode to generate bipolar switching behavior. Hirose's 1976 study used Ag-photodoped amorphous As_2S_3 with Ag annihilation as the 'I' sheet in a lateral MIM device to discover switching behavior induced by silver dendrite development and annihilation [43]. Subsequently, it was shown that GeSe could be applied to NVM by serving as the vertical ion conductor in the MIM system [44]. The synthesis of resistive cycling GeSe nanostructures that are as tiny as 20 nm is shown in Figure 7.11. In such systems, the cations made up of metal migrate in the direction of the cathode and form a metallic filament that grows in the direction of the anode, which activates the switch. The oxidation of the electrochemically activated anode and the dissolution of metal atoms into the ionic conductor guarantee continuous electrochemical deposition. The anode and cathode are experiencing chemical reactions [45].

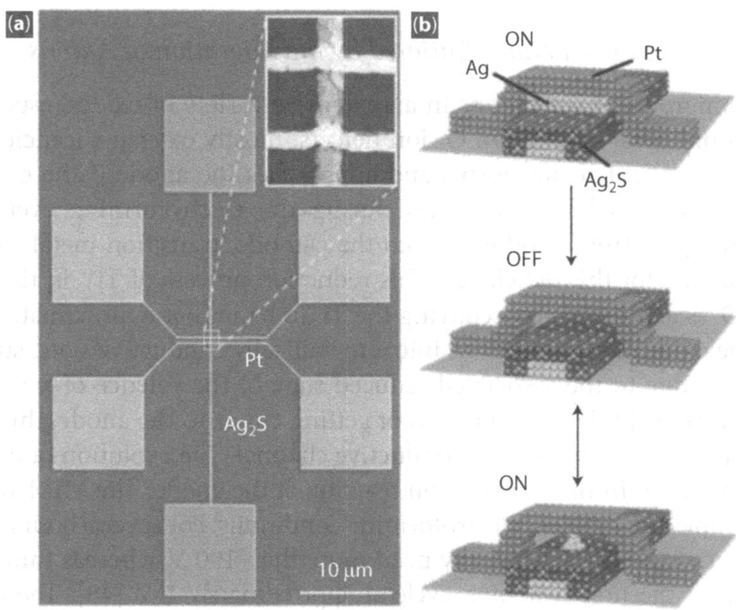

Figure 7.11 The cross-sectional view illustrates a vertical MIM switch that employs a solid electrolyte conducting Ag^+ ions. Within the GeSe layer, an electrochemically formed silver filament is established to activate the switch. This cross-sectional diagram corresponds to the apparatus for which the curve of I-V shown in the inset was measured [45].

$$Ag^{+}(Ag_2S) + e^{-} \rightarrow Ag \text{ (Reduction)}$$

$$\text{and } Ag \rightarrow Ag^{+}(Ag_2S) + e^{-} \text{ (Oxidation).}$$

Filament annihilation and switch deactivation are the results of the breakdown of metal atoms near the edge of the metal filament caused by a change in bias voltage polarity. The switch should theoretically last an infinite amount of time since the chemical reaction in the MIM system is non-damaging. Sometimes one of the "metal" electrodes in the MIM structure is composed of solid electrolytes. Such a MIM switch has been developed for one of the electrodes using an ionic and an electronic mixed conductor. In this setup, a metal filament connects the mixed conductor electrode and the counter electrode by forming in the absence of air nano-gap, which serves as an insulated layer. The switching time exponentially shortens with an enhancement in the switching bias voltage, and faster switching occurs with a reduction the size of the metal filament [46].

7.3.2.2 Redox Reactions Initiated by the Migration of Anions

More migratory than cations in an extensive variety of oxides, especially transition metal oxides, are O_2 ion defects, mostly oxygen vacancies. An oxygen-deficient region forms and approaches the anode if the cathode prevents ion exchange activities during the electroforming event. By absorbing electrons produced from the cathode, transition metal cations compensate for this deficiency. This reduction process of Ti^{4+} in titanates or TiO_2 is analogous to occupying the Ti 3d band. For approximately $n > 1.5$, the oxide often transitions into a metallically conductive state, such as $TiO_{(2-n)/2}$, due to the associated reduced state of the valence of transition metal cations [47]. In the process of getting closer to the anode, this "virtual cathode" finally forms a conductive channel. The evolution of oxygen gas may result from the oxidation reaction at the anode. The MIM system determines the different electroforming conditions. For several hours, macroscopic single crystals usually need more than 100 V, whereas thin films can finish the first switching cycle at approximately 1 V [48]. The entire charge governs the electroforming process. Regional redox processes in the virtual cathode and anode once electroforming is complete provide bipolar switching, which creates or breaks the conductive contact. By adjusting the resistance of the system to intermediate values in response to the charge transfer that takes place during switching, future memory cells might be able to generate multi-bit storage. The evidence for a filamentary switching

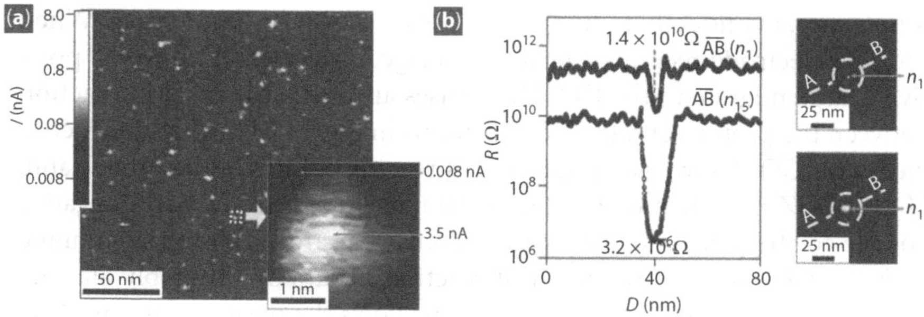

Figure 7.12 The conductance characteristics of individual $SrTiO_3$ dislocations. After a mild heat reduction, the conducting profile of a non-doped $SrTiO_3$ individual surface of the crystalline obtained by C-AFM is displayed in panel (a). Panel (b) shows a line scan over the selected point (where D is the distance along AB) showing how much conductance was increased after the AFM tip was applied with a negative voltage. Right: maps showing the conductivity of chosen area both before and after electroformation [54].

process can be observed in nanoscale confined TiO_2 [49], TiO_2 thin films [50], VO_2 thin films [51], Nb_2O_5 thin films and nanocrystalline Ta_2O_5 [52], and axial $SrZrO_3$:Cr thin films [53]. Undoped $SrTiO_3$ thin films and single crystals have the same dislocations and conductive filaments. A 1-2 nm wide region at the departure shift exhibits several orders of magnitude enhancement in conductivity is shown in Figure 7.12. The dislocation can be turned on and off by periodically scanning with an appropriately biased AFM tip. In the ON state, there are metallic features to the conductivity. Complete sections that may rapidly change between an ON and OFF state by conductive-tip atomic force microscopy (C-AFM) illustrate the fibrous nature of the conductivity in ultrathin epitaxial $SrTiO_3$ films. Conductivity is restricted to the regions dividing islands with a diameter of around 100 nm for certain materials, such as thin films of (La,Sr)MnO_3 [54].

7.3.3 Transistor

By employing the electrochemical process within ions in an ionic conductor and a substrate at its heterointerface, a transistor with a controlled bandgap has been created. There has been a report of a transistor utilizing graphene oxide (GO) in conjunction with oxygen atoms in the material graphene lattice. By adjusting the oxygen content, GO functions as a semiconductor with a changeable bandgap. The device fabrication involves depositing GO on the substrate, overlaying it with an yttria-stabilized zirconia (YSZ) solid-state electrolyte and forming three-terminal platinum electrodes for electrical conductivity measurements as shown in Figure 7.13. When the gate electrode receives a positive voltage, protons that possess a positive charge migrate to the GO. This causes an electrochemical interaction between the protons and the oxygen atoms in the GO. Oxygen atoms coupled with GO desorb as a result of this reaction [41]. On the other hand, H_2O in YSZ is broken down electrochemically by introducing a negative voltage to the gate, and the oxygen atoms that are produced recombine with the GO. This electrochemical reaction occurs by reversibly altering the polarity of the applied voltage, which allows the oxygen concentration of the GO to be modulated. Reducing the oxygen content in conjunction with GO permits bandgap modification. The regulated bandgap of GO can be achieved by adjusting the voltage across the gate and source electrodes, as demonstrated by the experimental results shown in Figure 7.14. The good operation of the transistor is confirmed by the bandgap fluctuation, which corresponds to a switching characteristic [55]. As such, this phenomenon is considered to have practical implications for the creation of sensors and functional devices. By combining a Nafion proton-conducting polymer electrolyte with a WO_3 thin film grown on a $LaAlO_3$ (100) substrate, a neuromorphic transistor made solely of solid-state components was created. The creation of an all-solid-state neuromorphic transistor is necessary for the advancement of next-generation technologies for communication and information, particularly those involving neural networks that are artificial. This transistor is made out of an electrolyte that conducts protons and a thin layer of WO_3. The drain current showed an impressive 4-order-of-magnitude rise as a result of the proton-migrated redox reaction of WO_3 thin film. Gate management of WO_3 redox reactions efficiently regulated the learning and forgetting characteristics of the device. This regulation is caused by the structural separation of the current

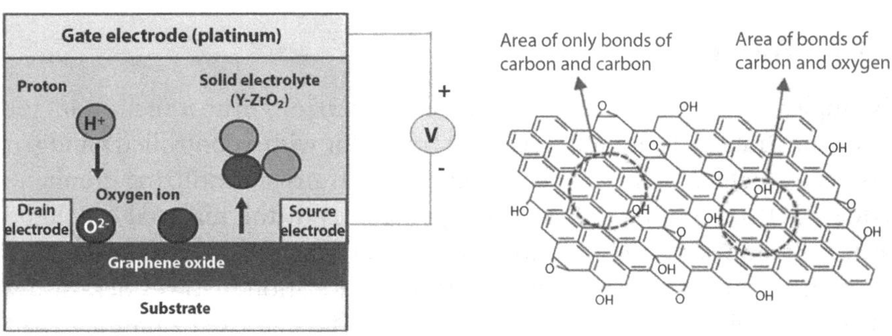

Figure 7.13 A transistor utilizing graphene oxide that offers adjustable bandgap, along with the crystal structure of GO [41].

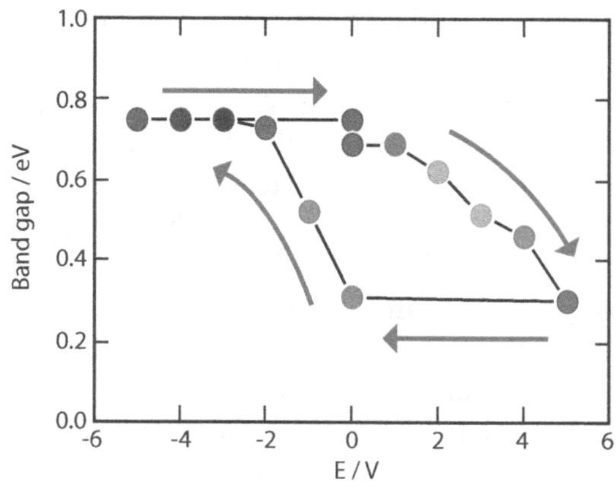

Figure 7.14 The connection between gate voltage (VG) and GO bandgap [55].

of the transistor scanning route and pulse delivery route. This innovative technique may lead to the advancement of low-energy, flexible neuromorphic devices. A tunable neuromorphic transistor was created by utilizing the redox process of WO_3 films that were thin. Exact control of the redox behavior was achieved by isolating the pulse applying route from the current sensing path. An individually tunable neuromorphic function was obtained by combining WO_3, a proton-electron mixed conductive oxide, with Nafion, a proton-conducting polymer electrolyte. Figure 7.15 displays

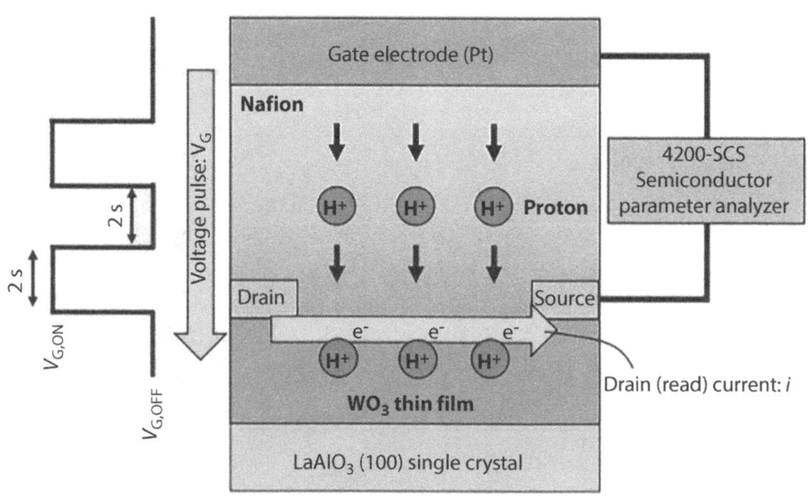

Figure 7.15 Visualization of a neuromorphic transistor constructed using a thin film of WO_3 and an electrolyte composed of proton-conducting Nafion [56].

the constructed neuromorphic transistor. An atomically flat surface of $LaAlO_3$ single crystal was coated with a 100 nm-thick WO_3 thin layer *via* radio frequency sputtering. With a constant drain voltage of 0.1 V, Figure 7.16 represents the correlation within the drain current and V_G; and illustrates the electrical conduction characteristic of the transistor [56]. The transistor exhibits a significant increase in the drain current from the sub-microampere region to the hundred microampere level, which is likely caused by proton migration in Nafion. This extraordinary 4-magnitude variation is seen at V_G higher than 1.2 V. When W^{6+} in WO_3 is doped with electrons to the W 5d orbital, it frequently reduces to a lesser oxidation state than 6^+. WO_3 has a large proton insertion/removal capability ($x < 0.5$) [57]. The roles of long-term potentiation and short-term memory were explained using the nanoionics principle. The current flowing through the drain showed a 4-order-of-magnitude rise because of the phase transition driven on by oxygen deprivation and proton insertion/desertion as a result of the redox process of WO_3 thin film. This approach is expected to yield neuromorphic devices that are adaptable and characterized by low power consumption. Figure 7.17 illustrates the fluctuation of current over time with the application of V_G maintained at 0.5 or -0.5 V [56].

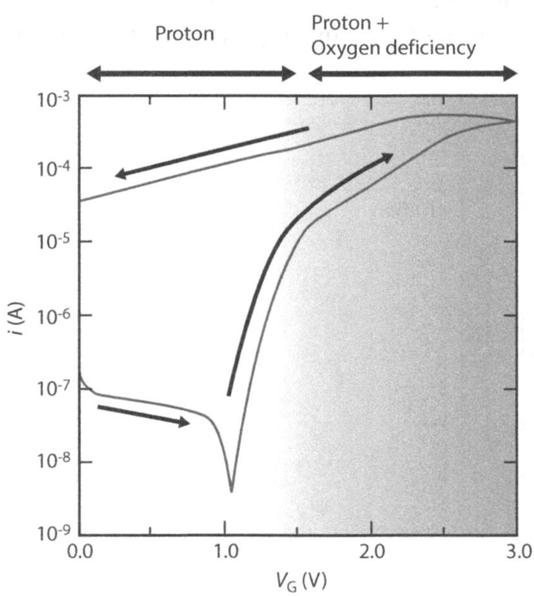

Figure 7.16 I-V properties of a transistor in a Nafion gated by protons transport. The rate at which the V_G was swept was 4 mV/s [56].

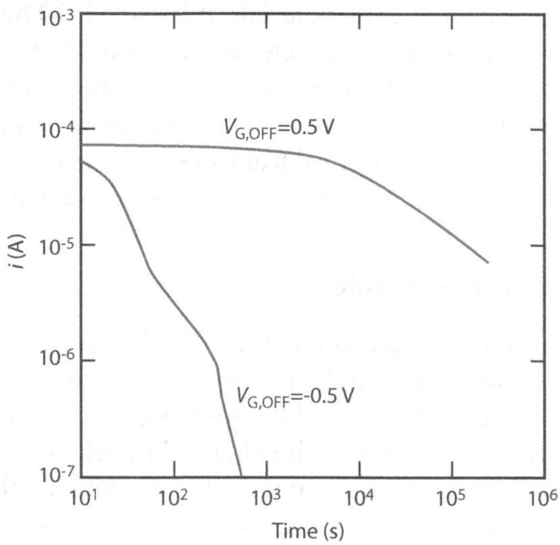

Figure 7.17 Change in current over time for the transistor while maintaining the applied V_G at either 0.5 or −0.5 V [56].

Enhanced retaining qualities have been established in a sodium-based nano-ionic sensory transistor. An all-solid-state Na-ion synaptic transistor is one suggested remedy for the low retention problem seen in LST. Analysis revealed that retention instability is mostly caused by the high ionic diffusivity in ionic synaptic transistors. Furthermore, near-ideal synaptic behaviors, have been made possible *via* material engineering. An Al_2O_3 barrier sheet was added to a LST with ultralow conductance to improve synaptic functions. The impact of an Al_2O_3 barrier layer was examined in an all-solid-state inorganic LST with a Li_3PO_4 electrolyte/ WO_x medium combinations. By sputtering a thin Al_2O_3 interfacial layer, near-ideal synaptic activity could be achieved in the ultralow conductivity region while abrupt ion migration could be limited. Cyclic voltammetry studies revealed that optimal ionic diffusion and accessibility were essential features for attaining perfect synaptic behavior. The retention qualities were significantly improved by the controlled ion migration facilitated by the Al_2O_3 barrier [58].

7.4 Application of Nanoionics in Optoelectronics

Nanoionics, which focuses on the influence and specification of ions at the nanoscale, holds the potential for various utilization within optoelectronics.

It enriches the characteristics of materials utilized in LED by managing the ion movement within them, creates elegant electrolytes with enhanced ion mobility, thereby improving the execution and consistency of dye-sensitized solar cells (DSSCs), and devises optical switches in which ion movement initiates modifications to the optical features of the material. The following are potential execution of nanoionics in the field of optoelectronics.

7.4.1 Light Emitting Diode

Nanoionics can be employed to enhance the charge transport efficiency within LED materials. Through the manipulation of ions at the nanoscale, the mobility of charge carriers can be improved, leading to more effective electron and hole transport within the LED and thereby increasing its overall efficiency. Nanoionics can also contribute to stabilizing the materials in LEDs, preventing degradation mechanisms, and extending the lifespan of LED. This has the potential to result in LED lighting solutions that are not only more efficient but are also longer-lasting and reliable. An example study scrutinized the augmented light extraction efficiency of LEDs through the development of two types of ZnO nanomaterials. The trials involved two categories of nanostructures, specifically one-dimensional nanorods and two-dimensional nanosheets, fostered directly atop LED surfaces. The initiation of surface patterning on LEDs with ZnO nanorods heightened surface dissipation and escape cone, pioneering a 30% improvement in light output power contrasted to traditional LEDs without ZnO nanostructures [59]. However, LEDs with nanosheet structures featured lower light output proficiency due to upgraded intrinsic reflection and light absorption in ZnO nanosheets. Despite this, the use of ZnO nanorods shows assurance for improving the outcome power of LEDs, with the added benefit of not severely altering the deterioration of electrical properties in LED devices. Nanoionics facilitates precise manipulation of the properties of phosphor materials utilized in LEDs, providing enhanced color-tuning capabilities. This development enables the production of LEDs with an improved color rendering and a wider spectrum of colors, a critical feature for applications requiring accurate color representation. Univalent alkali metal ions, such as Na^+, Li^+, and K^+ play a noteworthy role in this process due to their precise ionic radii and reduced oxidation state.

These ions can transform the provincial site symmetry of materials comprising rare earth ions, resulting in an overall enhancement in luminescence behavior. Among these alkali metal ions, Li^+ is particularly eminent in this scenario. It is extensively deployed due to its versatile contributions, including supporting charge trap population, acting as a flux, influencing the creation of oxygen vacancy, altering the crystal field, and affecting the scattering rate of Li^+ ions in the anchor matrix [60]. Recent developments have highlighted the role of Li^+ ions as luminescence enrichment for Eu^{3+} in materials like Y_2O_3, $CaSiO_3$, and Gd_2O_3 [61, 62]. Luminescence improvement of Dy^{3+} by Li^+ co-doping was described in ZnO and MgO [63], $CaMoO_4$ [64], $Ca_3B_2O_6$ [65], Ba_2GdNbO_6 [66], and YPO_4 [67]. Li-doped ZrO_2:Dy^{3+} and ZrO_2 doped with Dy^{3+} have been studied and optical properties have been analyzed. A standard algorithm was used to calculate the color coordinates as portrayed in the inset of Figure 7.18 [68]. The coordinated color temperature (CCT) of the phosphors was determined using the McCamy [69] concrete formula:

$$CCT = -437n^3 + 3601n^2 - 6861n + 5514.31$$

Here, n represents the reciprocal slope line. The color appearance diagram illustrates the versatility of the emission color in ZrO_2:Dy^{3+} phosphors co-doped Li^+, as shown by the yellow points.

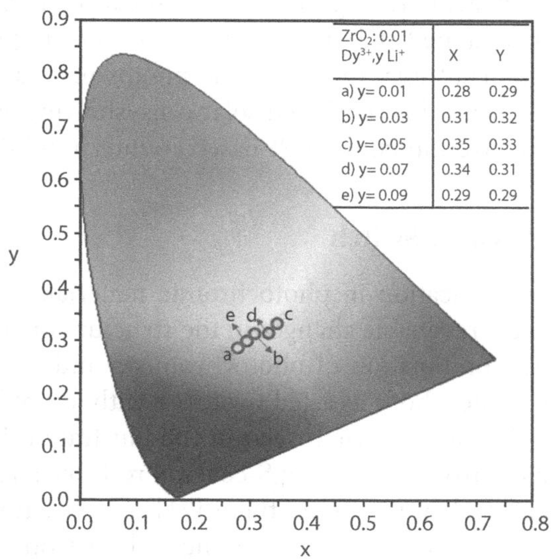

ZrO₂: 0.01 Dy³⁺,y Li⁺	X	Y
a) y= 0.01	0.28	0.29
b) y= 0.03	0.31	0.32
c) y= 0.05	0.35	0.33
d) y= 0.07	0.34	0.31
e) y= 0.09	0.29	0.29

Figure 7.18 Color coordinates calculated for the phosphors [68].

7.4.2 Solar Cell

The electrochemical stability of the electrolyte is a pivotal factor in determining the overall capability of solar cells. Nanoionics offers a means to tailor the properties of electrolyte materials, aiming for enriched electrochemical stability to prevent deterioration and assure prolonged functionality of organic solar cells. A noteworthy instance involves the use of $RbAg_4I_5$ in combination with β-AgI, which illustrates a record-breaking ionic conductivity. Within AgI, Ag^+ ions occasionally traverse back and forth within pairs of face-shared tetrahedra, overcoming high potential barriers. In superionic conductors, the I_4 tetrahedra contributes faces to at least two neighboring polyhedra, allowing Ag^+ ions to travel along structural channels throughout the entire crystal with minimal potential barriers [70]. Nanoionics methodologies can be employed to customize nanomaterials utilized in the photoelectrode of DSSCs. This may involve optimizing the structure, composition, and ion conductivity of materials to enhance charge transport and collection, ultimately leading to an improved overall efficiency of the solar cell. Some of the research focused on the formation of a titanium porous film-supported $NiCo_2S_4$ nanotube and its application as a counter electrode alternative to fluorine-doped tin oxide/Pt (FTO/Pt) in quantum-dot-sensitized solar cells of CdSe/CdS. The advanced porous counter electrode, created using a two-step hydrothermal process followed by acid etching, demonstrated excellent electrocatalytic characteristics, heightened stability, and improved loading capacity. This resulted in a significant enrichment in the photovoltaic achievement of the fabricated quantum-dot-sensitized solar cells, showing an enrichment compared to the conventional FTO/Pt-based counterpart [45].

7.4.3 Photo Assisted Switch

Nanoionics finds application in photochromic materials, wherein exposure to light triggers reversible changes in the structure or composition of the material. This combination of photochromism and ion mobility can be exploited to create photo-assisted switches with variable states controlled by light. The atomic switch seen in the illustration is made up of a Pt-covered Ag electrode and an Ag_2S-covered rod-like structure, with a gap of approximately 1 nm between them [71]. Ag structures are grown using Ag_2S, which functions as a hybrid ionic and electronic conductor in a specific manner. Ag^+ in the Ag_2S undergoes reduction when a positive bias is subjected to the Ag_2S/Ag electrode, allowing electrons from the Pt electrode to penetrate in. As a result, Ag atoms precipitate onto the Ag_2S

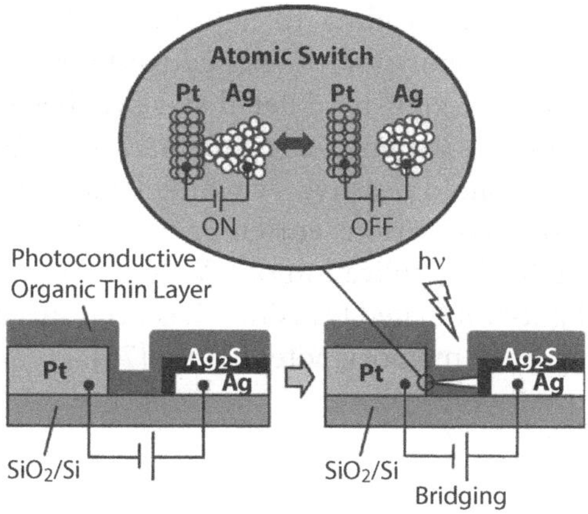

Figure 7.19 Diagrammatic representation of how an atomic switching is formed using photoassistance [73].

surface, creating an atomic bridge that connects the two electrodes and turns on the atomic switch. Then again, by providing a negative voltage to the Ag_2S/Ag electrode, the Ag structures are reduced, which shuts off the atomic switch. Utilizing a metal atomic bridge results in a smaller device and lower resistance, which makes this technology advantageous for certain uses such as programmable switches [72]. It has been introduced and demonstrated a novel category of atomic switch termed the "photo-assisted atomic switch." This innovative device leverages a photoconductive organic material, exterminating the need for an exceptionally small gap distance. In Figure 7.19, a schematic illustration of the photo-assisted atomic switch is shown. This recently developed gadget contains a small layer of a photoconductive organic molecule across the nanogap, a feature common to traditional atomic switches. Ag stubble does not extend when a forward bias is given to the Ag_2S/Ag electrode in low light conditions. Nonetheless, a photocurrent flows between the platinum and Ag_2S/Ag electrodes when light illumination takes place and a forward bias is provided to Ag_2S/Ag. This fills the gap and allows an Ag stubble to grow. The switch functions like a regular atomic switch after the connection is established [73].

7.4.4 High-Performance Optical Sensors

Nanoionics can be effectively used to create high-performance optoelectronic sensors. Specifically, the ionic conductor of photo-sensitive nitrogen-doped graphene QDs (N-GOQDs) exhibits n-p* electron shifts under UV excitation. The introduction of nitrogen doping further assists these electron transitions, leading to light-induced additional conductance [74]. A diffusive memristor, developed based on N-GOQDs shown in Figure 7.20 illustrated light-modulated synaptic behaviors. This clarifies the structure of the sentence and enhances readability. The device displays enduring threshold resistive switching behavior, attributed to the migration of Ag^{+} ions and various enrichments facilitated by assisted UV illumination. Notably, the application of UV light enhances post-synaptic current, and the light-stimulated transition has been successfully achieved from short-to-long-term memory potentiation [75].

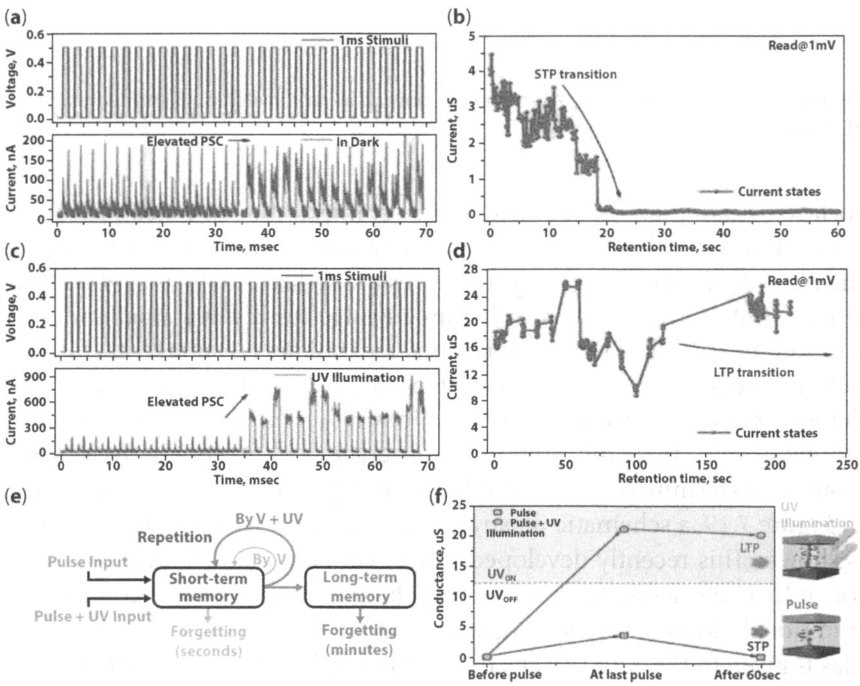

Figure 7.20 The shift from short term potentiation to long term potentiation with the aid of UV illumination in the synapse device is depicted as follows.: (a) Applying uniform short-duration spikes to the synapse device based on NGOQDs, along with the corresponding post synaptic current response (b) Weak retaining traits after stimuli-induced low-current post synaptic current, indicating a short-term potentiation transition in the sensor; (c) Applying the same brief pulses to the synapse device while keeping it under continuous UV light, which causes an increased post-synaptic current response; (d) Significant retention properties following greater current-elevated post synaptic current with UV aid, indicating long-term potentiation shift of the device; (e) Display of the model of memory schematic tailored for the NGOQDs-based optical synaptic device; and (f) the device's conductivity ranges before the pulse, towards the end of the pulse, in the dark, and under UV light, respectively [75].

7.5 Future Perspectives and Challenges

Nanoionics is expected to play a crucial role in advancing electronic devices, leading to faster, more efficient, and smaller components. Ensuring compatibility with existing technologies, scalability, and manufacturing processes will be essential for widespread adoption. Nanoionic materials have shown potential for high-density and ultra-fast memory technologies. However, achieving long-term stability, endurance, and addressing material degradation issues will be critical for practical implementation. Integration of nanoionics into optoelectronic devices could lead to improved performance and energy harvesting capabilities, developing materials that efficiently combine electronic and optical functionalities, and addressing issues like heat dissipation and stability under light exposure, controlling ion movement precisely, reducing energy consumption, and achieving compatibility with conventional electronic components. Nanoionic materials could revolutionize energy storage devices, enhancing energy density and charge/discharge rates, ensuring safety, addressing cycle life limitations, and optimizing the manufacturing processes for large-scale energy storage systems. Nanoionics may contribute to the development of flexible and wearable electronics with improved durability, maintaining performance under mechanical stress, ensuring flexibility in material design, and addressing potential health and environmental concerns. Collaborative efforts with other nanotechnologies could result in multifunctional devices with unprecedented capabilities, achieving seamless integration, overcoming compatibility issues, and optimizing synergies between different nanotechnologies. Nanoionic sensors may offer high sensitivity and selectivity for environmental monitoring; calibration for real-world conditions, ensuring long-term stability, and addressing potential interference from external factors. Ensuring the stability of nanoionic materials over extended usage and environmental conditions; scaling up nanoionic technologies for mass production while maintaining performance and cost-effectiveness; bridging the gap between disciplines to encourage collaboration between materials science, electronics, and optoelectronics; establishing standards for nanoionic devices to facilitate compatibility and interoperability in various applications; assessing and mitigating potential environmental impacts associated with the production and disposal of nanoionic materials; navigating regulatory processes for the approval of nanoionic devices in different applications and industries; as research in nanoionics progresses, addressing these challenges will be crucial for realizing the full potential of technologies of nanoionics in electronics and optoelectronics. The interdisciplinary nature of the field requires collaboration between scientists, engineers, and policymakers to ensure responsible and effective development and implementation.

7.6 Conclusions

In conclusion, the exploration of nanoionics in the realms of electronics and optoelectronics reveals a promising landscape with transformative potential. Investigation into nanoionic materials and devices has illuminated pathways toward advanced electronic technologies, memory devices, optoelectronic applications, and beyond. The convergence of ionics and nanotechnology opens avenues for innovation, pushing the boundaries of what is achievable in these fields. The fundamental understanding of ion dynamics within nanostructures has enabled the development of novel electronic components and memory technologies. Nanoionics has emerged as a key player in the pursuit of faster, more efficient, and scalable electronic devices. The potential for high-density, ultra-fast memory systems is a testament to the impact of nanoionics on data storage technologies. Optoelectronic applications stand to gain significantly from the integration of nanoionics, promising improved device performance and energy harvesting capabilities. The synergy between electronic and optical functionalities holds great potential for addressing challenges in modern optoelectronic devices. The potential to create robust and flexible devices opens new possibilities for applications in various industries. Collaboration with other nanotechnologies presents exciting prospects for multifunctional devices with unprecedented capabilities. The integration of diverse nanotechnological approaches could lead to breakthroughs in device functionalities and performance. Despite the optimism surrounding nanoionics, several challenges must be addressed. In essence, the application of nanoionics in electronics and optoelectronics marks a transformative journey, offering unprecedented opportunities and challenges. The journey into the nanoionics frontier continues, promising a future where electronic and optoelectronic technologies reach.

References

1. Despotuli, A.L. and Nikolaichik, V.I., A step towards nanoionics. *Solid State Ionics*, 60, 275–278, 1993.

2. Maier, J., Nanoionics: Ion transport and electrochemical storage in confined systems. *Nat. Mater.*, 4, 805–815, 2005.
3. Maier, J., Defect chemistry and ionic conductivity in thin films. *Solid State Ionics*, 23, 59–67, 1987.
4. Maier, J., Ionic transport in nano-sized systems. *Solid State Ionics*, 175, 7–12, 2004.
5. Maier, J., Ionic conduction in space charge regions. *Prog. Solid State Chem.*, 23, 171–263, 1995.
6. Schoonman, J., Nanostructured materials in solid state ionics. *Solid State Ionics*, 135, 5–19, 2000.
7. Gomes, A.R.Q., Cruz, J.N., Castro, A.L.G., Cordovil Brigido, H.P., Varela, E.L.P., Vale, V.V., Carneiro, L.A., Ferreira, G.G., Percario, S., Dolabela, M.F., Participation of Oxidative Stress in the Activity of Compounds Isolated from Eleutherine Plicata Herb. *Molecules*, 28, 5557, 2023.
8. Tuller, H.L., Ionic conduction in nanocrystalline materials. *Solid State Ionics*, 131, 143–157, 2000.
9. Lee, J.H., Park, J.H., Dongale, T.D., Kim, T.G., Vacancy-modulated self-rectifying characteristics of NiO_x/Al_2O_3-based nanoscale ReRAM devices. *J. Alloys Compd.*, 821, 153247, 2020.
10. Sahoo, S., Manoravi, P., Prabaharan, S.R.S., Titania Based Nano-ionic Memristive Crossbar Arrays: Fabrication and Resistive Switching Characteristics. *Nanosci. Nanotechnol.-Asia*, 9, 486–493, 2018.
11. Dash, C.S., Sahoo, S., Prabaharan, S.R.S., Resistive switching and impedance characteristics of $M/TiO_{2-x}/TiO_2/M$ nano-ionic memristor. *Solid State Ionics*, 324, 218–225, 2018.
12. Lee, M.J., Lee, C.B., Lee, D., Lee, S.R., Chang, M., Hur, J.H., Kim, Y.B., Kim, C.J., Seo, D.H., Seo, S., Chung, U.I., Yoo, I.K., Kim, K., A fast, high-endurance and scalable non-volatile memory device made from asymmetric Ta_2O_{5-x}/TaO_2-xbilayer structures. *Nat. Mater.*, 10, 625–630, 2011.
13. Hatem, F.O., Kumar, T.N., Almurib, H.A.F., A SPICE Model of the Ta_2O_5/TaO_x Bi-Layered RRAM. *IEEE Trans. Circuits Syst. I Regul. Pap.*, 63, 1487–1498, 2016.
14. Sun, B., Zhang, X., Zhou, G., Zhang, C., Li, P., Xia, Y., Zhao, Y., Effect of Cu ions assisted conductive filament on resistive switching memory behaviors in $ZnFe_2O_4$-based devices. *J. Alloys Compd.*, 694, 464–470, 2017.
15. Zhou, G., Yang, X., Xiao, L., Sun, B., Zhou, A., Investigation of a submerging redox behavior in Fe_2O_3 solid electrolyte for resistive switching memory. *Appl. Phys. Lett.*, 114, 1635061-5, 2019.
16. Hota, M.K., Bera, M.K., Maiti, C.K., Switching mechanism in au nanodot-embedded Nb_2O_5 memristors. *J. Nanosci. Nanotechnol.*, 14, 3538–3544, 2014.
17. Sahoo, S., Conduction and switching behavior of e-beam deposited polycrystalline Nb_2O_5 based nano-ionic memristor for non-volatile memory applications. *J. Alloys Compd.*, 866, 158394, 2021.

18. Hu, S.G., Wu, S.Y., Jia, W.W., Yu, Q., Deng, L.J., Fu, Y.Q., Liu, Y., Chen, T.P., Review of nanostructured resistive switching memristor and its applications. *Nanosci. Nanotechnol. Lett.*, 6, 729–757, 2014.
19. Van De Burgt, Y., Lubberman, E., Fuller, E.J., Keene, S.T., Faria, G.C., Agarwal, S., Marinella, M.J., Alec Talin, A., Salleo, A., A non-volatile organic electrochemical device as a low-voltage artificial synapse for neuromorphic computing. *Nat. Mater.*, 16, 414–418, 2017.
20. Ramos, I.N.F., da Silva, M.F., Lopes, J.M.S., Cruz, J.N., Alves, F.S., do Rego, J.A.R., da Costa, M.L., de Assumpção, P.P., Barros Brasil, D.S., Khayat, A.S., Extraction, Characterization, and Evaluation of the Cytotoxic Activity of Piperine in Its Isolated form and in Combination with Chemotherapeutics against Gastric Cancer. *Molecules*, 28, 5587, 2023.
21. Park, M., Zhang, X., Chung, M., Less, G.B., Sastry, A.M., A review of conduction phenomena in Li-ion batteries. *J. Power Sources*, 195, 7904–7929, 2010.
22. Nagai, J., Optical And Electrical Properties Of Amorphous Li_xWO_3 Films. *Opt. Mater. Technol. Energy Effic. Sol. Energy Convers. VII*, vol. 1016, p. 28, 1989.
23. Fuller, E.J., El Gabaly, F., Léonard, F., Agarwal, S., Plimpton, S.J., Jacobs-Gedrim, R.B., James, C.D., Marinella, M.J., Talin, A.A., Li-Ion Synaptic Transistor for Low Power Analog Computing. *Adv. Mater.*, 29, 1–14, 2017.
24. Yiyang, L., Fuller, E.J., Asapu, S., Agarwal, S., Kurita, T., Yang, J.J., Talin, A.A., Low-Voltage, CMOS-Free Synaptic Memory Based on Li_xTiO_2 Redox Transistors. *ECS Meet. Abstr.*, MA2019-02, 1227–1227, 2019.
25. Nikam, R.D., Kwak, M., Lee, J., Rajput, K.G., Hwang, H., Controlled Ionic Tunneling in Lithium Nanoionic Synaptic Transistor through Atomically Thin Graphene Layer for Neuromorphic Computing. *Adv. Electron. Mater.*, 6, 19011001–10, 2020.
26. Tang, J., Bishop, D., Kim, S., Copel, M., Gokmen, T., Todorov, T., Shin, S., Lee, K.T., Solomon, P., Chan, K., Haensch, W., Rozen, J., ECRAM as Scalable Synaptic Cell for High-Speed, Low-Power Neuromorphic Computing. *Tech. Dig. - Int. Electron Devices Meet. IEDM*, 2018-December, pp. 13.1.1–13.1.4, 2018.
27. Bishop, D.M., Solomon, P., Kim, S., Tang, J., Tersoff, J., Todorov, T., Copel, M., Collins, J., Lee, K.T., Shin, S., Haensch, W., Rozen, J., Time-resolved Conductance in Electrochemical Systems for Neuromorphic Computing. *Ext. Abstr. 2018 Int. Conf. Solid State Devices Mater.*, 2020.
28. Lee, J., Nikam, R.D., Lim, S., Kwak, M., Hwang, H., Excellent synaptic behavior of lithium-based nano-ionic transistor based on optimal $WO_{2.7}$ stoichiometry with high ion diffusivity. *Nanotechnology*, 31, 235203, 2020.
29. Long, S., Lian, X., Cagli, C., Perniola, L., Miranda, E., Liu, M., Suñé, J., A model for the set statistics of rram inspired in the percolation model of oxide breakdown. *IEEE Electron Device Lett.*, 34, 999–1001, 2013.
30. Liu, S., Liu, R., Chen, Y., Ho, S., Kim, J.H., So, F., Nickel oxide hole injection/transport layers for efficient solution-processed organic light-emitting diodes. *Chem. Mater.*, 26, 4528–4534, 2014.

31. Sun, Y., Jiang, Y., Sun, X.W., Zhang, S., Chen, S., Beyond OLED: Efficient Quantum Dot Light-Emitting Diodes for Display and Lighting Application. *Chem. Rec.*, 19, 1729–1752, 2019.
32. Ji, W., Shen, H., Zhang, H., Kang, Z., Zhang, H., Over 800% efficiency enhancement of all-inorganic quantum-dot light emitting diodes with an ultrathin alumina passivating layer. *Nanoscale*, 10, 11103–11109, 2018.
33. Yang, X., Zhang, Z.H., Ding, T., Wang, N., Chen, G., Dang, C., Demir, H.V., Sun, X.W., High-efficiency all-inorganic full-colour quantum dot light-emitting diodes. *Nano Energy*, 46, 229–233, 2018.
34. Tay, Y.Y., Tan, T.T., Liang, M.H., Boey, F., Li, S., Specific defects, surface band bending and characteristic green emissions of ZnO. *Phys. Chem. Chem. Phys.*, 12, 6008–6013, 2010.
35. Mashford, B.S., Stevenson, M., Popovic, Z., Hamilton, C., Zhou, Z., Breen, C., Steckel, J., Bulovic, V., Bawendi, M., Coe-Sullivan, S., Kazlas, P.T., High-efficiency quantum-dot light-emitting devices with enhanced charge injection. *Nat. Photonics*, 7, 407–412, 2013.
36. Tsuchiya, T., Tsuruoka, T., Terabe, K., Aono, M., *In situ* and nonvolatile photoluminescence tuning and nanodomain writing demonstrated by all-solid-state devices based on graphene oxide. *ACS Nano*, 9, 2102–2110, 2015.
37. Li, H. and Nogami, M., Pore-controlled proton conducting silica films. *Adv. Mater.*, 14, 912–914, 2002.
38. Kim, S., Nanostructured Solid-State Electrochemical Systems (Applications and Challenges). *Electrochem. Soc. Interface*, 15, 28–30, 2006.
39. Liang, C.C., Conduction Characteristics of the Lithium Iodide-Aluminum Oxide Solid Electrolytes. *J. Electrochem. Soc.*, 120, 1289, 1973.
40. Sata, N., Eberman, K., Eberl, K., Maier, J., Mesoscopic fast ion conduction in nanometre-scale planar heterostructures. *Nature*, 408, 946–949, 2000.
41. Terabe, K., Tsuchiya, T., Yang, R., Aono, M., Nanoionic devices enabling a multitude of new features. *Nanoscale*, 8, 13873–13879, 2016.
42. Wagner, C., Physical chemistry of ionic crystals involving small concentrations of foreign substances. *J. Phys. Chem.*, 57, 738–742, 1953.
43. Hirose, Y. and Hirose, H., Polarity-dependent memory switching and behavior of Ag dendrite in Ag-photodoped amorphous As2S3 films. *J. Appl. Phys.*, 47, 2767–2772, 1976.
44. Calvo, R.J., Pennington Corner: Just Around the Corner. *Electrochem. Soc. Interface*, 8, 9–9, 1999.
45. Terabe, K., Hasegawa, T., Nakayama, T., Aono, M., Quantized conductance atomic switch. *Nature*, 433, 47–50, 2005.
46. Kund, M., Beitel, G., Pinnow, C.U., Röhr, T., Schumann, J., Symanczyk, R., Ufert, K.D., Müller, G., Conductive bridging RAM (CBRAM): An emerging non-volatile memory technology scalable to sub 20nm. *Tech. Dig. - Int. Electron Devices Meet. IEDM, 2005*, pp. 754–757, 2005.

47. Baiatu, T., Waser, R., Härdtl, K.-H., dc Electrical Degradation of Perovskite-Type Titanates: III, A Model of the Mechanism. *J. Am. Ceram. Soc.*, 73, 1663–1673, 1990.
48. Watanabe, Y., Bednorz, J.G., Bietsch, A., Gerber, C., Widmer, D., Beck, A., Wind, S.J., Current-driven insulator-conductor transition and nonvolatile memory in chromium-doped $SrTiO_3$ single crystals. *Appl. Phys. Lett.*, 78, 3738–3740, 2001.
49. Ogimoto, Y., Tamai, Y., Kawasaki, M., Tokura, Y., Resistance switching memory device with a nanoscale confined current path. *Appl. Phys. Lett.*, 90, 1435151–3, 2007.
50. Choi, B.J., Jeong, D.S., Kim, S.K., Rohde, C., Choi, S., Oh, J.H., Kim, H.J., Hwang, C.S., Szot, K., Waser, R., Reichenberg, B., Tiedke, S., Resistive switching mechanism of TiO_2 thin films grown by atomic-layer deposition. *J. Appl. Phys.*, 98, 0337151–10, 2005.
51. Beaulieu, R.P., Sulway, D.V., Cox, C.D., The detection of current filaments in VO_2 thin-film switches using the scanning electron microscope. *Solid State Electron.*, 16, 428–429, 1973.
52. Pinto, R., Filamentary switching and memory action in thin anodic oxides. *Phys. Lett. A*, 35, 155–156, 1971.
53. Rossel, C., Meijer, G.I., Brémaud, D., Widmer, D., Electrical current distribution across a metal-insulator-metal structure during bistable switching. *J. Appl. Phys.*, 90, 2892–2898, 2001.
54. Szot, K., Speier, W., Bihlmayer, G., Waser, R., Switching the electrical resistance of individual dislocations in single-crystalline $SrTiO_3$. *Nat. Mater.*, 5, 312–320, 2006.
55. Tsuchiya, T., Terabe, K., Aono, M., *In situ* and non-volatile bandgap tuning of multilayer graphene oxide in an all-solid-state electric double-layer transistor. *Adv. Mater.*, 26, 1087–1091, 2014.
56. Tsuchiya, T., Jayabalan, M., Kawamura, K., Takayanagi, M., Higuchi, T., Jayavel, R., Terabe, K., Neuromorphic transistor achieved by redox reaction of WO_3 thin film. *Jpn. J. Appl. Phys.*, 57, 04FK01, 2018.
57. Glemser, O. and Naumann, C., Kristallisierte Wolframblauverbindungen; Wasserstoffanaloga der Wolframbronzen H_xWO_3. *ZAAC - J. Inorg. Gen. Chem.*, 265, 288–302, 1951.
58. Lee, K., Kwak, M., Choi, W., Lee, C., Lee, J., Noh, S., Lee, J., Lee, H., Hwang, H., Improved synaptic functionalities of Li-based nano-ionic synaptic transistor with ultralow conductance enabled by Al_2O_3 barrier layer. *Nanotechnology*, 32, 275201, 2021.
59. Nguyen, H.P.T., Arafin, S., Piao, J., Cuong, T.V., Nanostructured Optoelectronics: Materials and Devices. *J. Nanomater.*, 2016, 1–2, 2016.
60. Liu, B., Gu, M., Liu, X., Ni, C., Wang, D., Xiao, L., Zhang, R., Effect of Zn^{2+} and Li^+ codoping ions on nanosized Gd_2O_3:Eu^{3+} phosphor. *J. Alloys Compd.*, 440, 341–345, 2007.

61. Hari Krishna, R., Nagabhushana, B.M., Nagabhushana, H., Chakradhar, R.P.S., Sivaramakrishna, R., Shivakumara, C., Thomas, T., Auto-ignition based synthesis of Y_2O_3 for photo- and thermo-luminescent applications. *J. Alloys Compd.*, 585, 129–137, 2014.
62. Gu, F., Li, C., Cao, H., Shao, W., Hu, Y., Chen, J., Chen, A., Crystallinity of Li-doped MgO:Dy^{3+} nanocrystals *via* combustion process and their photoluminescence properties. *J. Alloys Compd.*, 453, 361–365, 2008.
63. Gu, F., Wang, S.F., Lu, M.K., Zhou, G.J., Xu, D., Yuan, D.R., Structure evaluation and highly enhanced luminescence of Dy^{3+}-doped ZnO nanocrystals by Li^+ doping *via* combustion method. *Langmuir*, 20, 3528–3531, 2004.
64. Guan, L., Jia, G., Yang, B., Li, X., Jin, L., Yang, Z., Fu, G., Synthesis and optical properties of Dy^{3+}, Li^+ doped CaMoO4 phosphor. *J. Rare Earths*, 29, 540–543, 2011.
65. Yu, C.C., Liu, X.M., Yu, M., Lin, C.K., Li, C.X., Wang, H., Lin, J., Enhanced photoluminescence of Ba_2GdNbO_6: Eu^{3+}/Dy^{3+} phosphors by Li^+ doping. *J. Solid State Chem.*, 180, 3058–3065, 2007.
66. Sun, X.Y., Zhang, J.C., Liu, X.G., Lin, L.W., Enhanced luminescence of novel $Ca_3B_2O_6$:Dy^{3+} phosphors by Li^+-codoping for LED applications. *Ceram. Int.*, 38, 1065–1070, 2012.
67. Parchur, A.K., Prasad, A.I., Rai, S.B., Ningthoujam, R.S., Improvement of blue, white and NIR emissions in YPO_4:Dy^{3+} nanoparticles on co-doping of Li^+ ions. *Dalton Trans.*, 41, 13810–13814, 2012.
68. Prakashbabu, D., Ramalingam, H.B., Krishna, R.H., Nagabhushana, B.M., Shivakumara, C., Munirathnam, K., Ponkumar, S., A potential white light emitting cubic ZrO_2:Dy^{3+}, Li^+ nano phosphors for solid-state lighting applications. *J. Lumin.*, 192, 496–503, 2017.
69. McCamy, C.S., Correlated color temperature as an explicit function of chromaticity coordinates. *Color Res. Appl.*, 17, 142–144, 1992.
70. Raleigh, D.O., Structural Principles for Silver Halide Superionic Conductors. *J. Electrochem. Soc.*, 124, 1157–1160, 1977.
71. Deng, J., Wang, M., Song, X., Yang, Z., Yuan, Z., Ti porous film-supported $NiCo_2S_4$ nanotubes counter electrode for quantum-dot-sensitized solar cells. *Nanomaterials*, 8, 251, 2018.
72. Miracle, D.B., Overview No. 104 The physical and mechanical properties of NiAl. *Acta Metall. Mater.*, 41, 649–684, 1993.
73. Hino, T., Tanaka, H., Hasegawa, T., Aono, M., Ogawa, T., Photoassisted formation of an atomic switch. *Small*, 6, 1745–1748, 2010.
74. Abbas, H., Ali, A., Jung, J., Hu, Q., Park, M.R., Lee, H.H., Yoon, T.S., Kang, C.J., Reversible transition of volatile to non-volatile resistive switching and compliance current-dependent multistate switching in IGZO/MnO RRAM devices. *Appl. Phys. Lett.*, 114, 0935031-5, 2019.
75. Ali, M., Sokolov, A., Ko, M.J., Choi, C., Optically excited threshold switching synapse characteristics on nitrogen-doped graphene oxide quantum dots (N-GOQDs). *J. Alloys Compd.*, 855, 157514, 2021.

8

Challenges and Opportunities in Nanoionics: Towards Breakthrough Applications

Saranya J.*, Selvakumar V. S., Suganthi S., T. Helan Vidhya and Dinesh K.

Department of Electronics and Communication Engineering, Rajalakshmi Engineering College, Thandalam, Tamil Nadu, India

Abstract

Nanoionics, a burgeoning interdisciplinary field at the confluence of nanotechnology and ionics, offers unprecedented insights into ion behaviour within nanoscale constraints. This exploration, delving deep into the world of atomic and molecular interactions, holds transformative potential across diverse domains, prominently in energy storage, sensors, and Nano electronics. Nanoionics' essence lies in comprehending ion transport and manipulation within nanoscale landscapes, encompassing thin films, nanoparticles, and nanostructured interfaces. This understanding is the bedrock for revolutionary technologies, poised to revolutionize energy storage, substance detection, and data processing. In energy storage, nanoionics propels advancements in solid-state batteries and fuel cells through nanostructured materials, boosting performance and stability. This field also pioneers the development of sensitive sensors and innovative memory devices like resistive random-access memory (ReRAM). Nanoionics, characterized by its amalgamation of nanoscale precision and ionic exploration, stands as a harbinger of transformative advancements across industries, promising a future shaped by science and technology. This article provides deep insights on the myriad challenges that confront the field of nanoionics and showcase the complex interplay of materials, processes, and technologies at the nanoscale, highlighting scalability hurdles and safety concerns.

Keywords: Nanoionics, nano electronics, nanoscale, energy storage, sensor, nanostructured materials, resistive random access memory, nanotechnology

**Corresponding author*: saranya.j@rajalakshmi.edu.in

Inamuddin, Tariq Altalhi, Mohammad Luqman and Jorddy Neves Cruz (eds.) Nanoionics: Fundamentals and Applications, (195–232) © 2025 Scrivener Publishing LLC

8.1 Introduction

Nanoionics, an innovative and rapidly advancing field at the intersection of nanotechnology and ionics, explores the complex behaviour and manipulation of ions within nanoscale environments, unlocking the potential of materials at the atomic and molecular levels. This exploration is crucial for understanding the unique ion movement and interaction in confined spaces, leading to significant technological advancements in various sectors, particularly in energy storage, sensors, and nano electronics [1]. The core of nanoionics includes the study of transportation of ions and its modulation within nanoscale regions, encompassing the investigation of ion conduction mechanisms using different materials and structures such as thin films, nanoparticles, and nanostructured interfaces [2]. This deep understanding is pivotal for developing advanced technologies that can revolutionise how we store and utilise energy, detect various substances, and process information. Nanoionics research has been crucial in the development of high-performance solid-state batteries and fuel cells in the field of energy storage. These devices rely on solid electrolytes to enhance their ionic conductivity and stability. The integration of nanostructured materials in electrolytes has shown promising results, significantly improving their performance and efficiency [3]. This advancement is not just limited to batteries but extends to the development of more efficient and cleaner energy sources like fuel cells, where nanoionics is contributing to overcoming challenges related to ion transport and material stability under operational conditions. The field also makes substantial contributions to sensor technology, where nanoionic materials, with their high surface area and unique ionic properties, offer enhanced sensitivity and faster response times. This makes them ideal for detecting a wide range of chemical and biological substances with higher accuracy and efficiency than traditional methods.

Moreover, nanoionics is at the forefront of developing innovative memory devices like resistive random-access memory (ReRAM) [4]. These devices exploit the movement of ions within nanostructures to create more efficient, scalable, and faster memory storage solutions, which are crucial in the era of big data and advanced computing. As nanoionics continues to grow and evolve, it stands as a beacon of scientific research and innovation, promising revolutionary applications across multiple industries [5]. The ongoing exploration in understanding ionic phenomena at the nanoscale is not only reshaping our existing knowledge but also paving the way for technological innovations that could significantly impact our daily lives and the future of various technologies. Nanoionics is a monument to the potential of interdisciplinary research and its role in pushing the boundaries of science and technology, from producing more sensitive sensors and energy storage systems to build the next generation of electronic products [6]. In recent years, substantial research efforts have been directed towards further understanding and enhancing nanoionic mechanisms. For instance, researchers at MIT and Stanford University have made significant strides in developing advanced solid-state batteries leveraging nanoionic principles. Their work focuses on creating solid electrolytes with higher ionic conductivities and stability, which could potentially revolutionise the battery industry [7].

Another notable area of research is in the development of nanoionic-based memristors and resistive random-access memory (ReRAM) devices. Institutions like the University of California and the Max Planck Institute are exploring how nanoionic mechanisms can be harnessed for creating more efficient data storage solutions [8]. These devices utilise ion movement in nanostructures to alter electrical resistance, enabling non-volatile memory that combines the speed of RAM with the permanence of flash storage.

Furthermore, research in nanoionic sensors has gained momentum, with teams from universities like Oxford and the National University of Singapore delving into creating highly sensitive and selective nanoionic sensors [9]. These sensors are being developed for a variety of uses, including early illness identification in medical settings and environmental pollution detection.

8.2 Mechanism Behind Nanoionics

Ion transport mechanisms at the nanoscale are the main focus in the area of nanoionics, which carefully combines the ideas of ionics and nanotechnology [10]. This section delves into the fundamental processes that govern nanoionic mechanisms. At this scale, ions behave differently due to the increased surface-to-volume ratio and quantum effects, leading to unique ion conduction properties.

The nanoionic mechanism fundamentally revolves around the movement of ions through a medium, which could be a solid electrolyte or a nanostructured material. This movement is often driven by an external stimulus such as an electric field, temperature gradient, or chemical potential difference [11]. The transport of ions through nanomaterials is governed by several factors, including the size and shape of the nanoparticles, the nature of the ion-conducting pathways, and the interaction of ions with the surface of the material [12].

In solid-state nanoionics, ion transport occurs through solid electrolytes with nanoscale features. These materials are designed to have high ionic conductivity while maintaining electronic insulation. This is achieved through the careful manipulation of material composition and structure at the nanoscale, creating pathways that facilitate rapid ion movement. For instance, nanostructuring can produce certain lattice patterns or an increase in defects in the electrolyte, which act as conduits for accelerated ion transport [13].

The efficiency of ion transport in nanoionic systems is also influenced by the interface between different materials. At the nanoscale, interfaces can significantly alter the properties of materials, often leading to enhanced ionic conduction. This is particularly important in composite electrolytes, where the interface between different phases can be engineered to optimise ion transport.

The unique mechanisms of ion transport in nanoionics have opened up a plethora of applications across various fields. In energy storage, nanoionic materials have been instrumental in the development of more efficient and higher capacity batteries [14]. The enhanced ion transport in these materials allows for faster charging and discharging cycles, thus making them ideal for applications requiring rapid energy delivery, like electric vehicles and portable electronic devices.

In the field of sensors, nanoionic materials, owing to their high sensitivity to changes in their ionic environment, are used to detect a wide range of substances. These sensors are capable of detecting minute changes in concentration of gases or liquids, making them invaluable in environmental monitoring, medical diagnostics, and industrial process control [15].

8.3 Significance of Nanomaterials in Nanoionics

In the realm of nanoionics, the use of nanomaterials represents a paradigm shift, heralding new possibilities in material science and engineering. This section will explore the multifaceted world of nanomaterials within the context of nanoionics, illuminating how these tiny yet powerful materials are redefining the boundaries of science and technology as illustrated in Figure 8.1. Nanoionics, at its core, is concerned with the movement and behaviour of ions at the nanoscale. The field stands at the crossroads of

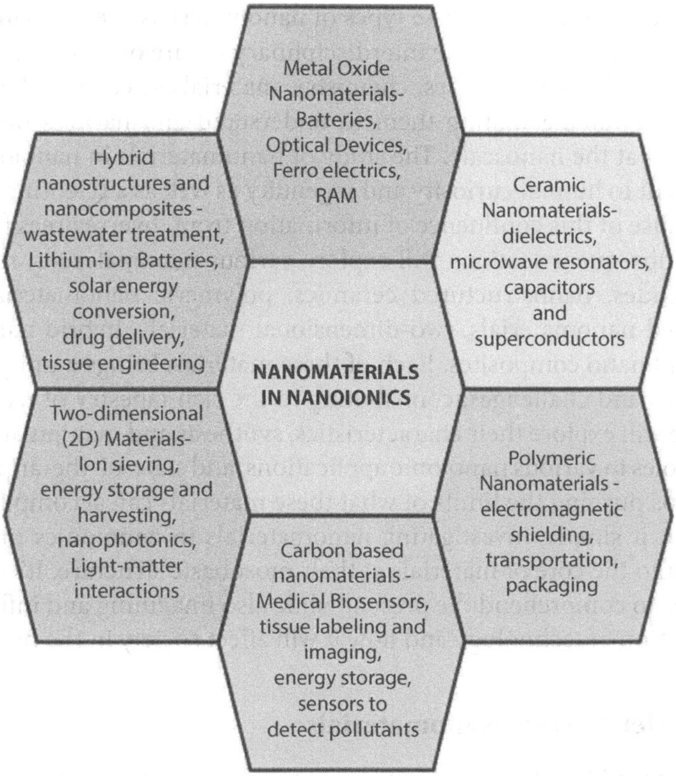

Figure 8.1 Usage of nanomaterials in various applications.

traditional ionics and nanotechnology, merging the principles of ionic conductivity with the unique properties that emerge at the nanoscale [16]. Their dimensions, often only a few nanometres across, bestow them with extraordinary physical, chemical, and electrical properties that are not observable in their bulk counterparts. This is the realm where quantum effects and increased surface area come into play, leading to enhanced reactivity and novel phenomena which is pivotal for nanoionic applications.

Nanoionics offer a new dimension of control over ionic transport and electrochemical reactions, which is crucial for the development of advanced technologies. From energy storage in batteries and capacitors to sensing devices and next-generation electronics, nanomaterials are reshaping the landscape of these applications. They put forth improvements in efficiency, speed, and scalability, which paves way to entirely new functionalities that were previously unimaginable [17].

As we delve into the diverse types of nanomaterials used in nanoionics, it's essential to appreciate the interdisciplinary nature of this field. It draws upon principles from physics, chemistry, materials science, and electrical engineering, amalgamating them to understand and harness the potential of ions at the nanoscale. The study of nanomaterials in nanoionics is a monument to human curiosity and ingenuity as well as a scientific endeavour because of this confluence of information from diverse areas [18].

The subsequent sections will explore various nanomaterials, including metal oxides, nanostructured ceramics, polymeric nanomaterials, carbon-based nanomaterials, two-dimensional materials, hybrid nanostructures, and nano composites. Each of these materials brings a unique set of properties and challenges, contributing to the rich tapestry of nanoionics [19]. We will explore their characteristics, synthesis and manipulation processes, roles in various nanoionic applications, and state-of-the-art research that keeps pushing the limits of what these materials can accomplish.

To put it simply, investigating nanomaterials in nanoionics involves a voyage into the core of materials at their most basic structure. It's a voyage that aims to comprehend the present while also imagining and influencing the direction of technology and how it will affect society in the future [20].

8.3.1 Metal Oxide Nanomaterials

Metal oxides like titanium dioxide (TiO_2) and zirconium dioxide (ZrO_2) have emerged as significant players in nanoionics. Their wide bandgap, stability, and strong ionic conductivity make them perfect for use in solid-state batteries and sensors, among other applications. Recent advancements in doping these oxides with elements like lithium have shown enhanced ionic transport properties, making them even more efficient. Li-doped TiO_2, for example, has shown significantly improved lithium-ion mobility, which is important for high-performance battery applications [21].

8.3.2 Ceramic Nanomaterials

Nanostructured ceramics, especially those based on perovskite structures, have attracted considerable attention in nanoionics. These materials, characterised by their unique crystal structures, offer tailored pathways for ion transport, making them highly effective as electrolytes in fuel cells and batteries [22]. Recent research has focused on refining the composition and structure of these ceramics to produce greater ionic conductivities at lower temperatures, boost their energy efficiency, and make them suitable for commercial use.

8.3.3 Polymeric Nanomaterials

The incorporation of polymers into nanoionics represents a revolutionary advancement, providing adaptability and adjustable characteristics. In solid polymer electrolytes, polymers such as polyethylene oxide (PEO) integrated with nanoscale ionic conductors have demonstrated encouraging outcomes [23]. Recent research has focused on refining the composition and structure of these ceramics to create greater ionic conductivities at lower temperatures, boost their energy efficiency, and make them suitable for commercial uses.

8.3.4 Carbon-Based Nanomaterials

Nanomaterials such as carbon, graphene, and carbon nanotubes (CNTs), have completely changed the field of nanoionics [24]. Their exceptional electrical conductivity, high surface area, and mechanical strength make them excellent candidates for electrode materials in energy storage devices. Recent research has explored the functionalisation of these materials to improve their ionic interaction and compatibility with electrolytes. For example, the functional groups in graphene oxide give sites for ion adsorption and transport improving the efficiency of batteries and super capacitors. Because of its small size, large surface area, low cost, and beneficial non-covalent interactions, pristine functionalized graphene oxide (GO) demonstrated great water dispersibility and maintained its high colloidal stability, making it a potential contender for biological applications [16, 75].

8.3.5 Two-Dimensional (2D) Materials

The exploration of 2D materials like transition metal dichalcogenides (TMDs) has opened new avenues in nanoionics. These materials, often only a few atoms thick, exhibit unique ionic transport properties due to their confined dimensions and high surface-to-volume ratio. Recent studies have focused on leveraging these properties for applications in ion-selective membranes and nanoscale transistors, where precise ion control is crucial [25].

8.3.6 Hybrid Nanostructures

Hybrid nanostructures, combining different types of nanomaterials, present a versatile approach in nanoionics. These structures can offer the best of both worlds, – for instance, combining the mechanical strength of carbon nanotubes with the high ionic conductivity of metal oxides [26]. Recent developments in this area include creating composite materials for solid electrolytes that exhibit enhanced ionic conductivity and mechanical stability, essential for durable and efficient energy storage systems [27].

8.3.7 Nanocomposites

Nanocomposites, particularly those involving ionic liquids and nano-fillers, have gained traction in nanoionics for their unique electrochemical

Table 8.1 Feature analysis on nanomaterials for nanoionic applications.

S. no.	Features	Description	Examples of nanomaterials
1	High Surface Area	Electrochemical processes and ionic conductivity are improved by increased surface area.	Nanoparticles, nanowires, nanosheets
2	Reduced Dimension	The rate of ionic transport is accelerated by shorter ion diffusion routes.	Nanotubes, nanorods, nanofilms
3	High Electronic Conductivity	The overall functionality of nanoionic devices is improved by good electronic conductivity.	Carbon nanotubes, doped nanomaterials
4	Phase Stability	For long-term performance, stable phases under a range of electrochemical circumstances can be provided by nanomaterials.	Nano-stabilized ceramics, perovskite nanostructures
5	Electromagnetic Properties	In nanoionic applications such as sensors and actuators, certain capabilities can be improved via tunable electromagnetic characteristics.	Ferromagnetic nanoparticles, piezoelectric nanomaterials

properties [28]. A lot of study has been done on these composites for application in battery separators and electrolytes. In order to increase ionic conductivity and thermal stability, recent developments have focused on improving the interaction between the nano-fillers and ionic liquids. The exploration and development of these nanomaterials in nanoionics are continuously evolving, driven by the need for more efficient, sustainable, and high-performing materials. Globally, research organisations and businesses are devoting substantial resources to comprehending the basic characteristics of these materials and turning them into useful applications. These cutting-edge nanomaterials will power the development of nanoionics, which will usher in a new age in material science and engineering and hold enormous promise for revolutionary innovations in the fields of electronics, energy, and other fields. Table 8.1 presents various features of nanomaterials which can be used widely for nanoionic applications.

8.4 Energy Storage Applications

Energy storage and conversion stand as critical pillars in the modern world, directly impacting everything from portable electronics to large-scale power grids. Nanoionics, an area of study that combines advanced material science with the complexities of ion transport at the nanoscale, plays a major role in this context. It introduces groundbreaking approaches to enhance the efficiency, capacity, and life span of energy storage and conversion systems. This section will explore the vital role of nanoionics in revolutionising these systems, offering insights into the current state of the art and envisioning future possibilities [9]. At the heart of this revolution lies the unique ability of nanoionic materials to manipulate ion transport and storage at the atomic level. These materials, engineered with precision, exhibit extraordinary electrical, chemical, and physical properties that are leveraged to optimise energy storage and conversion processes. From high-performance batteries to efficient fuel cells, nanoionics is reshaping the landscape of energy technologies, promising a future with more sustainable and efficient energy solutions [29].

The significance of nanoionics in energy storage is particularly evident in the development of advanced batteries. Here, nanoionic materials contribute to enhancing ionic conductivity, reducing charge transfer resistance, and improving overall battery performance. Similarly, in the realm of energy conversion, nanoionics is instrumental in developing more effective fuel cells and photovoltaic cells, where efficient ion transport is key to maximising energy conversion efficiency [30].

The area of nanoionics is still facing significant challenges that require ongoing attention from academics and engineers, including material stability, scalability, and cost-effectiveness [31]. As we delve deeper into the specific roles of nanoionics in energy storage and conversion, this section will also highlight the innovative research and technological advancements being made to overcome these challenges.

In summary, the intersection of nanoionics with energy storage and conversion represents a dynamic and rapidly evolving area of research. It holds not just the promise of improved current technologies but also the potential for completely new paradigms in how we store and convert energy. This exploration is not only a testament to human ingenuity but also a crucial step towards a more sustainable and energy-efficient future [32].

The application of nanoionics in energy storage, particularly in battery technology, is a domain of significant innovation. Nanoionic materials have revolutionised battery design, leading to the creation of batteries that are more efficient, have higher energy densities, and offer longer life spans [6].

Superconductors are another area where nanoionics is making substantial contributions. Nanoionic materials in supercapacitors enhance energy storage capacity by facilitating higher ion storage at the electrode/electrolyte interface. This results in supercapacitors with higher energy densities, opening up new possibilities for their use in applications requiring rapid energy delivery and high-power density [5].

In addition to lithium-ion batteries and supercapacitors, nanoionics is also instrumental in developing other advanced energy storage systems such as sodium-ion batteries, solid-state batteries, and redox flow batteries. These technologies leverage nanoionic materials to overcome the limitations of traditional batteries, such as safety concerns and lower energy densities [33].

In the field of energy conversion, nanoionics plays a crucial role in enhancing the efficiency of fuel cells and photovoltaic cells. Fuel cells, which convert chemical energy into electrical energy, benefit significantly from nanoionic materials. These materials improve ion transport within the fuel cell, thereby increasing its efficiency and power output. Nanostructured electrolytes in fuel cells reduce the operating temperature and enhance the ionic conductivity, making them more suitable for a broader range of applications [34].

Photovoltaic cells, or solar cells, have also seen advancements through the incorporation of nanoionic materials. These materials can be used to create thin-film solar cells with enhanced light absorption and conversion efficiency. By manipulating the bandgap of these materials at the nanoscale, it's possible to optimise the absorption of solar energy, leading to more efficient solar cells.

Despite the significant advancements, several challenges remain in fully harnessing the potential of nanoionics in energy storage and conversion. One of the primary challenges is the stability of nanoionic materials under operational conditions. Long-term stability and durability are crucial for

Table 8.2 Feature analysis on various nanomaterials for energy storage applications.

S. no.	Energy storage applications	Role of nanoionics	Examples of nanomaterials
1	Lithium-Ion Batteries	Improving the rates of charge and discharge and ionic conductivity. Enhancing the performance of electrode materials by nanostructuring them. Supplying solid electrolyte interphases that are stable (SEI).	Nanosized LiFePO4, LiCoO2, Nano-Si anodes
2	Fuel Cells	Improving ionic transport in the electrolyte membrane. Enhancing catalytic activity of electrodes.	Proton-conducting nanoparticles, Perovskite oxides (LSMO)
3	Hybrid Supercapacitors	Combining battery-like energy density with capacitor-like power density. Enhancing electrode materials for better performance.	Li2S/graphene nanocomposites, Carbon-based nanomaterials
4	Zinc-Air Batteries	Improving the kinetics of the oxygen reduction/evolution reaction (ORR/OER). Enhancing materials for electrodes and electrolytes.	ZnO nanoparticles, Graphene-based materials

commercial applications, and ongoing research is focused on enhancing the robustness of these materials [35]. Another challenge is scalability and cost-effectiveness. While nanoionic materials offer superior performance, scaling up their production in a cost-effective manner is essential for widespread adoption. Research is ongoing to develop more cost-effective synthesis and fabrication methods for these materials [36].

Looking towards the future, the field of nanoionics in energy storage and conversion is poised for significant growth. Advancements in nanotechnology and material science are expected to yield new nanoionic materials with even better performance and stability [5]. The integration of artificial intelligence and machine learning in materials research could accelerate the discovery of new nanoionic materials and their applications in energy systems.

In summary, the field of study that lies at the confluence of energy storage and conversion and nanoionics is dynamic and fast developing. It has the potential to completely transform the way we convert and store energy, and it will be essential in the shift to more efficient and sustainable energy systems. Table 8.2 presents the applications of nanomaterials which can be used for energy storage applications.

8.5 Emerging Electronics

In the ever-evolving landscape of technology, nanoionics stands out as a transformative field, ingeniously blending the principles of nanotechnology and ionics to redefine the future of electronics. At its core, nanoionics is the study of ion transport and reactions on a nanometre scale, a realm where the conventional rules of ionics morph under the influence of quantum mechanics and surface phenomena [30]. The complex ion dance in nanomaterials presents a plethora of opportunities for the creation of electrical devices that surpass the capabilities of conventional electronics while also being faster and more efficient.

The significance of nanoionics in the realm of electronics stems from its ability to manipulate ion behaviour in ways that can dramatically alter the electrical properties of materials. In nanoionic materials, ions exhibit unique behaviours due to their interactions at the nanoscale, leading to enhanced ionic mobility, high reactivity, and novel electronic properties. These characteristics are pivotal in crafting cutting-edge electronic devices, from high-density memory storage to ultra-sensitive sensors and beyond [37].

As the demand for smaller, more powerful, and energy-efficient electronic devices continues to soar, nanoionics offers a pathway to meet these challenges head-on. By exploiting the unique properties of ions at the nanoscale, nanoionic-based electronics can transcend the limitations of conventional materials and designs [38]. This not only means enhanced performance in current applications but also paves the way for innovative technologies that were once deemed impractical or impossible.

The fusion of nanoionics with electronics is not just a scientific endeavour; it represents a radical shift in how we conceive and develop electronic devices. It brings together interdisciplinary expertise from physics, chemistry, materials science, and electrical engineering, creating a synergistic platform for innovation [6]. From the development of energy-efficient batteries and memory devices to the creation of new forms of sensors and energy harvesting systems, nanoionics is at the forefront of the next technological revolution.

In essence, the introduction of nanoionics into the field of electronics marks the dawn of a new era. It heralds the advent of devices that are more adaptable, efficient, and capable of tackling the complex challenges of the modern world [16]. As we delve deeper into this section, we will explore the fundamental principles of nanoionics, the groundbreaking materials involved, their diverse applications in electronics, and the exciting new frontiers they are opening up.

The fundamental principles of nanoionics in electronics revolve around the intricate interactions of ions at the nanoscale. In this domain, ions exhibit unique transport and reaction behaviours that are substantially different from those observed in bulk materials [39]. These differences arise due to the quantum effects and high surface-to-volume ratios characteristic of nanoscale structures. Understanding these behaviours is crucial for harnessing the potential of nanoionics in electronic applications [17].

At the nanoscale, ion transport is influenced by the dimensions and properties of the materials involved. Nanoionic materials, such as thin films and nanostructured electrodes, provide unique pathways for ion movement, offering enhanced conductivity and electrochemical reactions. This is primarily because ions in nanostructures interact more intimately with their environment, leading to faster and more efficient transport mechanisms. This phenomenon is central to developing various electronic devices, from batteries to sensors and memristors.

Moreover, the ion-electron interactions at the nanoscale play a significant role in defining the electronic properties of materials [10]. In nanoionics, these interactions can be precisely controlled to tailor the material's electrical conductivity, charge storage capacity, and other relevant properties.

This control is achieved through careful design and engineering of the nanostructures, allowing for the customization of electronic devices to suit specific needs and applications.

The interplay of ions and electrons in nanostructures also leads to unique phenomena like ionic gating, where the application of a voltage can modulate the material's electronic properties. This principle is integral to the functioning of nanoionic transistors and switches, which are key components in modern electronic circuits [40].

The progression of nanoionics in electronics is profoundly influenced by the development and integration of various nanoionic materials. These materials are pivotal in enhancing the performance and expanding the functionalities of electronic devices. Key among these are metal oxides, carbon-based nanomaterials, and emerging 2D materials, each offering distinct properties and advantages. The variety of nanoionic materials and their tailored properties are driving the advancement of electronic devices. These materials enable the development of faster, more efficient, and more durable electronic components, heralding a new age in the field of electronics powered by the principles of nanoionics [41]. The diverse and powerful properties of nanoionic materials have led to their widespread application across various segments of electronics. These materials are revolutionising traditional electronic devices, offering enhanced performance, greater efficiency, and new capabilities.

8.5.1 Memory Devices

The development of memory devices, particularly non-volatile memory like Resistive Random Access Memory (ReRAM), is the most important use of nanoionics in electronics [38]. ReRAM utilises the resistance change in nanoionic materials to store and read information, offering a promising alternative to traditional flash memory. This technology benefits from the high-speed ion transport and the ability to precisely control the distribution of ions in nanoionic materials, leading to faster, more scalable, and more energy-efficient memory devices. Moreover, the potential for 3D stacking of ReRAM cells promises significantly increased data storage densities.

8.5.2 Sensors

Nanoionic materials are also instrumental in the evolution of sensor technology. Leveraging their high sensitivity to environmental changes and their ability to selectively interact with specific molecules, nanoionic sensors are being developed for a wide array of applications. These include environmental monitoring, where they can detect pollutants at very low concentrations, and healthcare, where they are used in the detection of biomarkers for various diseases [42]. The high surface area and reactive sites of nanoionic materials enable these sensors to provide rapid and accurate responses, which is critical in these applications [43].

8.5.3 Energy Harvesting Devices

In the field of energy harvesting, nanoionics is facilitating the development of more efficient and versatile devices. In thermoelectric and piezoelectric generators, respectively, nanoionic materials are employed to transform thermal and mechanical energy into electrical energy. The unique characteristics of nanoionic materials, such as their high electrical conductivity and capacity to function under a variety of mechanical and temperature circumstances, greatly increase the efficiency of these operations [12].

8.6 Challenges in Nanoionics Technology

Nanoionics, while heralding a new era in material science and technology, confronts a myriad of challenges that span across the spectrum of research, development, and practical application, each interwoven with complexities and requiring multifaceted solutions. At the core of these challenges is the intricate process of synthesising nanoionic materials, where the precision and control required at the atomic or molecular level often necessitate the use of advanced, sometimes prohibitively expensive techniques such as chemical vapour deposition or molecular beam epitaxy [29]. This precision is vital to ensure that materials possess the desired properties and functionalities, but it significantly escalates the cost and complexity of production. Furthermore, the fabrication of devices that incorporate these nanoionic materials, particularly on a large scale, poses additional hurdles. These processes must not only be precise but also compatible with existing manufacturing technologies, and this need for compatibility often limits the choice of materials and techniques, potentially impeding innovation [44]. Another major challenge is scalability, which is pivotal for transitioning nanoionic technologies from laboratory prototypes to mass-produced, commercially viable products; scaling up production involves not just the replication of processes but also maintaining consistency and quality, ensuring that the unique properties of nanoionic materials are preserved at larger scales, which is a task fraught with technical difficulties and often requires substantial financial investment. In addition to manufacturing challenges, the stability and durability of nanoionic materials under operational conditions remain pressing concerns, especially in applications that involve harsh environments or require long-term reliability [45]. For instance, in energy storage applications like batteries and supercapacitors, nanoionic materials must withstand repeated charging and discharging cycles without significant degradation, and achieving this level of durability is a complex challenge that involves not just the materials themselves but also the interfaces and interactions within the device. The integration of nanoionic devices with existing technological infrastructures also presents a significant challenge, necessitating innovations not only in the materials but also in system design and architecture; this is particularly pertinent in fields like electronics and energy, where nanoionic devices must seamlessly interface with existing circuits and systems. Economic factors also play a critical role, as the development and implementation of nanoionic technologies must be cost-effective to be viable for widespread adoption; this necessitates not only reducing the costs of materials and production but also ensuring that the end products offer sufficient advantages over existing technologies to justify any additional expenses. Furthermore, there is the challenge of environmental impact and sustainability, as the production and disposal of nanoionic materials and devices must be considered in the context of their ecological footprint [11]. This is increasingly important in a world where environmental consciousness is rising, and technologies are often evaluated not just on their performance but also on their sustainability. Safety and health concerns are also paramount, especially given that the nanoscale manipulation of materials can sometimes lead to unexpected biological interactions; therefore, rigorous testing and standardisation are essential to ensure that nanoionic materials and devices are safe for widespread use. In addition to these technical and economic challenges, there is also the hurdle of public perception and acceptance, particularly in applications that directly impact consumers; the introduction of new technologies often meets with scepticism, and thus effective communication and demonstration of the benefits and safety of nanoionic technologies are crucial for their acceptance. Lastly, the field of nanoionics, being highly interdisciplinary, requires a convergence of expertise from various domains such as chemistry, physics, materials science, and engineering, and this necessitates a collaborative approach to research and development [9] fostering such interdisciplinary collaborations, while essential, can be challenging due to differences in terminology, methodologies, and perspectives.

In the realm of nanoionics, recent years have seen a focus on addressing and understanding various challenges, especially in the synthesis and functional properties of nanoionic materials. In 2021, research shed light on the defect chemistry of transition metal oxide thin films, which are crucial in a range of applications. A novel spectroscopic ellipsometry approach was applied to La1-xSrxFeO3-d (LSF) thin films, enabling a systematic description of defect chemistry in these materials for the first time. This was significant for emerging low-temperature applications, as understanding and controlling point defects in such oxides is essential for their functional properties [17].

Another focus area has been on mixed ionic electronic conductors (MIEC), pivotal in clean energy technologies and industrial chemical processes. Advances in tuning surfaces and interfaces of MIEC fluorite and perovskite oxides for applications like solid oxide fuel cells were highlighted. This included insights on improving oxygen reduction reaction kinetics and fuel oxidation reaction kinetics through heterointerfaces in epitaxial thin films and vertically aligned nanocomposites [46]. Additionally, recent results in grain boundary engineering and strain engineering for the development of advanced electrodes were discussed.

In the context of grain boundaries, research in 2019 delved into Sr-doped LaMnO3±δ thin films, showing how grain boundaries can strongly influence electronic and oxygen mass transport properties. This study highlighted the role of cation concentration modifications at interfaces and their impact on material nature, unravelling unique defect chemistry and electronic properties of nanomaterials.

Furthermore, earlier research in 2017 explored space charge storage in composites from the point of view of point defect thermodynamics. This involved considering contacts of two phases for synergistic dissociative storage in space charge zones, relevant for applications like supercapacitors and batteries. This work aimed to connect mass storage with outer partial pressure or cell voltage, reflecting theoretical charge/discharge curves in battery cells [7].

These studies represent a snapshot of the ongoing efforts in the field of nanoionics, where understanding material properties at the nanoscale is key to advancing applications in energy storage, conversion, and electronics [47]. The focus on space charge procedures, interface and grain boundary engineering, and defect chemistry highlights the diverse and intricate character of the problems in nanoionics.

8.7 Sustainability and Ethical Considerations in Nanoionics

In the complex and rapidly evolving field of nanoionics, sustainability and ethical considerations play a pivotal role in shaping research, development, and application. The integration of nanoionic technologies into various aspects of our lives, from healthcare and energy to environmental protection, demands a deep understanding of their long-term impacts on both the environment and society [18]. The environmental implications of nanoionics, while potentially beneficial due to the minimised use of resources and the ability to replace larger, more environmentally damaging equipment, also bring forth concerns regarding the long-term ecological impact of nanomaterials. This is particularly relevant in the context of alternative energy sources such as nano-photovoltaic panels, where a balance must be struck between the higher costs of such technologies and their environmental benefits, as opposed to cheaper, traditional energy sources that may have a more detrimental environmental impact. Ethical considerations in nanoionics extend beyond environmental impact, touching upon profound societal and humanistic concerns [48]. One of the critical areas of focus is the health and safety of workers in industries dealing with nanoionics, particularly those exposed to nanoparticles. The history of technology development shows that often, the harmful effects on workers' health are recognized only after substantial damage has occurred, underscoring the need for proactive risk assessment and the establishment of strict safety guidelines in workplaces handling nanotechnologies.

Furthermore, the advancement of medical nanotechnologies, such as the COVID-19 nano vaccines, highlights ethical issues related to trust, equity, and global health. Inequities in the development, distribution, and accessibility of these technologies can lead to mistrust and potential public health hazards, especially in lower- and middle-income countries (LMICs) that have historically been marginalised in the global healthcare landscape [49]. The ethical responsibility of the global community is to harness these advancements in nanomedicine to address the pressing health needs of vulnerable populations, particularly in the developing world, while preventing the creation of a 'nano-divide'. This includes ensuring equitable access to life-saving nanomedical advances and designing these technologies with equity as both a moral principle and a policy goal [50].

These challenges span environmental impact, worker safety, global health equity, and societal acceptance, necessitating a balanced approach that ensures the benefits of nanotechnology are realised without compromising environmental integrity, human health, and global equity. Addressing these challenges is not only a scientific and technological endeavour but also a moral and ethical imperative, requiring collaborative efforts from researchers, policymakers, industry leaders, and the global community at large. The path forward in nanoionics must be navigated with a keen awareness of these multifaceted challenges, ensuring that the advancements in this field are sustainable, equitable, and beneficial for all segments of society [51]. In the last few years, the field of nanoionics has faced numerous sustainability and ethical challenges that have evolved and diversified, mirroring the growth and expansion of nanotechnology itself. These challenges have been shaped not only by technological advancements but also by societal and environmental concerns. One of the major ethical challenges highlighted in recent years relates to the potential risks associated with nanomaterials. There has been a growing debate on whether synthetic nanomaterials could cause harm to health and the environment. This concern is akin to earlier apprehensions about technologies like nuclear energy and parallels drawn with historical cases such as asbestos. The debate has emphasised the importance of the precautionary principle in evaluating the risks associated with nanoionics, particularly when considering the potential long-term impacts on health and the environment.

The ethical discourse has gradually shifted from more speculative concerns about nano-enhanced humans and immortality, more commonly associated with transhumanism, to more grounded issues such as nanoparticle risks and sustainable development. This shift signifies a normalisation of nano ethics, integrating it into broader frameworks of technology and risk ethics [52]. Furthermore, the field has seen a significant development in the understanding of the ethical implications of nanoionics in the workplace, particularly concerning the exposure of workers to nanoparticles. Historically, the adverse health effects of new technologies were often recognized only after substantial harm had occurred. Therefore, there is a recognized need for proactive measures in assessing and controlling the risks associated with exposure to nanoparticles in the workplace, emphasising the ethical responsibility towards workers' health and safety.

On a broader scale, nanoionics and nanotechnology have been increasingly linked to global health equity issues. For instance, the development and implementation of medical nanotechnologies, such as nano vaccines, have raised concerns about trust and equity, especially in lower- and middle-income countries. These concerns stem from historical inequities and the fear of a widening 'nano-divide' between those with access to advanced technologies and those without [53]. There is a call for the global community to adopt ethical practices that ensure the equitable distribution of benefits from nanomedical advances, focusing on addressing the health needs of vulnerable populations. In conclusion, as nanoionics continues to advance, it's imperative to address these ethical and sustainability challenges with a multi-faceted approach. This includes careful consideration of environmental impacts, worker safety, global health equity, and broader societal implications. The field must navigate these challenges with an awareness that goes beyond scientific and technological considerations, embracing ethical and moral responsibilities to ensure that the advancements in nanoionics are sustainable, equitable, and beneficial for all segments of society [54].

8.8 Cross-Disciplinary Opportunities

Cross-disciplinary collaborations in nanoionics represent a crucial convergence of diverse scientific disciplines, methodologies, and perspectives, fostering a holistic approach to understanding and manipulating ionic phenomena at the nanoscale. These collaborations, by combining expertise from various fields, facilitate long-range epistemic exploration, a vital aspect of classic convergence science. This approach is especially effective in domains where complex challenges necessitate a blend of theoretical understanding, practical demand, and technological capabilities. Successful collaborations often involve teams that integrate knowledge across multiple domains, leveraging collective intelligence and diverse perspectives. This model of teamwork not only increases the likelihood of impactful research but also mitigates the uncertainties inherent in exploring new interdisciplinary interfaces. The growing emphasis on such collaborative efforts reflects a strategic shift towards harnessing the full potential of diverse expertise in addressing complex scientific and societal challenges [55].

Physicists contribute their understanding of quantum mechanics and electrodynamics, crucial for grasping the behaviour of ions in nanomaterials, while chemists bring expertise in synthesis and characterization of these materials, essential for developing innovative nanoionic applications. In nanoionics research, engineers play a pivotal role with their expertise in experimental design, ensuring that the unique challenges of nanoscale phenomena are effectively addressed. Their work involves designing sophisticated experimental setups and devices, essential for exploring and understanding ion transport and electrochemical processes at the nanoscale. Engineers' contributions are vital in ensuring these studies are conducted with optimal performance and efficiency [56]. Their innovative approaches and designs are instrumental in overcoming the complexities inherent in nanoionics research, enabling more precise and accurate investigations into the behaviour and properties of materials at the nanoscale. Through their ingenuity, engineers facilitate significant advancements in the field, contributing to deeper insights and more effective applications of nanoionics. Materials scientists bridge the theoretical and practical aspects, focusing on how materials can be engineered to optimise their nanoionic properties, and collaborating closely with engineers who are instrumental in the actual design and fabrication of nanoionic devices like batteries, sensors, and memory storage units. Moreover, computational modelling and experimental research form a symbiotic relationship in nanoionics. Computational scientists provide simulations and models predicting ion behaviour, which guide and inform experimental research. These models are vital for understanding complex phenomena that are challenging to observe directly, such as ion transport mechanisms and interface interactions within nanostructures [57]. Experimentalists, in turn, validate and refine these models with empirical data, creating a dynamic feedback loop that propels the field forward. Collaborations in nanoionics also extend to industry and policy making, crucial for translating scientific discoveries into marketable technologies and ensuring that these technologies are developed responsibly and sustainably. Industry partnerships focus on scaling up production, improving cost-effectiveness, and ensuring the reliability and durability of nanoionic devices, while policymakers and regulatory bodies address ethical, environmental, and safety concerns, aligning research and development with societal needs and standards [58].

Interdisciplinary education and training in nanoionics are also gaining emphasis, preparing the next generation of scientists and engineers for effective cross-disciplinary collaboration. Educational programs now often provide a broad knowledge base across multiple scientific disciplines, coupled with skills in teamwork and communication, essential for working in diverse teams. While these collaborations offer immense benefits, they are not without challenges, such as differences in terminology, methodology, and research culture across disciplines. Overcoming these challenges requires open communication, mutual respect, and a commitment to interdisciplinary learning. The integration of diverse disciplines in nanoionics is not just a pathway to technological advancement but also a means to foster innovation, leading to breakthroughs that could significantly impact technology and society.

In short, cross-disciplinary collaborations in nanoionics are fundamental to the field's advancement, leveraging the strengths of various disciplines and fostering innovation through combined expertise [59]. These collaborations ensure that nanoionic technologies are developed in a manner that is responsible, sustainable, and aligned with the broader goals of societal welfare and environmental stewardship.

8.9 Educational Outreach and Knowledge Transfer

Nanoionics, a cutting-edge field at the intersection of nanotechnology and ionics, is pivotal in advancing various technologies, from energy storage to bioengineering. The crux of this field lies in manipulating ions at the nanoscale, leading to groundbreaking applications in batteries, fuel cells, sensors, and more. Given its complexity and potential, educational outreach and knowledge transfer in nanoionics are vital for fostering a deeper understanding and appreciation of this science among both the academic community and the public. This outreach is not just about disseminating information; it's about engaging diverse audiences, from school students to industry professionals, in a way that sparks curiosity and drives innovation. Initiatives like interactive workshops, public lectures, and hands-on demonstrations play a crucial role in demystifying the science behind nanoionics, making it more accessible and relatable. These programs are essential for inspiring the next generation of scientists and engineers who will continue to explore the boundaries of what's possible in nano-scale ionics [60].

The burgeoning field of nanoionics, where nanoscale manipulation of ions unlocks myriad technological possibilities, necessitates a robust regulatory framework and global cooperation. This imperative stems from unique challenges inherent in nanoionic technologies, including potential environmental, health, and safety risks, alongside broader ethical and societal implications. The safe and responsible development and application of these technologies call for regulatory guidelines and standards, informed by a comprehensive understanding of nanoionics' diverse applications spanning from energy storage to biomedicine. Initiating an effective regulatory framework for nanoionics begins with a thorough risk assessment of nanomaterials and nanoionic devices, considering their entire lifecycle from synthesis to disposal or recycling [61]. Identifying risks at each stage is vital for formulating safety guidelines that nurture innovation. This process demands rigorous scientific research and collaborative efforts among researchers, industry experts, and regulatory bodies. The dynamic nature of nanoionics also necessitates a regulatory approach that can adapt to new discoveries and technologies without compromising safety standards. Regularly updated regulatory guidelines, informed by the latest scientific and technological advancements, ensure ongoing relevance and effectiveness in protecting public health and the environment, while supporting the field's growth.

Global cooperation is crucial in nanoionics regulation, given its transnational research and development nature. Universally accepted standards and guidelines can facilitate international collaboration, trade, and knowledge exchange. Such a global framework helps harmonise regulatory approaches, mitigates barriers to research collaborations, and nurtures a worldwide market for nanoionic products. International organisations and regulatory agencies must collaborate, pooling diverse expertise to address the ethical and societal issues associated with nanoionic technologies and ensuring alignment with global values. The regulatory framework must also address intellectual property (IP) rights. Given nanoionics' innovation-driven nature and significant commercial potential, clear and equitable IP regulations are essential [59]. This involves balancing patent and trademark protection with issues surrounding data sharing, open access, and ethical technology usage. Such policies must foster both innovation and accessibility.

Public engagement and transparency are key in the nanoionics regulatory process. This engagement is instrumental in building public trust, dispelling misconceptions about risks and benefits. Transparent communication about regulatory processes and criteria further bolsters this trust, underpinning an informed environment for nanoionics development and application. Education and training are critical for effectively implementing regulatory standards in nanoionics. This includes educating scientists and engineers about regulations and keeping regulatory bodies updated on field advancements. These initiatives build a knowledgeable community capable of safely and responsibly navigating nanoionics' complexities [62].

An inclusive regulatory framework is essential, considering the diverse impacts of nanoionic technologies across regions and communities. This inclusivity ensures universal access to nanoionics' benefits while addressing disproportionate risks or ethical concerns in different socio-economic or cultural contexts. Developing a comprehensive regulatory framework and fostering global cooperation are paramount for the safe and responsible advancement of nanoionics. Informed by rigorous risk assessment, adaptable to field advancements, and harmonised at a global level, this framework should integrate IP rights, public engagement, education, and inclusivity elements. As nanoionics evolves, a collaborative and dynamic approach to regulation is key to harnessing its potential while safeguarding public health, the environment, and societal values [63].

The universal nature of the challenges posed by nanoionics underscores the importance of global cooperation. Nanoionic materials and technologies, due to their nanoscale dimensions, present risks that are not yet fully understood. Global cooperation enables the pooling of resources and expertise for extensive research and risk assessments, leading to effective, science-based regulatory standards. Global cooperation is also critical for addressing the ethical, legal, and social implications of emerging nanoionic technologies. Collaborative international discussions and agreements on privacy, security, and ethical usage are vital. International forums can facilitate these discussions, ensuring that diverse cultural and ethical perspectives are incorporated into guidelines and policies.

A globally coordinated IP framework is essential in nanoionics, protecting inventors' rights while ensuring broad access to new technologies. This balance encourages continued investment in research and development and fosters equitable technological access. Global cooperation extends to human capital development. Cross-border educational programs, research initiatives, and exchange programs are crucial in building a skilled workforce knowledgeable in nanoionics. Sharing best practices in education and training uplifts the global standard of research and development in this field [64]. Furthermore, international cooperation is vital for mobilising resources for nanoionics research and development, particularly in resource-limited regions. International partnerships and funding programs ensure equitable resource distribution, maximising nanoionics' potential to address global challenges like energy security, environmental sustainability, and healthcare. Addressing nanoionics' challenges on a global scale can also lead to the development of international emergency response protocols, essential for preparedness and resilience in potential nanoionic-related crises.

Global cooperation in nanoionics is not merely about tackling challenges; it's about seizing opportunities. Collaborative efforts can accelerate groundbreaking technological development, effectively address global issues, and stimulate economic growth. International collaborations can yield innovative solutions unattainable by individual countries, leveraging the global community's diverse strengths and perspectives. In conclusion, nanoionics' advancement necessitates a regulatory framework supported by global cooperation. This cooperation is crucial for managing nanoionic technologies' multifaceted challenges and unlocking their immense opportunities. Through shared research, harmonised regulations, collaborative ethical discussions, educational initiatives, and resource mobilisation, the global community can ensure nanoionics' development is safe and beneficial for all [55]. As nanoionics continues to evolve, fostering international collaboration and unity will be essential for navigating its complexities and realising its full potential for humanity's betterment.

8.10 Significance of Nanoionics in Industrial Revolution

The field of nanoionics stands at the precipice of a revolution, poised to significantly influence various industries through its commercialization. Nanoionic technologies, which harness the movement of ions at the nanoscale, offer groundbreaking potential in numerous applications. The journey from laboratory breakthroughs to commercial products, however, involves navigating through a labyrinth of challenges and opportunities. This section examines the potential for the commercialization of nanoionic technologies and their integration into various industries, highlighting the transformative impact they could have on the global market. The allure of nanoionics lies in its versatility and the breadth of its potential applications. In the energy sector, for instance, nanoionic materials are pivotal in developing more efficient and higher-capacity batteries, fuel cells, and supercapacitors. This is particularly relevant in the context of the growing demand for renewable energy sources and the need for efficient energy storage solutions. The commercialization of nanoionic-based energy storage systems could revolutionise the automotive industry, particularly in electric vehicles (EVs), by providing batteries that are lighter, charge faster, and hold more power than current models [65].

In the field of electronics, nanoionics paves the way for smaller, faster, and more energy-efficient components. The miniaturisation of electronic components using nanoionic materials can lead to the development of advanced semiconductors, memory devices, and transistors. This technological leap is vital for the continuous scaling down of electronic devices and the enhancement of their performance, opening up new avenues in consumer electronics, telecommunications, and computing. In the semiconductor industry, collaborations between nanoionics researchers and technology giants have yielded impressive results. Intel's partnership with academic institutions for the development of nanoionic transistors is a case in point. These transistors use nanoionic materials to achieve faster switching speeds and lower power consumption. Intel's investment in this technology has led to a 10% improvement in transistor performance, significantly impacting the semiconductor market, which, according to a report by Mordor Intelligence, is expected to reach USD 129 billion by 2025 [66].

Healthcare is another domain where nanoionics can make a substantial impact. Nanoionic materials can be used in targeted drug delivery systems, biosensors, and diagnostic devices. The ability to manipulate ions at the nanoscale enables the development of highly sensitive diagnostic tools and efficient drug delivery mechanisms, potentially revolutionising treatments and diagnostics in fields like oncology, neurology, and cardiology. The healthcare sector has also seen fruitful partnerships in the realm of nanoionics. A notable collaboration between pharmaceutical companies and nanoionic material researchers has led to the development of targeted drug delivery systems. These systems use nanoionic carriers to deliver drugs more efficiently to specific body sites, enhancing therapeutic effectiveness while minimising side effects. For instance, a collaboration between Novartis and MIT on nanoionic drug delivery systems has resulted in a 20% increase in drug targeting efficiency, as reported in a study published in the Journal of Controlled Release [67]. This innovation is significant in the pharmaceutical industry, which, according to a Grand View Research report, is forecasted to reach USD 1.17 trillion by 2021.

In the semiconductor industry, collaborations between nanoionics researchers and tech giants have yielded impressive results. Intel's partnership with academic institutions for the development of nanoionic transistors is a case in point. These transistors use nanoionic materials to achieve faster switching speeds and lower power consumption. Intel's investment in this technology has led to a 10% improvement in transistor performance, significantly impacting the semiconductor market, which, according to a report by Mordor Intelligence, is expected to reach USD 129 billion by 2025 [68]. In environmental applications, partnerships between nanoionic material developers and environmental agencies have led to the creation of advanced sensors and filtration systems. These technologies, utilising nanoionic materials, offer enhanced sensitivity and specificity in monitoring and treating environmental pollutants. A collaboration between a leading nanotechnology company and the Environmental Protection Agency (EPA) in the U.S. has resulted in the development of nanoionic-based sensors that demonstrate a 30% increase in pollutant detection sensitivity. This advancement is crucial in the global environmental monitoring market, which, according to a Markets and Markets report, is expected to reach USD 19.56 billion by 2025 [69].

The commercialization and industry integration of nanoionic technologies have been markedly accelerated by successful industry partnerships.

These collaborations, exemplified in the fields of electric vehicles, semiconductors, healthcare, and environmental technology, have led to significant advancements and economic growth. The statistics from these successful partnerships provide compelling evidence of their impact, underscoring the importance of collaborative efforts in bringing nanoionic innovations to the market. As the field of nanoionics continues to evolve, fostering such strategic partnerships will be key to unlocking its full potential across diverse industries.

8.11 Innovation and Future Prospects

The field of nanoionics, a nexus of nanotechnology and ionics, is burgeoning with innovations that promise transformative changes across a spectrum of industries, characterised by key findings and future prospects that are as diverse as they are impactful. Central to these advancements is the development of advanced nanoionic materials for batteries, a leap that has revolutionised energy storage capabilities; these materials significantly enhance the energy density and charging speed of batteries, presenting a future where electric vehicles can travel longer distances and recharge in a fraction of the current times, a prospect that aligns well with the burgeoning global push towards sustainable transportation [6]. In the realm of consumer electronics and computing, nanoionics has facilitated the creation of smaller, faster, and more energy-efficient semiconductors, thereby enabling the continual miniaturisation of electronic devices while simultaneously boosting their performance, a trend that is vital in an era increasingly dominated by portable, high-performance gadgets [70]. The healthcare industry, too, stands at the cusp of a nanoionic revolution, with the development of targeted drug delivery systems and highly sensitive diagnostic tools; these innovations, based on the precise manipulation of ions at the nanoscale, promise to usher in new paradigms in personalised medicine, particularly in the treatment of chronic diseases like cancer, where targeted therapy is crucial. Furthermore, the environmental sector is witnessing the emergence of nanoionic-based sensors and filtration systems, offering unprecedented sensitivity and specificity in the detection and treatment of pollutants, a development that is of paramount importance in the current age of environmental consciousness and could potentially reshape global strategies in environmental monitoring and remediation. These technological advancements in nanoionics are underpinned by fundamental breakthroughs in material science, particularly in the synthesis and manipulation of nanomaterials, which have opened new frontiers in the understanding and application of ion transport at the nanoscale. The ability to engineer materials at this level has not only expanded the potential applications of nanoionics but also presented new challenges, particularly in the realms of manufacturing scalability and integration into existing industrial frameworks [71]. Looking to the future, the prospects of nanoionics are vast and varied; the field is poised to make significant contributions to renewable energy, particularly in improving the efficiency and reducing the costs of solar cells and fuel cells. In the telecommunications sector, nanoionic materials are expected to play a crucial role in the development of next-generation communication devices, offering faster data transmission rates and higher storage capacities. The burgeoning field of wearable technology and the Internet of Things (IoT) also stands to benefit immensely from nanoionic advancements, with the development of more efficient, flexible, and durable sensors and devices. The potential of nanoionics in biomedicine extends beyond drug delivery and diagnostics; research is ongoing into the use of nanoionic materials in tissue engineering and regenerative medicine, offering the possibility of novel treatments for a range of diseases and injuries [72]. In the face of these promising prospects, the field of nanoionics is not without its challenges. Issues related to the toxicity and environmental impact of nanomaterials need to be addressed, necessitating rigorous safety assessments and the development of sustainable production methods. Furthermore, the integration of nanoionic technologies into existing systems and industries requires not only technical compatibility but also economic viability, a hurdle that must be overcome through continued research, development, and collaboration between academia, industry, and regulatory bodies.

The future of nanoionics, a field at the intersection of nanotechnology and ionics, is teeming with potential for groundbreaking advancements across various sectors. In energy storage, the development of nanoionic materials is expected to revolutionise battery technology, significantly enhancing energy density and charging speeds, which is critical for the proliferation of electric vehicles and the effective harnessing of renewable energy sources. In electronics, nanoionics is paving the way for further miniaturisation and enhanced performance of components, promising more efficient and powerful computing devices, advanced smartphones, and sophisticated wearable technologies. The healthcare industry anticipates transformative changes with nanoionic applications in drug delivery systems, enabling targeted and efficient treatments with minimal side effects, crucial in cancer therapy and chronic disease management [73]. Additionally, the potential for developing advanced biosensors and diagnostic tools through nanoionics heralds a new era of personalised and preventive healthcare. Environmental applications are equally promising, with nanoionic materials poised to enhance the sensitivity and accuracy of pollutant monitoring, improving environmental control and remediation strategies. However, these prospects are not without challenges; scalability, manufacturability, stability, and environmental impacts of nanomaterials remain significant hurdles [74]. Addressing these challenges requires collaborative efforts among researchers, industry experts, and policymakers, along with the establishment of regulatory frameworks and intellectual property protections. Education and public engagement will also play a crucial role in fostering a skilled workforce and an informed public ready to adopt these advanced technologies. In essence, the field of nanoionics stands at a pivotal point, with the promise of technological revolutions that could address some of the most pressing global challenges, including sustainable energy, advanced healthcare, and environmental protection, heralding a future that is more efficient, sustainable, and health-focused [55].

Conclusion

As we conclude this extensive exploration of nanoionics, it is clear that this field, situated at the confluence of nanotechnology and ionics, represents a significant frontier in modern scientific research and technological innovation. The profound understanding and manipulation of ions at the nanoscale, which is the hallmark of nanoionics, has opened up a plethora of possibilities across various sectors, signalling a paradigm shift in how we approach material science, energy solutions, electronic devices, and healthcare technologies.

In the realm of energy storage and conversion, nanoionics stands as a beacon of innovation, offering transformative solutions that are poised to revolutionise this domain. The development of advanced nanoionic materials for batteries and supercapacitors has already shown remarkable progress, enhancing energy density, efficiency, and charging speeds. These advancements are not only pivotal for the future of electric vehicles and renewable energy systems but also for the broader goal of achieving sustainable and efficient energy management on a global scale. The prospects in this area are bolstered by continuous research efforts aimed at overcoming challenges related to material stability, scalability, and cost-effectiveness, promising an era of energy storage solutions that are more robust, eco-friendly, and accessible.

In the sphere of electronics, nanoionics heralds a new age of miniaturisation and enhanced functionality. The unique ion transport mechanisms harnessed in nanoionic materials have led to the development of smaller, faster, and more energy-efficient electronic components, pushing the boundaries of what is possible in semiconductor technology, data storage, and sensor development. This progress is crucial in an era where the demand for high-performance, portable, and energy-efficient electronic devices is ever-increasing. The integration of nanoionics into electronics is not only a testament to technological ingenuity but also a reflection of the field's ability to adapt and evolve in response to emerging needs and challenges. The significance of nanoionics in the evolution of future technologies cannot be overstated. In energy storage, the field has revolutionised the development of batteries and supercapacitors, leading to devices with enhanced capacities, faster charging times, and greater efficiencies. These advancements are not only crucial for the burgeoning electric vehicle market but also for the broader adoption of renewable energy sources, as efficient energy storage is key to offsetting the intermittency of sources like solar and wind power. The potential of nanoionics in this sector holds the promise of a more sustainable and energy-efficient future, aligning with global efforts to combat climate change and reduce reliance on fossil fuels. In the electronics industry, nanoionics has emerged as a game-changer. By enabling the miniaturisation of components and enhancing their performance, nanoionic technologies are driving the evolution of smaller, faster, and more powerful electronic devices. This is particularly important in an era where the demand for advanced computing power and portable electronics is constantly increasing. The implications of these advancements extend beyond consumer electronics, impacting sectors such as telecommunications, data storage, and even space exploration, where efficient and reliable electronic devices are critical.

In healthcare, the impact of nanoionics is profound and far-reaching. The field has introduced groundbreaking approaches to drug delivery systems and diagnostic tools, harnessing the precision and efficiency afforded by nano-scale ion manipulation. These advancements promise a future where medical treatments are more targeted, less invasive, and more effective, particularly in the management and treatment of chronic diseases and conditions. The potential of nanoionics in revolutionising healthcare extends beyond treatment to diagnostics and personalised medicine, offering tools and techniques that are more sensitive, accurate, and patient-specific.

However, the journey of nanoionics towards these promising horizons is not devoid of challenges. Issues related to the scalability of production, the long-term stability of nanoionic materials, and their integration into existing industrial and technological infrastructures are significant hurdles that must be addressed. Moreover, the environmental impact and sustainability of these materials, along with their safety and health implications, present ongoing concerns that require continuous attention and research. Addressing these challenges necessitates a multifaceted approach, involving collaborative efforts among scientists, engineers, policymakers, and industry leaders.

The future prospects of nanoionics are intrinsically linked to its ability to continue evolving and adapting in response to these challenges. The integration of interdisciplinary knowledge and expertise is paramount in this endeavour. The convergence of physics, chemistry, materials science, engineering, and other related fields within the domain of nanoionics has been a key driver of its progress and will continue to be essential for its future advancements. This interdisciplinary approach not only enriches the research and development within nanoionics but also fosters innovation by bringing together diverse perspectives and methodologies.

In addition to scientific and technological advancements, the field of nanoionics also encompasses important ethical, societal, and policy considerations. As nanoionic technologies become more integrated into various aspects of our lives, it is imperative to consider their long-term impacts on society and the environment. This includes addressing questions of sustainability, equity in access and distribution, and the ethical implications of emerging nanoionic applications. The development of comprehensive regulatory frameworks, informed by a deep understanding of nanoionics and its potential impacts, will be crucial in ensuring that these technologies are developed and applied responsibly and ethically.

Educational outreach and knowledge transfer in nanoionics are also vital components of its future trajectory. Cultivating a deeper understanding and appreciation of nanoionics among the academic community, industry professionals, and the general public is essential for fostering a supportive environment for innovation and application. Engaging diverse audiences through interactive programs, workshops, and public lectures will not only demystify the science behind nanoionics but also inspire the next generation of scientists and engineers who will propel the field forward.

In conclusion, nanoionics stands at a critical juncture, with its advancements poised to bring about significant transformations across multiple industries and sectors. The exploration and understanding of ionic phenomena at the nanoscale have opened new doors to technological innovations that have the potential to significantly impact our daily lives and the future of various technologies. From creating more efficient energy storage systems and sensitive sensors to developing the next generation of electronic devices, nanoionics is a testament to the power of interdisciplinary research and its role in driving forward the frontiers of science and technology. As the field continues to grow and evolve, it holds the promise of revolutionising how we store and utilise energy, detect various substances, and process information, making it a key player in shaping the technological landscape of the future.

References

1. Schoonman, J., Nanoionics. *Solid State Ionics*, 157, 319, 2003.
2. Maier, J., Nanoionics: Ion transport and electrochemical storage in confined systems. *Nat. Mater.*, 4, 805, 2005.
3. Waser, R. and Aono, M., Nanoionics-based resistive switching memories. *Nat. Mater.*, 6, 833, 2007.
4. Maier, J., Nanoionics: Fundamentals and Applications, in: *21st Century Nanosci. – A Handb.*, 8, 2020.
5. Chen, Y., Gerdes, K., Paredes Navia, S.A., Liang, L., Hinerman, A., Song, X., Conformal Electrocatalytic Surface Nanoionics for Accelerating Higherature Electrochemical Reactions in Solid Oxide Fuel Cells. *Nano Lett.*, 19, 8767, 2019.
6. Zhang, J., Liu, W., Dai, J., Xiao, K., Nanoionics from Biological to Artificial Systems: An Alternative Beyond Nanoelectronics. *Adv. Sci.*, 9, 2200534, 1–17, 2022.
7. Sangwan, V.K. and Hersam, M.C., Neuromorphic nanoelectronic materials. *Nat. Nanotechnol.*, 15, 517, 2020.
8. Hoffmann, M., Schmitt, P., Wittemeier, S., Schaller, F., Shaporin, A., Stockel, C., Geneis, V., Forke, R., Hedayat, C., Hilleringmann, U., Kuhn, H., Zimmermann, S., Storing MEMS Interfaces Without Electrical Auxiliary Energy for Long-Time Monitoring. *Proc. IEEE Int. Conf. Micro Electro Mech. Syst.*, 522, January 2023.
9. Bo, Z., Xu, C., Huang, Z., Chen, P., Yan, G., Yang, H., Yan, J., Cen, K., Ostrikov, K.K., Photo-electric capacitive deionization enabled by solar-driven nano-ionics on the edges of plasma-made vertical graphenes. *Chem. Eng. J.*, 422, 130156, 2021.
10. Dittmann, R., Menzel, S., Waser, R., Nanoionicmemristive phenomena in metal oxides: the valence change mechanism. *Adv. Phys.*, 70, 155, 2021.
11. Islam, M.A., Bouldin, J., Yang, J., Han, S.D., Electrochemical Sodiation Mechanism in Magnetite Nanoparticle-Based Anodes: Understanding of Nanoionics-Based Sodium Ion Storage Behavior of Fe_3O_4. *ACS Appl. Mater. Interfaces*, 14, 50773, 2022.
12. Sepúlveda, P., Muga, I., Sainz, N., Rojas, R.G., Ossandón, S., Nanoionics from a quantum mechanics point of view: Mathematical modeling and numerical simulation. *Comput. Methods Appl. Mech. Eng.*, 407, 115926, 2023.

13. Kumar, M., Lim, J., Park, J.Y., Kim, S., Seo, H., Electric-field-induced healing of inanimate topographies: Multistate resistive switching and nano-sized artificial synapse functionality. *Appl. Surf. Sci.*, 530, 147190, 2020.
14. Kim, T.H., Kim, S., Park, B.G., Improved rectification characteristics by engineering energy barrier height in TiOx-based RRAM. *Microelectron. Eng.*, 237, 111498, 2021.
15. Sun, B., Xiao, M., Zhou, G., Ren, Z., Zhou, Y.N., Wu, Y.A., Non–zero-crossing current-voltage hysteresis behavior in memristive system. *Mater. Today Adv.*, 6, 100056, 2020.
16. Zhan, H., Xiong, Z., Cheng, C., Liang, Q., Liu, J.Z., Li, D., Solvation-Involved Nanoionics: New Opportunities from 2D Nanomaterial Laminar Membranes. *Adv. Mater.*, 32, 15214095, 1–47, 2020.
17. Terabe, K., Tsuchiya, T., Tsuruoka, T., A Variety of Functional Devices Realized by Ionic Nanoarchitectonics, Complementing Electronics Components. *Adv. Electron. Mater.*, 8, 2100645, 2022.
18. Tsuchiya, T., Terabe, K., Aono, M., Nanoionic Devices for Physical Property Tuning and Enhancement. *At. Switch*, p. 161, 2020.
19. Zotov, A.V., Sirotkin, V.V., Il'in, A.I., Trofimov, O.V., Borisenko, D.N., Kolesnikov, N.N., Tulin, V.A., Multilevel memristive structures based on bismuth selenide microcrystals. *Chaos, Solitons Fractals*, 143, 110542, 2021.
20. Rao, A., Kumar, G.S., Roy, S., Rajesh, A.T., Devatha, G., Pillai, P.P., Turn-On Selectivity in Inherently Nonselective Gold Nanoparticles for Pb2+ Detection by Preferential Breaking of Interparticle Interactions. *ACS Appl. Nano Mater.*, 2, 5625, 2019.
21. Yang, C.P. and Guo, Y.G., Nanostructures and nanomaterials for lithium metal batteries, in: *Nanostructures Nanomater. Batter. Princ. Appl.*, p. 159, 2019.
22. Wang, L., Zhang, Y., Zhang, P., Wen, D., Physically Transient, Flexible, and Resistive Random Access Memory Based on Silver Ions and Egg Albumen Composites. *Nanomaterials*, 12, 3061, 2022.
23. Barabash, M.L., Gibby, W.A.T., Guardiani, C., Smolyanitsky, A., Luchinsky, D.G., McClintock, P.V.E., Origin and control of ionic hydration patterns in nanopores. *Commun. Mater.*, 2, Article ID: 65, 1–9, 2021.
24. Selamneni, V., Bokka, N., Adepu, V., Sahatiya, P., Carbon Nanomaterials for Emerging Electronic Devices and Sensors, in: *Carbon Nanomater. Electron. Devices Appl.*, p. 215, 2021.
25. NSM, J., SBRS, A., Ahmad, N., Sodium-ion nanoionic hybrid solid electrolyte: Extended study on enhanced electrical and electrochemical properties. *Solid State Ionics*, 377, 115882, 2022.
26. Quill, T.J., LeCroy, G., Halat, D.M., Sheelamanthula, R., Marks, A., Grundy, L.S., McCulloch, I., Reimer, J.A., Balsara, N.P., Giovannitti, A., Salleo, A., Takacs, C.J., An ordered, self-assembled nanocomposite with efficient electronic and ionic transport. *Nat. Mater.*, 22, 362, 2023.

27. Saeed, S.W., Norby, T., Bjørheim, T.S., First-Principles Analyses of Nanoionic Effects at Oxide-Oxide Heterointerfaces for Electrochemical Applications. *J. Phys. Chem. C*, 124, 14072, 2020.
28. Ullah, M.K., Ahmad, N., Ud-Din Khan, S., Alvi, F., Abbas, G., Rafique, A., Ali, A., Ahmad, M.A., Raza, R., Structural and electrochemical studies of microwave sintered nanocomposite electrolytes for solid oxide fuel cells. *Int. J. Hydrogen Energy*, 44, 10964, 2019.
29. Banerjee, W., Kim, S.H., Lee, S., Lee, D., Hwang, H., An Efficient Approach Based on Tuned Nanoionics to Maximize Memory Characteristics in Ag-Based Devices. *Adv. Electron. Mater.*, 7, 4, 2100022, 2021.
30. Al-Harahsheh, M., AlJarrah, M., Alrebaki, M., Mayyas, M., Nanoionic exchanger with unprecedented loading capacity of uranium. *Sep. Purif. Technol.*, 238, 116423, 2020.
31. Wu, M.C., Chen, J.Y., Ting, Y.H., Huang, C.Y., Wu, W.W., A novel high-performance and energy-efficient RRAM device with multi-functional conducting nanofilaments. *Nano Energy*, 82, 105717, 2021.
32. Zhang, M., Zhao, P., Li, P., Ji, Y., Liu, G., Jin, W., Designing Biomimic Two-Dimensional Ionic Transport Channels for Efficient Ion Sieving. *ACS Nano*, 15, 5209, 2021.
33. Nikam, R.D., Kwak, M., Lee, J., Rajput, K.G., Hwang, H., Controlled Ionic Tunneling in Lithium Nanoionic Synaptic Transistor through Atomically Thin Graphene Layer for Neuromorphic Computing. *Adv. Electron. Mater.*, 6, 2, 1901100, 2020.
34. Lu, J., Xu, H., Yu, H., Hu, X., Xia, J., Zhu, Y., Wang, F., Wu, H.A., Jiang, L., Wang, H., Ultrafast rectifying counter-directional transport of proton and metal ions in metal-organic framework-based nanochannels. *Sci. Adv.*, 8, Vol 8, 14, 1–10, 2022.
35. Zhu, X., Lee, S.H., Lu, W.D., Nanoionic Resistive-Switching Devices. *Adv. Electron. Mater.*, 5, 9, 1900184, 2019.
36. Nayak, A.K. and Sasmal, A., Recent advance on fundamental properties and synthesis of barium zirconate for proton conducting ceramic fuel cell. *J. Clean. Prod.*, 386, 135827, 2023.
37. Habasaki, J., Classification of Nanostructured Materials and Effects of Nano-Sizing, in: *Mol. Dyn. Nanostructures Nanoionics*, p. 11, 2020.
38. Shaporin, A., Stockel, C., Melzer, M., Schaller, F., Forke, R., Zimmermann, S., Kuhn, H., Optimal design of piezoelectric MEMS for vibration monitoring system with nanoionics zero-energy memory elements. *2023 24th Int. Conf. Therm. Mech. Multi-Physics Simul. Exp. Microelectron. Microsystems, EuroSimE 2023*, 2023.
39. Despotuli, A.L. and Andreeva, A.V., Structure-Dynamic Approach of Nanoionics, in: *21st Century Nanosci. – A Handb.*, 9, 2020.
40. Sanna, S., Firodaliso, E.M., Kasama, T., Castelli, I.E., Esposito, V., Effect of high oxygen deficiency in nano-confined bismuth sesquioxide. *JPhys Energy*, 2, 24010, 2020.

41. Gadermaier, B., Resch, L., Pickup, D.M., Hanghofer, I., Hanzu, I., Heitjans, P., Sprengel, W., Würschum, R., Chadwick, A.V., Wilkening, H.M.R., Influence of defects on ionic transport in $LiTaO_3$ – A study using EXAFS and positron annihilation lifetime spectroscopy. *Solid State Ionics*, 352, 115355, 2020.
42. Feng, Y., Li, X., Ji, D., Tian, J., Peng, Q., Shen, Y., Xiao, Y., Functionalised penetrating peptide-chondroitin sulphate□gold nanoparticles: Synthesis, characterization, and applications as an anti-Alzheimer's disease drug. *Int. J. Biol. Macromol.*, 230, 123125, 2023.
43. Huang, Y., Huynh, T.T., Sun, L., Hu, C.H., Wang, Y.C., Rogers, B.E., Mirica, L.M., Neutral Ligands as Potential 64Cu Chelators for Positron Emission Tomography Imaging Applications in Alzheimer's Disease. *Inorg. Chem.*, 61, 4778, 2022.
44. Park, J.M., Hwang, H., Song, M.S., Jang, S.C., Kim, J.H., Kim, H., Kim, H.S., All-Solid-State Synaptic Transistors with Lithium-Ion-Based Electrolytes for Linear Weight Mapping and Update in Neuromorphic Computing Systems. *ACS Appl. Mater. Interfaces*, 15, 47229, 2023.
45. Sarfraz, M.H., Zubair, M., Aslam, B., Ashraf, A., Siddique, M.H., Hayat, S., Cruz, J.N., Muzammil, S., Khurshid, M., Sarfraz, M.F., Hashem, A., Dawoud, T.M., Avila-Quezada, G.D., Abd_Allah, E.F., Comparative analysis of phyto-fabricated chitosan, copper oxide, and chitosan-based CuO nanoparticles: antibacterial potential against Acinetobacter baumannii isolates and anti-cancer activity against HepG2 cell lines. *Front. Microbiol.*, 14, 1188743, 2023.
46. Habasaki, J., Nanostructures in Nanoionics and Colloidal Chemistry: Overview and Problems, in: *Mol. Dyn. Nanostructures Nanoionics*, p. 49, 2020.
47. Younis, A., Solution processed laterally grown Zinc Oxide microstructures for next generation computing devices. *Physica E*, 126, 114475, 2021.
48. Park, M., Jeon, B., Park, J., Kim, S., Memristors with Nociceptor Characteristics Using Threshold Switching of Pt/HfO2/TaOx/TaN Devices. *Nanomaterials*, 12, 4206, 2022.
49. Shaikh, M.T.A.S., Nguyen, T.H.V., Jeon, H.J., Prasad, C.V., Kim, K.J., Jo, E.S., Kim, S., Rim, Y.S., Multilevel Reset Dependent Set of a Biodegradable Memristor with Physically Transient. *Adv. Sci.*, 11, 4, 2306206, 1–11, 2023.
50. Alam, F., Jalal, A.H., Forouzanfar, S., Karabiyik, M., RabieiBaboukani, A., Pala, N., Flexible and Linker-Free Enzymatic Sensors Based on Zinc Oxide Nanoflakes for Noninvasive L-Lactate Sensing in Sweat. *IEEE Sens. J.*, 20, 5102, 2020.
51. Das, S., Chen, A., Marinella, M., Beyond CMOS. *Proc.-2021 IEEE Int. Roadmap Devices Syst. Outbriefs, IRDS 2021*, 2021.
52. Hum. Rights Doc. Online, the-elections-in-argentina-one-step-toward-democracy-cosponsored-with-the-international-human-rights-law-group-dec-1-1983-27-pp.

53. Yao, Q. and Zhu, C., Advanced Post-Potassium-Ion Batteries as Emerging Potassium-Based Alternatives for Energy Storage. *Adv. Funct. Mater.*, 30, 2005209, 49, 2020.
54. Xiao, K., Wan, C., Jiang, L., Chen, X., Antonietti, M., Bioinspired Ionic Sensory Systems: The Successor of Electronics. *Adv. Mater.*, 32, 2000218, 1–9, 2020.
55. Wang, Z., Wang, L., Nagai, M., Xie, L., Yi, M., Huang, W., Nanoionics-Enabled Memristive Devices: Strategies and Materials for Neuromorphic Applications. *Adv. Electron. Mater.*, 3, 1600510, 7, 2017.
56. Yang, Y. and Huang, R., Probing memristive switching in nanoionic devices. *Nat. Electron.*, 1, 274, 2018.
57. Nessel, J.A., Lee, R.Q., Mueller, C.H., Kozicki, M.N., Ren, M., Morse, J., A novel nanoionics-based switch for microwave applications. *IEEE MTT-S Int. Microw. Symp. Dig.*, 1051, 2008.
58. Ariga, K., Ito, M., Mori, T., Watanabe, S., Takeya, J., Atom/molecular nanoarchitectonics for devices and related applications. *Nano Today*, 28, 100762, 2019.
59. Kim, J., Ahmed, T., Nili, H., Yang, J., Jeong, D.S., Beckett, P., Sriram, S., Ranasinghe, D.C., Kavehei, O., A Physical Unclonable Function with Redox-Based Nanoionic Resistive Memory. *IEEE Trans. Inf. Forensics Secur.*, 13, 437, 2018.
60. Valov, I. and Waser, R., Physics and Chemistry of Nanoionic Cells, in: *Resist. Switch*, p. 253, 2016.
61. Karimi, B., Tavakolian, M., Mansouri, F., Vali, H., Nanopalladium on Magnetic Ionic Nanoparticle Network (MINN) as an Efficient and Recyclable Catalyst with High Ionic Density and Dispersibility. *ACS Sustain. Chem. Eng.*, 7, 3811, 2019.
62. Indiveri, G., Linn, E., Ambrogio, S., ReRAM-Based Neuromorphic Computing, in: *Resist. Switch*, p. 715, 2016.
63. Chadwick, A.V. and Savin, S.L.P., Structure and dynamics in nanoionic materials. *Solid State Ionics*, 177, 3001, 2006.
64. Despotuli, A. and Andreeva, A., Method of uniform effective field in structure-dynamic approach of nanoionics. *Ionics (Kiel)*, 22, 1291, 2016.
65. Habasaki, J., Leon, C., Ngai, K.L., Nanoionics. *Top. Appl. Phys.*, 132, 277, 2017.
66. Nagata, T., Haemori, M., Chikyow, T., Combinatorial synthesis of Cu/(TaxNb1-x) 2O5 stack structure for nanoionics-type ReRAM device. *ACS Comb. Sci.*, 15, 435, 2013.
67. Spring, J., Sediva, E., Hood, Z.D., Gonzalez-Rosillo, J.C., O'Leary, W., Kim, K.J., Carrillo, A.J., Rupp, J.L.M., Toward Controlling Filament Size and Location for Resistive Switches via Nanoparticle Exsolution at Oxide Interfaces. *Small*, 16, 16136829, 1–20, 2020.

68. Pervov, V.S., Ovchinnikova, S.I., Medvedeva, A.E., Makhonina, E.V., Kireeva, N.V., Nanoionics: Principles of ceramic materials fabrication for electrochemical power generation. *Inorg. Mater.*, 52, 83, 2016.
69. Garbayo, I., Baiutti, F., Morata, A., Tarancón, A., Engineering mass transport properties in oxide ionic and mixed ionic-electronic thin film ceramic conductors for energy applications. *J. Eur. Ceram. Soc.*, 39, 101, 2019.
70. Gupta, S.K., Gupta, A.K., Yadav, R.K., Badatya, S., Gupta, M.K., Tailoring of the electrochemical stability and dielectric behaviour of inorganic matrix-based ionogels. *Phys. Scr.*, 97, 65812, 2022.
71. Koh, S.G., Koide, T., Morita, T., Kinoshita, K., Ionic liquid-loaded metal-organic framework system for nanoionic device applications. *Jpn. J. Appl. Phys.*, 60, SBBK10, 2021.
72. Subramani, K. and Mehta, M., Nanodiagnostics in microbiology and dentistry, in: *Emerg. Nanotechnologies Dent.*, p. 391, 2017.
73. Chokkareddy, R., Kanchi, S., Inamuddin, Altalhi, T.A., Smart Nanodevices for Point-of-Care Applications. *Curr. Anal. Chem.*, 18, 415, 2021.
74. Wagemaker, M. and Mulder, F.M., Properties and promises of nanosized insertion materials for li-ion batteries. *Acc. Chem. Res.*, 46, 1206, 2013.
75. Saranya, J., Sreeja, B.S., Padmalaya, G., Radha, S., Manikandan, T., Ultrasonic Assisted Cerium Oxide/Graphene Oxide Hybrid: Preparation, Anti-proliferative, Apoptotic Induction and G2/M Cell Cycle Arrest in HeLa Cell Lines. *J. Inorg. Organomet. Polym. Mater.*, 30, 2666, 2020.

9

Nanoscale Modeling and Simulation in Nanoionics: Insights into Material Behavior and Device Design

M. Rizwan[1]*, A. Ayub[2], I-S. Ilyas[2], K. Zaman[2] and G. Nabi[2]

[1]*School of Physical Sciences, University of the Punjab, Lahore, Pakistan*
[2]*Department of Physics, University of Gujrat, Gujrat, Pakistan*

Abstract

This cohesive exploration discusses advancements in electronic components, covering the finite element method (FEM) modeling of nanostructured superionic systems, the evolutionary trajectory of memristors, SPICE-based modeling, and the intricacies of resistor-switching devices. The chapter delves into the application of FEM in nanostructured superionic systems, showcasing the efficiency of this method in modeling enhanced conductivity, particularly through the space charge layer. It comprehensively outlines the evolution of memristors, detailing their conceptualization by Leon Chua and exploring various types such as resistive, spintronic, organic, and ferroelectric. This chapter emphasizes the pivotal role of memristors in neuromorphic systems, highlighting their memory and resistive properties that excel in data storage and processing. A SPICE-based memristor model is also discussed, focusing on electron transfer techniques in the Schottky-barrier area. The categorization of I-V curves and exploration of oxygen-deficient ions and electron-transferring phenomena deepen the understanding of memristor behavior in SPICE simulations. The discussion expands the exploration to resistor-switching devices in nano-ionic systems, emphasizing resistive-switching effects based on cations, anions, and their combination. The discussion introduces three categories of resistive-switching effects and underscores the significance of selector devices in enhancing device efficiency through controlled ion movement.

**Corresponding author*: rizwan.sps@pu.edu.pk

Inamuddin, Tariq Altalhi, Mohammad Luqman and Jorddy Neves Cruz (eds.) Nanoionics: Fundamentals and Applications, (233–262) © 2025 Scrivener Publishing LLC

***Keywords*:** Finite element method (FEM), nanostructured superionic systems, memristors, SPICE model, resistor-switching devices, ion movement, conductivity, neuromorphic systems

9.1 Introduction

In the ever-growing field of nanotechnology, the ability to manipulate matter at atomic and subatomic levels has become the cornerstone of innovation and research. One of the pivotal fields is the emerging field of nanoionics, which has enabled the innovation and creation of unique ionic materials and devices, that have modified properties, advanced stability, tunable character, and broader range [1]. Nanoionics mainly deals with the manipulation of interface properties in solid-state ionic materials. The properties of the ionic materials differ greatly at the interface in comparison to their bulk behavior owed to many effects such as defect segregation, structural order, and work function discrepancies. The mismatched work function that arises due to contact between ionic conductors with a higher-density semiconductor is greatly exploited in many anionic systems [2].

The integration of advanced modeling and simulation tools at the nanoscale level has provided an indispensable tool in the field of nanoionics, providing further insights into ionic materials and their behavior. The nanoscale chemical and physical behavior of materials defy conventional expectations. Nanoionics focuses on the transport and storage properties of ions, that are harnessed in storage devices, sensors, and more devices. To achieve nanoscale precision, traditional experiments face limitations caused by size, cost, and time. Nanoscale modeling and simulation serve as virtual laboratories for scientists to explore, and manipulate the behavior of nanoionic materials with great precision. Simulation methods such as density functional theory and molecular dynamics enable researchers to study interactions at an unimaginable scale [3]. Phenomena such as ion diffusion, charge transfer, and surface interactions can be studied in detail *via* these tools. Comprehension of these phenomena and properties makes tuning these properties easy for enhanced properties and better functioning nanoionic devices. Despite remarkable progress in modeling and simulation techniques of nanoionics, there are still challenges that persist.

Many nanoionic systems have been studied experimentally, but there is a serious lack of theoretical analysis of their properties, such as enhancement of conductivity of superionic composites *via* theoretical modeling. Ionic devices cannot vie with electronic devices with regards to speed due to the electrons' reduced mobility. But this does not hold at the nanoscale, and at nanoscale ionic processes can be increased substantially. Nanoscale devices that have controlled ionic processes have opened up applications beyond energy storage. Ionic resistive switching (RS) refers to the rearrangement of ions *via* the process of ion dispersion and the electrochemical phenomenon that leads to nanoscale conductive paths [4]. The ability to control these ionic processes will enable better devices with enhanced functionality and reliability. Such control can be achieved *via* simulation tools.

Solid-state ionic materials have a wide range of applications in modern high-end technologies. Modeling and simulations of properties of such solid-state nano-ionic materials have progressed to a great degree in a few years. Simulation models such as Materials Studios, Atomistix Toolkit, and COSMOL are used for nanoscience. Recently the field of nanoionics has focused on memristive devices. Memristors merge the processing and storage capabilities activated by an external electrical stimulus, mirroring the functions found in nerve structures. Research and development of memristors require theoretical tools for ionic transport modeling [5].

9.2 Modeling and Simulation Methods in Nanoionics

Despite extensive use of experimental methods in nanotechnology, the field of nanoionics particularly lacks well-developed simulation models and tools. Simulation techniques can provide new insights into the field of nanoionics and act as a bridge between theory and experiment [6]. Ionic motion is very important in nanoionic materials. To truly understand how ions transport in devices, we need detailed, real-time investigations at an atomistic level. Molecular dynamic simulations provide comprehension at the atomistic level, and comprehension of properties such as conductivity, and activation energy, but there are still some technological problems that pose limits to MD simulation tools widespread in the field of nanoionics.

9.2.1 Molecular Dynamic Simulations (MD)

MD simulations based on the *ab initio* method track the movement of atoms over time in a model system. The derived forces acting between atoms are large, so a numerical approach is employed. The trajectory is divided into many intervals, such that force remains constant in each interval. Molecular dynamics simulations can be combined with other algorithms to resize the simulation cell. The interaction forces can be based on quantum calculations, which are rendered more accurate in comparison to force fields. Current software that employs empirical force fields can predict properties with more accuracy. Based on what properties are to be explored, the simulation model should balance accurate forces and statistical quality. For ionic mobility in nanoionic materials, the balancing criteria are lattice constants and elastic moduli along with the defect inclusion energy [7]. The initial criteria for the evaluation of empirical force fields for the comprehension of ionic mobility may include the lattice constants, elastic moduli of single crystalline models, and the energy needed. Ionic mobility is related to the formation of vacancies and interstitial sites within the crystal lattice or Frenkel defects; therefore, the chemical potential associated with Frenkel defects is associated with the vacancy and interstitial chemical potential.

To evaluate this potential in MD, Boltzmann statistics is utilized. For the calculation of ionic conductivity, we can differentiate interstitial and

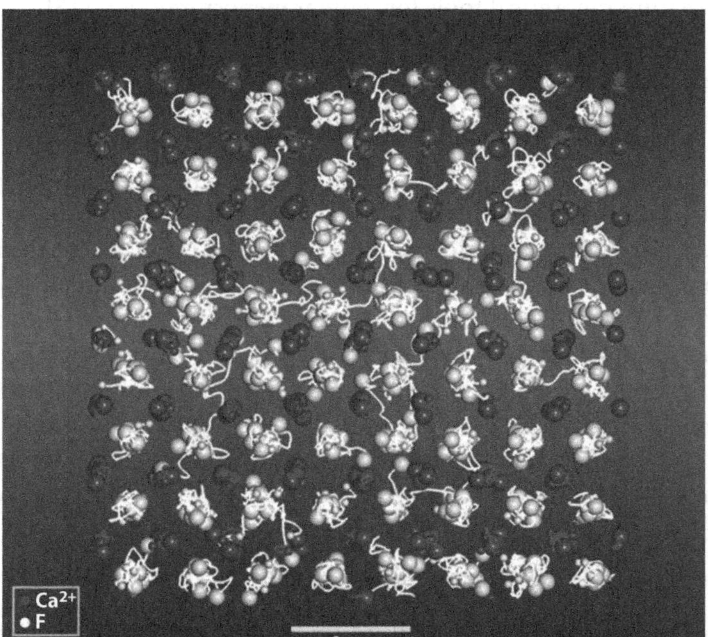

Figure 9.1 Illustration of an MD simulation depicting a Calcium fluoride (CaF_2) crystal at T=1800 K and room pressure. Solid lines delineate ion trails, representing a 10 ps timeframe. Calcium ions are represented in blue and fluoride ions are represented in white. While calcium ions remain fixed in their lattice locations, undergoing vibrations, F ions exhibit a dynamic design involving the creation and recombination of multiple Frenkel defects, coupled with the transport of vacancies as indicated by white lines [8].

vacancy mobility, which is crucial when interstitial and vacancy migration are considered. Supercell techniques are employed when single crystallines are studied in MD. Applying boundary conditions to such systems requires that the interaction occurring at interstitial and vacancy sites should be evaded. Such a problem can be avoided by making dimensions of the box to nm or more. Due to ion migration at these sites, Frenkel defects are unavoidable. But if the systems are very large, they can have the capability to accommodate Frenkel defects. The exploration of conductivity in MD depends on the system under study, such as in crystalline compounds; the nearest neighbor's analysis can assist in identifying interstitials and vacancies. Amorphous systems are characterized by ion-ion contact statistics. Figure 9.1, shows a nm-scale simulation model of CaF_2. The solid lines indicate the ionic pathways.

9.2.2 Charge Transport Model (CTM) for Nanoionic Memristors

Memristor or memory-resistor was discovered in 1971 [4]. Resistive switching in these devices provides an alternative to flash memory, which has limited performance in the nanoscale. The CTM offers a more accurate alternative to existing SPICE models by considering ionic transport mechanisms. CTM is based on ionic translocation and tunneling phenomenon [9]. The suggested memristor for simulation is based on Ag/TiO_2/ITO and is composed of an input voltage connected in series. There have been a lot of modeling efforts for memristors. In one of the first modeling efforts, memristors were considered to be a series of connections maintained between two resistors [10]. The proposed model had a memory resistance that depended on the width ratio of total dielectric thickness and the doped region. Windowing functions are used to accommodate the non-linear effect, but such simulation models still have limited performance.

Another memristor modeling with better performance is resistors connected in series along with an electron tunnel barrier, and the state variable is the Simmons tunnel barrier, which is ultimately the programming threshold. SPICE or simulation program that has an integrated circuit emphasis (SPICE) based model with a Simmons tunnel barrier has high accuracy but a large computation time as well. Many other models were developed with many varying factors to improve performance and accuracy [11].

The CTM model is based on TiO_2, and the conduction properties are based on the oxide. The model has a semiconductor and insulating side, the ionized atoms can travel to transition oxide under the action of the electric field, resulting in a change in ion concentration and thus conductivity. To comprehend charge transport in memristors, the vacancy drift is important.

9.2.3 Linear Drift Memristor Model

Despite existing for many years, the full potential of memristors, including their non-volatile nature, wasn't realized until 2008. Memristors have non-volatility, infinite dimensions, low power operation, and dynamic response, such diverse properties correspond to many applications. Many analytical models have been proposed to describe memristors that provide intrinsic dopant kinetic approximation in linear and non-linear behavior. Memristance correlates charge and flux as given in equation 9.1,

$$M = \frac{d\varnothing}{dq} \tag{9.1}$$

Voltage v and current I of the memristive system are given as

$$V = IR(x) \tag{9.2}$$

$$f(x, I) = \frac{dx}{dt} \tag{9.3}$$

Where, R(x)= instantaneous resistance (dependent on state variable x). x(state variable), is normalized doped region width related to its thickness, D by $x = \frac{\omega}{D}$. The current is given as derivative of the function, ω,

$$\begin{aligned} &= U_D = \frac{d\omega}{dt} = \mu E, \text{ and} \\ &\frac{\mu R_{on} I(t)}{D} = \mu E \end{aligned} \tag{9.4}$$

Where, U_D = drift speed at boundary

μ = mobility of dopant

E = electric field, present in the doped region in the company of current I(t)

R_{on} = resistance of the device when the device is activated.

In case, E has a relatively reduced value, this linear drift model can successfully represent the dynamics of memristors. This model is invalid beyond thickness less than the function ω. The projected model is supple for different memristor systems [12].

9.2.4 SPICE Model for Memristors

SPICE model is governed by electron passage in the Schottky barrier. Current in memristors is governed by tunneling, ohmic contact, and thermionic emission, The current produced *via* thermionic emission is very small compared to the one produced *via* tunneling. Therefore, to reduce complexity thermionic current is ignored [9]. Oxygen vacancies and electron transport decide the memristor current. The proposed SPICE model for Silver/Titanium dioxide/indium Tin Oxide or (Ag/TiO_2/ITO) is given in Figure 9.2.

As shown in Figure 9.2, the current-voltage sweep is portioned into 6 sections, grounded on voltage range and Schottky barrier current. The state variable for this model is given by

$$\frac{dn_L}{dt} = (V, N_L) \tag{9.5}$$

The state variable, n_L provides the oxygen decencies near the Schtokky region. The schematic diagram of the planned SPIC model is given in Figure 9.3.

The simulation using the spice model is given in Figure 9.4.

9.2.5 Structure-Dynamic Approach (SDA)

The problems related to electric double layers can be solved by various mathematical approaches for ionic liquids and electrolytes, but the same does

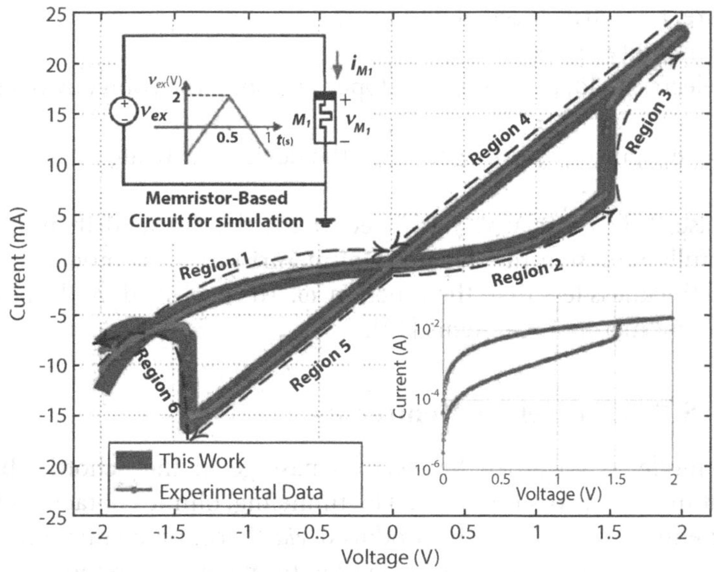

Figure 9.2 Illustration of the current-voltage (I-V) of the Ag/TiO_2/ITO memristor based on the charge transport mechanism (CTM model), demonstrating higher accuracy than previous studies. The top left corner shows the proposed circuit of the memristor. The memristor's current voltage curve is segmented into six parts, with each part associated with a different voltage region. The I-V curve is for a Voltage=± 2 V, frequency=1 Hz triangular wave input voltage. The I-V curve in the bottom right corner is the logarithmic scale I-V graph for positive values of voltages only [9].

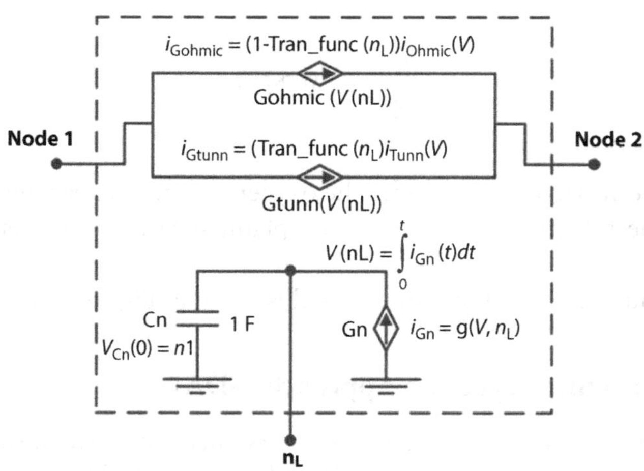

Figure 9.3 Schematic diagram of the SPICE model for a memristor. The model incorporates three nodes: Node 1 and Node 2 correspond to the +ve and -ve terminals of the device. Node n_L is used for imagining the number of O-vacancies in the neighborhood of the Schottky barrier [9].

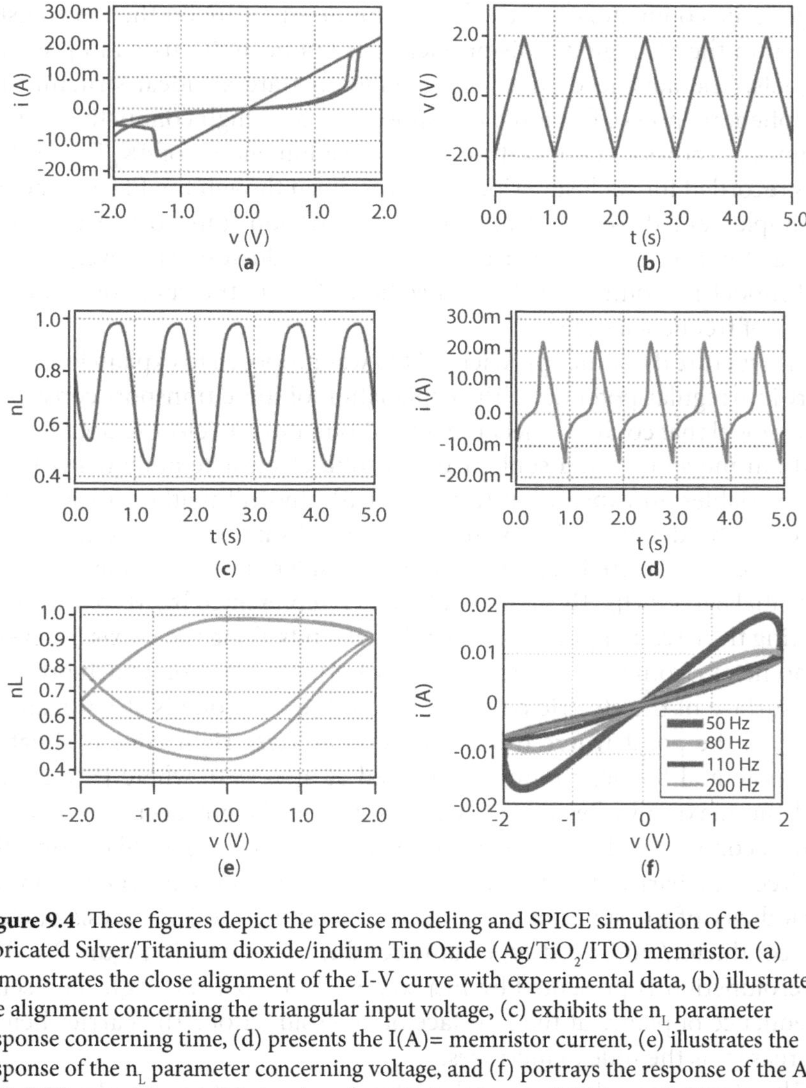

Figure 9.4 These figures depict the precise modeling and SPICE simulation of the fabricated Silver/Titanium dioxide/indium Tin Oxide ($Ag/TiO_2/ITO$) memristor. (a) demonstrates the close alignment of the I-V curve with experimental data, (b) illustrates the alignment concerning the triangular input voltage, (c) exhibits the n_L parameter response concerning time, (d) presents the I(A)= memristor current, (e) illustrates the response of the n_L parameter concerning voltage, and (f) portrays the response of the $Ag/TiO_2/ITO$ memristor model at various frequencies when a ± 1 V sinusoidal input voltage is applied [9].

not apply to solid nanosystem, and nanoionics, which have fast ionic transport. Comprehension of dynamics of ionic transport at the nanolevel could reveal frequency-capacitance properties of conductor hetero-junctions. A commonly used approach to process FIT devices is mainly based on diffusion and drift coefficients, such models are used in modeling memristors, which are nanoionic memory systems. The huge problem faced in superionic conductors is the effect of atomic structure on the ion translocation in the SC region. This problem has not been discussed in previous models. The advanced superionic conductors are an ideal structure for the phenomenon of fast ionic transport. A record high conductivity of 0.3 $Ohm^{-1}cm^{-1}$ at room temperature with activation energy of $4K_BT_{300}$ =0.1eV was recorded for such superionic devices. Even though the FIT theory can be implemented to obtain information about ion transport in the space charge (SC) region of advanced superionic conductors. However, a physical model is required for the realization of ionic transport on a variety range of frequencies [13].

A structure dynamic approach (SDA) was proposed to explain ion transport in irregular potentials. The explanation of ionic transport consists of (i) a model that could connect the relaxation rate of an electric double layer and ion movement in a sublattice of a solid electrolyte potential (ii) hidden variables, to explain ion transport (iii) physical-mathematical model based on hidden variables. A simple advanced superionic conductor/electric conductor layered heterojunction is considered with varying heights of potential barrier (η). The number of layers is represented by, M. $\eta(x)$, representing the irregular potential relief that extends along the x-axis to several nanometers from the electric conductor [14]. The position which prevents the surface of the electric conductor from the electrode is at x_o=0, and at $x=x_m$, is the point where the electric double layer right edge is situated. The potential barrier minima is situated at $\eta(x_j)=0$, where j=1, 2....M, and the maximum lies at $x_j - \Delta/2$. The barrier height decreases from x_0 and becomes equal to the barrier total height at x_m. just like along the x-direct, the barrier height (η), is a discrete function that depends on the periodicity of the superionic conductor. Structure model analysis of electric conductor/ advanced superionic conductor is given in Figure 9.5. The heterojunction is presented by EC/$\{x\}_j$, where j=1,2,....M. $\{x_j\}$, represents a sequence of planes at the interface of heterojunction. The barrier height decreases, as the index j increases.

The structural coherence at the interface implies that the planes, $\{X_j\}$, situated in the anionic sublattice of the heterojunction are parallel to the electric conductor surface planes and the normal of both planes meets with the FIT tunnel situated in heterojunction. The heterojunction of electric conductors and advanced superionic conductors can be grown epitaxially. The SDA method differs from other models by the existence of potential barrier minima inside each barrier layer, which decreases with increasing j. Figure 9.5, also compares a classic ionic crystal potential relief with the heterojunction potential [15]. The proposed model explains the structure and ion transport in advanced superionic conductors, such as α-AgI and

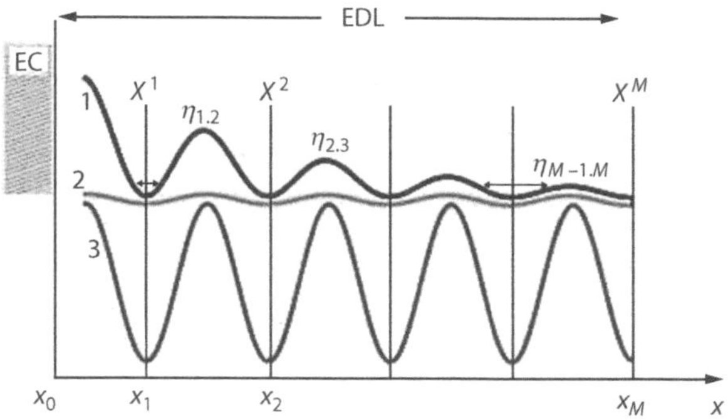

Figure 9.5 Schematic illustration of the crystal potential relief for electric conductor/ $\{X_j\}$ heterojunction (1) indicated by solid blue, advanced superionic conductor bulk (2) indicated by solid orange, and classical ionic conductor (3) indicated by solid brown [5].

α-$RbAg_4I_5$. In this case, the cation transport occurs at tetrahedral positions as illustrated in Figure 9.6 [5]. In advanced superionic conductors, the cations form a zigzag-like tunnel as a result of cation transport along the empty tetrahedral positions. The simulation results obtained by using SDA are given in Figure 9.7 [5].

9.2.6 Finite Element Method (FEM) Model

In a nanostructured superionic system, the conductivity is enhanced due to the space charge layer. The finite element method (FEM) is used to model such enhancement. FEM modeling can provide resolution up to the order of millions of elements and provides a very short computing time, as well as provides a general description of the effect of SCL on the conductivity of the superionic composite. This method can be an alternative to the quantum-based method, density functional theory, which has very exhausting computing time. The study employed simple cubic, face-centered cubic, and body-centered cubic geometries. The model equations were solved using a single unit cell of the nanocomposite. Nanoparticles having radius R were surrounded by a space charge layer, the ionic conductivity in the SCL layer showed an exponentially decreasing behavior from a maximum at the surface, σ_o and approaches the bulk value σ_b, that is far from the surface. The spatially varying conductivity, σ_j, is given as

$$\sigma_j = \sigma_b + (\sigma_o - \sigma_b)\frac{R}{r_j} e^{\left(-\frac{r_j - R}{\delta}\right)} \tag{9.6}$$

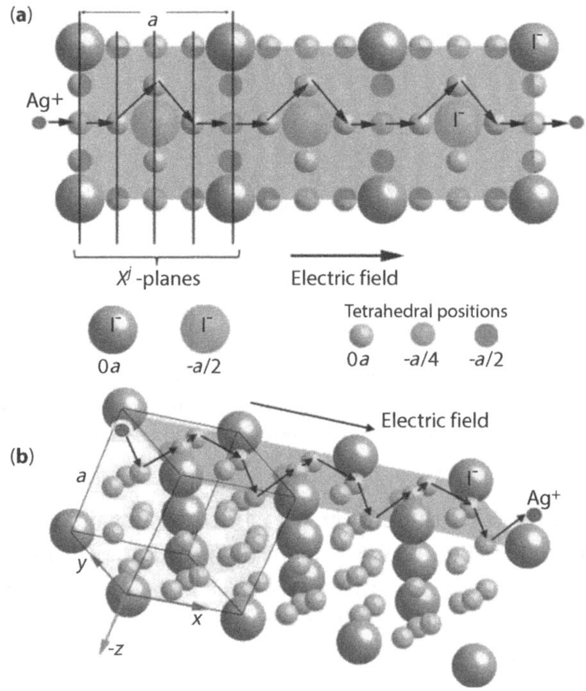

Figure 9.6 The arrangement of α-AgI/AdSIC (Symmetry=Im3m, lattice parameter, a = 0.509 nm) is characterized by a rigid iodine sub-lattice and tetrahedral locations that are accessible for the mobility of Ag+ cations. The potential movement of Ag^+ cations in response to an external electric field is depicted schematically with arrows in the <001> projection (Figure 9.6a) and within the crystal bulk (Figure 9.6b). These tetrahedral positions are situated on the crystallographic planes denoted as X_j. Different colors are used to signify distinct z-coordinates of I-ions and tetrahedral vacant positions concerning the drawing plane [5].

Where δ, is the spatial decay determining constant in the SCL is related to Debye length and has the order of nm. $\frac{R}{r_j}$, is the radial factor that is due to spherical geometry. The lower limit on conductivity increment in the FEM model is provided by assuming that nanoparticles had zero conductivity, and thus the nano-ionic system occupies volume and decreases the conductivity of the system. The insulation of nanoparticles is overturned by the space charge layer. The nanoionic composite has three domains, one where ionic conductivity is zero, second with improved conductivity, and third with reduced ionic conductivity. An upper limit is important as well,

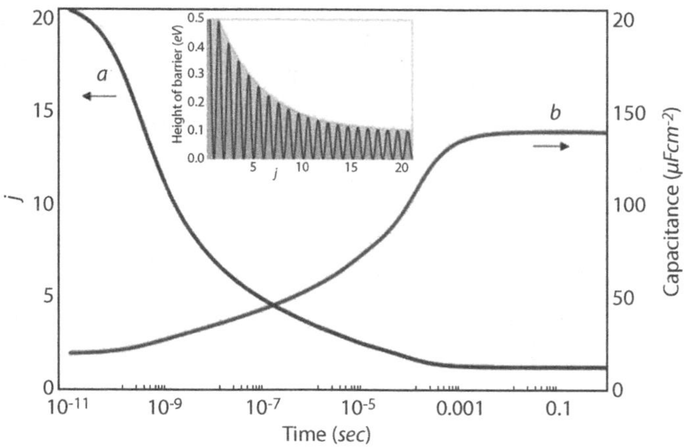

Figure 9.7 Galvanostatic charging of the electric conductor/{X_j} (j= 1,2,.. 21) blocking heterojunction. This figure analyzes the time evolution of the cations' center of mass and the system's capacitance density for a sequence of potential barriers [5].

ionic conductivity cannot increase without any limit, given packing limits to the carrier concentration and mobility. The simple unit cell used for FEM modeling along with the simulation results of effective conductivity obtained is given in Figures 9.8 and 9.9, respectively.

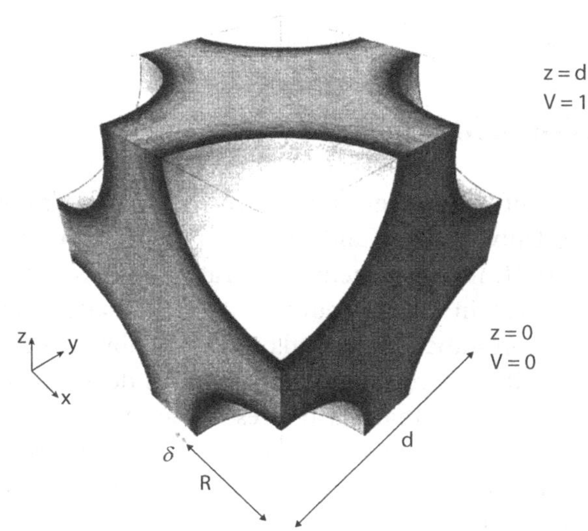

Figure 9.8 Nanoparticles arrangement of the cubic model unit cell with simple cubic nanoparticle [16].

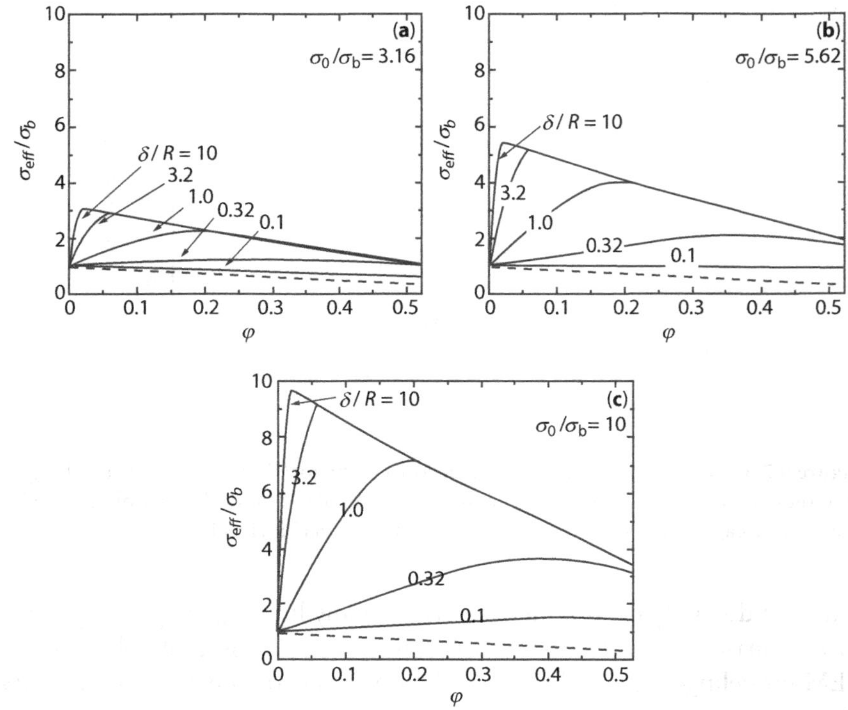

Figure 9.9 Effective conductivity of the simple cubic case illustrated as a function of the loading fraction for five values of $\frac{\delta}{R}$ at three values of the surface conductivity: (a) 3.16, (b) 5, and (c) 10. The dashed lines show the upper and lower limits [16].

9.3 Nanoionic Memristors

Memristors concept was first given by Leon Chua, who was working as a professor at the University of California in the Department of Electrical Engineering in 1971. Its experimental realization in May 2008 by Hewlett Packard as predicted in Nature journal attract, the attention of various researchers. Chua presumed that based on the equilibrium of the inductor, resistor, and capacitor, there should exist a fourth device that links magnetic flux and charge. After that, various researchers worked on the making of this device using various sorts of materials and structures. Biolek *et al.*, devised a drift-memristor carrying impurities and having non-linearity, through SPICE in 2009 [11, 17].

In 2010, Wei Lu suggested a memristor device at the nanoscale that can imitate synapse behavior in neuro-morphic systems.

In 2011, Researchers from Harvard University published about an important programmable-nano-wire circuits that can be used in nano-processing. Various developments in this field led to recent memristors and still because of their priceless importance much of the research is focused on it [18].

Memristor word comes from a combination of two words "Memory and Resistor". Their ability to memorize the state history makes them valuable and a prospective candidate for future research. It is a thin-layered electrical circuit component that varies its charge-sustaining resistance. Chua verified that it is a basic component in electrical circuits as its behavior cannot be copied by any of the other electronic components i.e., capacitor, inductor, or resistor [19].

A memristor is a basic electronic component belonging to the group of neuromorphic systems and having the capability of presenting a resistance based on the stored information about voltage and current already existing in the system in that cycle. These devices can store as well as process the given information compared to conventional devices which may only store data in the form of memories or can process that information using transistors [20].

Memristors can be classified into many groups like, two-terminal-vertical-stack memristors, two-terminal-planar memristors, and three-terminal-resistance-switches. Depending on the material of switching phenomena, these can be categorized into physical and chemical switches. Chemical switches are composed of ionic-switching material which comprises anionic or cationic materials, whereas both of these species are mobile and work on the basic principle of oxidation and reduction of materials and nanoionics. The electric field induces motion of the anions which indirectly changes the resistance of the material due to changes in the valence of metals like metallic oxides, nitrides, and so on [21]. Cationic memristors on the other side are composed of metals that are electrochemically active like silver, and copper which act as an atomic bridge between the electrodes. The other types of memristors which rely on Physical changes are made up of materials that have the capability of showing either the magnetic, phase-changing, ferroelectric, or electron-hole trapping phenomena. Memristors that rely on both the chemical and physical switching phenomena can also be found [22].

A memristor comprises two electrodes, which may be called an anode and a cathode. In between two electrodes lies a region, that is known as the switching region represented as 102. This region is made up of a conductive and an insulating material. Furthermore, in between the switching region and the second electrode (104), there lies a region known as the intermediate region represented as 103 and this intermediate region has electrical contact with both the region and electrode present on its sides that is an intermediate region and second electrode. This region contains metallic nanowires in a polymer matrix which play an essential role in nanoionic memristors because when we switch off the voltage, the memristor memorizes its current resistance and stores and utilizes that information when you again turn On that device either after one day or 1 year and this ability to memorize data is dependent on the nanomaterial used. So, we may conclude that memristor memory is dependent on the type of nanomaterial used in its fabrication [18].

9.3.1 Types of Memristors

A memristor usually consists of an insulating layer i.e., of TiO_2 sandwiched between two metallic layers usually made up of platinum exhibiting bipolar switching properties. Based on the type of material used and its switching phenomena it can be characterized into succeeding types:

- **Resistive-memristor**

In resistive memristors, the resistive material can switch its state connecting high-resistance (HRS) with low-resistance (LRS) when an external electrical signal is provided as input [23].

- **Spintronic-memristor**

This type of resistor varies their resistance by varying the spin direction of their electrons. To detect magnetic flux, magnetic tunneling junctions (MTJ) are used in industrial recording heads. For spin-torque-magnetic RAM and logic devices, it is the fundamental component. Figure 9.10, shows a schematic diagram of a spintronic memristor.

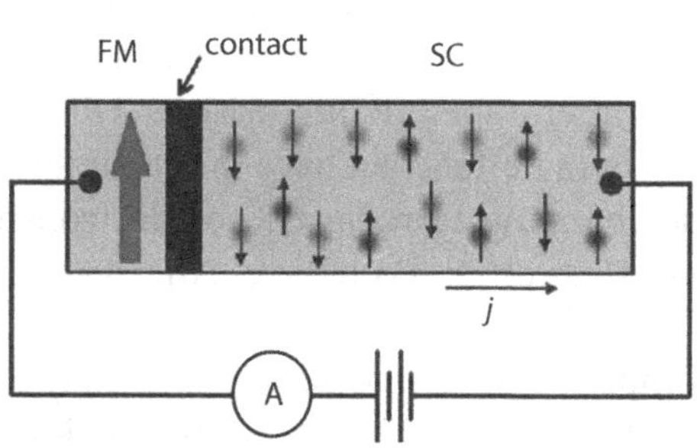

Figure 9.10 Schematic diagram of spintronic memristor. The flow of spin-up electrons from the semiconductor to the ferromagnet leads to a higher ratio of spin-down electrons near the junction [47].

• **Organic or Polymeric-memristor**
This type of memristor consists of a conducting part which plays a vital role. It consists of a papery polyaniline layer settled on an insulating base with the help of two electrodes. A thin layer of solid-electrolyte tampered with polyethylene-oxide (PEO) present only in the central part exhibits oxidation-reduction reactions. Only the central part of PANI present along with PEO as represented in the Figure 9.11 plays the role. To furnish it with the reference voltage a narrow strip of wire is dipped into the solid electrolyte. Variations in memory and conductivity are dependent on the redox reactions taking place in the operational zone, where PANI is shifted from an insulating state to an oxidized conducting state. From the framework of field-effect-transistors (FETs) the two electrodes are named Source and drain and the wire dipped in PEO is named as gate electrode [23].

• **Ferroelectric-memristor**
This type of memristor is related to ferroelectricity where two metallic electrodes are connected through an antiferroelectric material present between them. Two opposite spins can be regarded as "0" and "1" which are two binary bits, thus making Ferro-electric-random-access-memory (FeRAM). Due to the involatile nature of FeRAM, they can be broadly used in automobile components, smart cards, and various other memory-inserted applications [23]. The Internet of Things and AI are going to rule over the world, but these emerging technologies require faster speed with the help of nanoelectronics design. Much of the research based on nanoelectronics focuses on designing such systems that consume minimum power with faster speed and immense packing density. All of these requirements point towards a non-volatile memory device that may be considered as a future candidate, which may rule over existing memory devices and

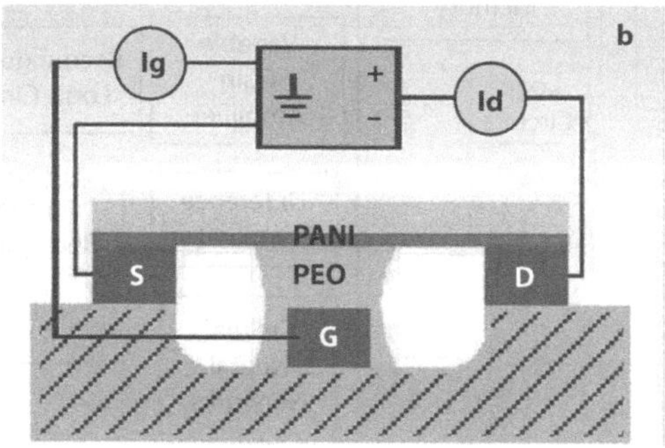

Figure 9.11 Schematic diagram of polymeric-memristor [47].

this is called a Memristor. It is well known as resistive RAM (Re-RAM). These devices form the basis of modern computing offering them high speed with smaller size and minimum energy required for operation [24].

Memristors offer various sorts of services including both volatile and involatile selector functions as well as memories, Digital to analog processing as well as data storing capacity, and sensor activities. It is their inherent property to imitate synaptic and neural functionalities and they may be appraised as important building blocks other than von Neumann sort of computing hardware. They show high stability against electromagnetic waves as well as high energy particles thus resulting in fewer errors. They may also work well in a broad range of temperatures from 4Kelvin to 600Kelvin. Their ability to work well in extreme conditions makes them the best component for various applications including space applications. Recently these devices have shown progress on a large scale in the areas of Integration on chips, downscaling, various sorts of arrays, and difficult computing architectures as well as face recognition [25]. Figure 9.12, shows modified applications of memristors.

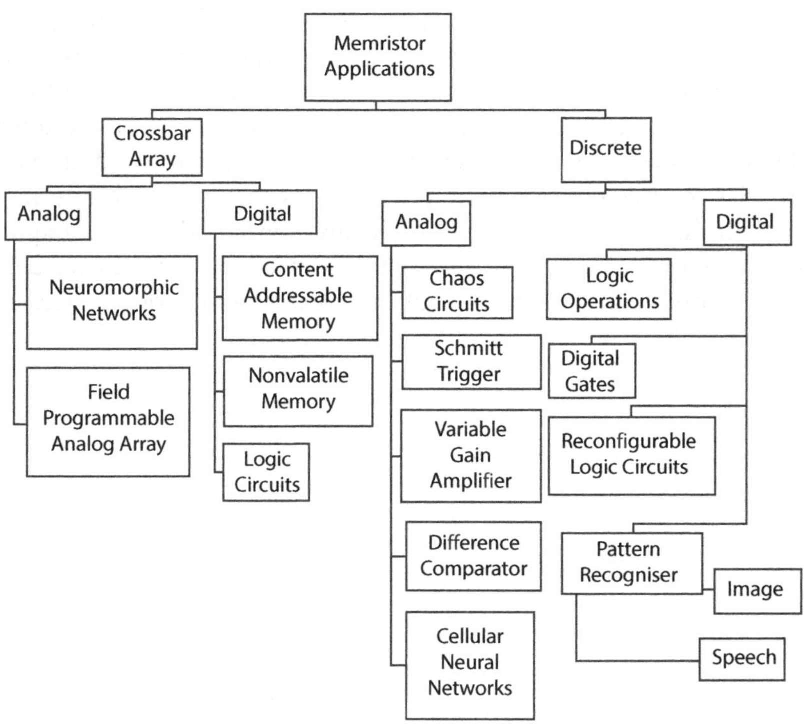

Figure 9.12 Future analog and digital applications of memristors [25].

SPICE represents 'Simulation-based Programming with Integrated-Circuit-Emphasis' and this is a libre-software that mimics active conditions of analog circuits. As this program provides analysis, this analysis uses experimental and theoretical results and equations based on Kirchoff's rules. ICs were designed by hand before SPICE [26].

Until the last three decades, memristors did not have any specific model in circuit simulation like SPICE, Using eqs. 9.7 and 9.8 respectively representing two specific sorts of memristors, charge and flux controlled were used [27].

$$V = \frac{\partial \varnothing}{\partial t} = \frac{\partial F}{\partial q}\frac{\partial q}{\partial t} = M(q)i \tag{9.7}$$

$$i = \frac{\partial q}{\partial t} = \frac{\partial G}{\partial \varnothing}\frac{\partial \varnothing}{\partial t} = W(\varnothing)V \tag{9.8}$$

The basic idea of SPICE based model relies on the e^- transferring technique in the Schottky-barrier area. Any of the process tunneling ohmic contact or thermionic emission or all of these together based on the presence of voltage and state may result in the presence of current in the memristor [20]. As the thermionic current present in the memristor is extremely small and does not affect the accuracy of results that much compared to the tunneling current this current can be ignored to solve the complexity of the circuit. The maneuver of oxygen-deficient ions and electron-transferring phenomena should be carefully observed to find the memristor current [27].

Based on the voltage and Schottky-barrier-currents, the I-V curve can be categorized into various regions as represented in Table 9.1.

The state variable can be found from

$$\frac{dn_L}{dt} = g(V, n_L) \tag{9.9}$$

Oxygen-vacancy-dynamics can be written as

$$g(V, n_L) = \frac{2\mu E_o}{L} \sin h \frac{|V|}{LE_o}\left[-stp(V)n_L + stp(-V)(1-n_L)\right] \tag{9.10}$$

Table 9.1 Categorization of I-V curve relying on the voltage & Schottky-barrier-currents [9].

Region	Voltage domain	Current-mechanism
1	Between -2 and 0	Tunneling Thermionic Emission
2	Between 0 and 1.5	Tunneling Thermionic Emission
3	Between 1.5 and 2	Ohmic-Tunneling Thermionic Emission
4	Between 0 and 2	Ohmic Tunneling
5	Between -1.4 and 0	Ohmic Tunneling
6	Between -2 and -1.4	Ohmic-Tunneling Thermionic Emission

We can also write

$$n_L = \frac{N_{Schottky}}{N_{Schottky} + N_{Rest}} \tag{9.11}$$

n_L des n_L describes the number of O-deficiencies close to the Schottky-barrier-area in the memristor. Equation 9.4 requires a capacitor with a specific capacitance, which is its ability to store electrical energy. The initial number of oxygen vacancies in the memristor needs to match the initial value stored in the capacitor, which is denoted by n_1. The dependent quantity current represented as G_n is equivalent to n_L depending on Voltage in equation 9.10. The oxygen deficiencies in the case of a memristor for forward and reverse bias are different and are higher in the case of forward biasing. The memristor changes itself from a higher to a lower state and in the case of switching, firstly ohmic current is dominant, and then the tunneling current becomes more effective. This can be represented by the equation 9.11:

$$\text{Trans-func} = \frac{1}{1 + e^{S\,Dec}} \tag{9.12}$$

Where S represents the parameters for state shifting. Based on the physical properties it may have different values.

$$\mathrm{Dec} = \mathrm{n_{th}} - \mathrm{n_L} \tag{9.13}$$

where n_{th} is the minimum value of oxygen deficiencies required for activation and this value is different for different metallic oxides. It is also dependent on the position of the electrode in the memristor device from the insulator. The memristor current can be found through eq. 9.14.

$$I_{mem} = \text{Trans-func}(n_L) i_{tunneling}(V) + 1 - \text{Trans-func}(n_L) i_{ohmic}(V) \tag{9.14}$$

Simulator internal limits should be carefully mentioned because they may affect simulation results but if by any chance this issue happens, a lower capacitance capacitor may be used to overcome the issue. By adjusting the parameters according to specific device's physical parameters, this code may also be used for VCM memristors. Based on their role, SPICE code parameters may be categorized into the following categories:

- Physical parameters: Depends on the physical attributes of memristors like thickness.
- Constant parameters: These parameters are general for every memristor and don't depend on any of its attributes like Boltzmann-constant.
- Setting Parameters: Depending on the experimental data, these parameters should be adjusted for every memristor like S, aM, and so on as used in the code.

Input Voltage, waveform, and frequency have a vital role in determining the results of the simulation.

Now, let's explore the different types of resistive switching devices based on the mobile ions involved.

9.4 Resistor-Switching Devices Design

In nanoionic devices, the speed of devices may be increased by applying higher fields. Controlled movement of ions has resulted in the development of various useful applications like ionic-resistive switching devices or memristors. The memristors are named so because of the resistance switching effect that plays the major role, as here, the resistance which is due to the dielectric layer present between two electrodes plays the major role. The functionality principle of these devices relies on the arrangement of diffused ions between two electrodes that affect the formation of a conductive layer or path uniting two electrodes. Controlling the formation of this conducting path through varying electrochemical processes taking place in the electrolytes, we may control the efficiency and working of these devices [9]. The switching phenomena in these devices play a vital role and ions are mostly used in this process. Based on the types of ions used, this effect may be categorized into three categories [28]:

9.4.1 A Cation-Based Resistive-Switching Effect

Abundantly used in electrochemical-metallization this process utilizes a switching-layer that plays the role of host for cations motion, uniting an inert metal electrode and an active metal electrode. If enough voltage is then applied to, the active-metal electrode, its atoms get oxidized into their respective positive ions. These positive ions then move to the inert electrode passing through the dielectric-insulating material present between the two electrodes, under the influence of an electric field. Under the influence of nucleation phenomena, these positive ions get contracted to neutral atoms, resulting in the formation of nanoclusters of the respective metal. This process continues and is accompanied by the development of metallic filament, in between two electrodes, which results in the switching of these devices from maximum-resistance-state (HRS) to minimum-resistance-state (LRS). This all happens in the "SET" command. Controlling the movement of ions and redox reaction rates, the geometry, and various other factors of the formed filament may be controlled. The rate of motion of these ions controls the direction of formation of these filaments, while the redox reaction rate determines the thickness of these formed filaments. Higher rates result in dense filaments [29].

If the device is set to a "RESET" state then the device changes itself from minimum or LRS to maximum or HRS, which happens because of the rupture of these filaments in which respective metallic atoms move back to the active electrode. Because of their higher stability, these ECM devices may be effectively used for memory-related applications.

Noble metals are mostly preferred for use as an inert metal electrode and filament mostly grows from Anode to Cathode. The crystal structure of materials may also play a role in increasing the efficiency of these devices. The rate of movement of ions may be controlled by crystal structure as in Lithium-Ion batteries [30].

9.4.2 B Anion-Based Resistive-Switching Effect

This effect is based on the oxide ions. Oxygen atoms or oxygen vacancies are moving which plays a role in the switching effect. These devices consist of two inert electrodes with a switching layer in between. This switching layer can be of two types: one with high oxygen content and another with low oxygen content. Oxygen-rich layer is composed of suboxides or metals that upon reaction with the layer may result in the formation of a sub-oxide junction. The O-deficient or switching layer on the other hand is composed of insulating material with extreme resistance. When enough voltage is solicited, the oxygen vacancies move from a layer having maximum oxygen to a layer with minimum oxygen going through the process of oxidization. These charged electron-deficient-oxygen-vacancies undergoing the process of reduction result in the formation of further oxygen-rich regions and increased conductivity [31]. Mostly higher oxygen vacancies result in higher conductivity as observed in the case of transition metals but based on the type of material the formed conducting filament may have oxygen-rich or oxygen-deficient atoms in excess. The joule-heating effect can also play a role in increasing the conductance of these devices. At higher temperatures higher conductivity is observed and upon lowering temperature, less joule-heating-effect results in higher resistance or lower conductance [32].

Much of the research has been done and many researchers are still working on finding interesting applications based on this switching phenomenon. As the controlled movement of ions in carbon-based-inorganic [33] may result in various applications including LEDs, Laser and so on [34].

9.4.3 Cation and Anion-Based Resistive-Switching Effect

Some devices including valence-charge-memory (VCM) devices instead of using cations or anions use both for switching purposes. In these devices, it has been observed that at the same time, metallic ions as well as oxygen vacancies travel from one electrode to the other turning the device, from H-R-S to L-R-S. On the other side, their reverse movement switches them back to HRS [35].

A memristive device, which is based on RS-switching phenomena is comparable to a synapse in the nervous system where its conductance determines how much input can be effectively converted into output. The conductance can be varied by requesting positive or negative pulses on the input side [36].

In memristive devices, short-term-memory (STM) and long-term-memory (LTM) utilize the concept of homosynaptic-plasticity [37] where synapses are input-dependent. Networks based on associative learning utilize the concept of heterosynaptic-plasticity, where synaptic variations not only depend on the input but also the performance of other neurons [38].

As mentioned earlier controlled movement of electrons and ions results in many of the future oriented applications.

9.5 Quantum-Point Contacts

In R-S devices, filaments made up of only a small number of atoms utilized in determining their width can be formed by careful administration of the programming input. So, the conductivity of these filaments becomes quantized when the size of the filament becomes comparable to electron fermi-wavelength resulting in the development of Quantum-point-contact (QPC) [39]. Conductance of QPC can be evaluated using the following equation:

$$G = n\,G_0 (G_0 = {}^{2\,e^2}\!/_{h})$$

Here in this equation "n" represents an integer, "h" denotes Planck's-constant and "e" denotes charge on an electron [40].

The quantization phenomena based on conductance have been observed in Cationic as well as Anionic-based R-S devices. Because of higher conductance, devices with these QPCs can be effectively used in high-density-memory applications. One of the major characteristics that enhance the importance of these devices is that these R-S devices based on QPC also enable us to study electrical attributes of various Nano-structures and develop circuits based on these properties, which was an impossible task before [41].

9.6 Magnetic Nanostructures

The motion of ions and their rearrangement, other than playing a role in determining the resistance may also affect determining the magnetization, like as administrating the motion of ferromagnetic-metallic-ions, in the insulating layer present between two electrodes, magnetic devices may be designed whose properties may be varied depending on the applied electric field. Magnetic tunneling junctions (MTJs) are one of the renowned examples of this concept. In these devices, the dielectric-switching layer acts as a tunneling layer, uniting two ferromagnetic electrodes. If the magnetization direction of two electrodes is oriented parallel to each other then it will show higher magneto-resistance (TMR) but if it is oriented anti-parallel it will show lower MTR [42].

Local magnetic domain structures can also be controlled by controlling the motion of these ions. This concept has been observed in $LiFe_5O_{8-x}$ films, where the removal of Li^+ ions resulted in an increase in magnetization as well as the growth of magnetic domains and vice versa. So, tuning the material composition, we may control the magnetic domains as well as the magnetization effect of magnetic devices, leading to new research prospects in this field [43].

Yang *et al.*, in their publication proved that the thickness of film can be efficiently controlled because of the TMR effect. AS by careful formation of Co filament in ZnO film in case of Co/ZnO/Fe sandwich, he observed that thickness of ZnO filament formed between Co and Fe electrodes can be controlled because, upon formation of a film with enough thickness, TMR effect becomes dominant [44].

9.7 Selector Devices

For repressing leakage current, selector devices are highly in demand along with R-S elements. For density-memory applications, these devices can be used along with R-S elements without using any transistors but these devices must have high linearity.

Z. Wang *et al.*, have researched that R-S devices relying on the development of AG filament show threshold-switching phenomena. Here, programming leads to the formation of filaments while on removal of the field, these particles immediately form nanoparticles and because of threshold-switching behavior, these devices may overcome the leakage current [45].

Xue *et al.* in their publication have also represented that VO_2 nanofilaments may be formed in V_2O_5 when an external field is applied and this transition behavior from metal to insulator can be used to build selector devices with minimum variations. Moreover, if we attach these selector devices with R-S devices in series, the leakage current can be minimized to a greater rate resulting in increased efficiency of the device [46].

9.8 Future Perspective

Nanoionic devices have changed the future of the world by effectively controlling the movement of ions. This effect can be effectively used in memristors which relies on resistance-switching phenomena. Besides RS switching devices which control resistance, the controlled movement of ions can be effectively utilized in the formation of magnetic nanostructures, selector devices, and quantum point contacts. These structures when used along with RS devices can increase the performance of devices with special reference to memories. The chapter provides a holistic perspective on advanced electronic components. Finite element method (FEM) modeling emerges as a powerful tool for understanding the conductivity of nanostructured superionic systems. Memristors, classified into various types, play a pivotal role in neuromorphic systems, offering advantages in memory and resistive properties. The SPICE-based model further deepens the understanding of memristor behavior, emphasizing the importance of electron transfer techniques. Resistor-switching devices, categorized by resistive-switching effects, showcase the potential for high-density-memory applications, with selector devices enhancing overall efficiency. A common theme across these chapters is the controlled movement of ions, influencing conductive paths and enabling threshold-switching phenomena.

References

1. Yamaguchi, S., Nanoionics-Present and future prospects. *Sci. Technol. Adv. Mater.*, *8*, 503, 2007.
2. Maier, J., Nanoionics: Ion transport and electrochemical storage in confined systems. *Nat. Mater.*, *4*, 805, 2005.
3. Despotuli, A.L. and Nikolaichik, V.I., A step towards nanoionics. *Solid State Ionics*, *60*, 275, 1993.
4. Chua, L.O., Memristor—The Missing Circuit Element. *IEEE Trans. Circuit Theory*, *18*, 507, 1971.

5. Despotuli, and Andreeva, Structure-dynamic approach in nanoionics. Modeling of ion transport on blocking electrode, 2013.
6. Habasaki, *Molecular Dynamics of Nanostructures and Nanoionics: Simulations in Complex Systems*, 2020.
7. Guo, X.X., Matei, I., Lee, J.S., Maier, J., Ion conduction across nanosized Ca F2 Ba F2 multilayer heterostructures. *Appl. Phys. Lett.*, *91*, 1013102, 2007.
8. Zahn, D., Molecular dynamics simulation of ionic conductors: Perspectives and limitations. *J. Mol. Model.*, *17*, 1531, 2011.
9. Amirsoleimani, A., Shamsi, J., Ahmadi, M., Ahmadi, A., Alirezaee, S., Mohammadi, K., Karami, M.A., Yakopcic, C., Kavehei, O., Al-Sarawi, S., Accurate charge transport model for nanoionic memristive devices. *Microelectron. J.*, *65*, 49, 2017.
10. Pickett, M.D., Strukov, D.B., Borghetti, J.L., Yang, J.J., Snider, G.S., Stewart, D.R., Williams, R.S., Switching dynamics in titanium dioxide memristive devices. *J. Appl. Phys.*, *106*, 074508, 2009.
11. Kolka, Z., Biolek, D., Biolkova, V., Improved model of TiO2 memristor. *Radioengineering*, *24*, 378, 2015.
12. Prodromakis, T., Peh, B.P., Papavassiliou, C., Toumazou, C., A versatile memristor model with nonlinear dopant kinetics. *IEEE Trans. Electron Devices*, *58*, 3099, 2011.
13. Despotuli, A. and Andreeva, A., Method of the uniform effective field in the structure-dynamic approach of nanoionics. *Ionics (Kiel)*, *22*, 1291, 2016.
14. Despotuli, and Andreeva, Computer modeling on subnanometer scale the ion-transport characteristics of the solid electrolyte/electronic conductor blocking heterojunctions. *Nano Microsyst. Tech. (Russian)*, 11, 15–23, 2012.
15. Andreeva, A.V. and Despotuli, A.L., Interface design in nanosystems of advanced superionic conductors. *Ionics (Kiel)*, *11*, 152, 2005.
16. Goodyer, C.E., Fish, J.S., Fehribach, J.D., O'Hayre, R., Bunge, A.L., Modeling space charge layer interaction and conductivity enhancement in nanoionic composites. *Electrochim. Acta*, *56*, 9295, 2011.
17. Mazumder, Kang, Waser, Memristors: devices, models, and applications. *Proc. IEEE*, *100*, 6, 1911–1919, 2012.
18. Goossens, Memristive Systems Using Nanoionic and Phase-Change Mechanisms.
19. Khrapko, Rusyn, Politansky, Investigation of the memristor nonlinear properties. *Inform. Autom. Pomiary w Gospod. Ochr. Śr.*, *8*, 1, 12–15, 2018.
20. Yakopcic, Taha, Subramanyam, Pino, Memristor SPICE model and crossbar simulation based on devices with nanosecond switching time, in: *The 2013 International Joint Conference on Neural Networks (IJCNN)*, IEEE, pp. 1–7, 2013.
21. Sun, Chen, Yan, The future of memristors: Materials engineering and neural networks. *Adv. Funct. Mater.*, *31*, 8, 2006773, 2021.
22. Jeong, Kim, Kim, Choi, Hwang, Memristors for energy-efficient new computing paradigms. *Adv. Electron. Mater.*, *2*, 9, 1600090, 2016.

23. Shahsavari, and Boulet, Memristor nanodevice for unconventional computing: review and applications, arXiv preprint arXiv:1703.00331, 2017.
24. Pershin, and Di Ventra, Spin blockade at semiconductor/ferromagnet junctions. *Phys. Rev. B*, *75*, 19, 193301, 2007.
25. Maan, Jayadevi, James, A survey of memristive threshold logic circuits. *IEEE Trans. Neural Netw. Learn. Syst.*, *28*, 8, 1734–1746, 2016.
26. Mestry, Dere, Hundiwale, Agashe, SPICE Modeling of Memristor and Memristive Systems: A Review.
27. Sharifi, and Banadaki, General spice models for memristor and application to circuit simulation of memristor-based synapses and memory cells. *J. Circuits Syst. Comput.*, *19*, 02, 407–424, 2010.
28. Lanza, Wong, Pop, Ielmini, Strukov, Regan, Larcher, Villena, Yang, Goux, Recommended methods to study resistive switching devices. *Adv. Electron. Mater.*, *5*, 1, 1800143, 2019.
29. Yang, Gao, Li, Pan, Tappertzhofen, Choi, Waser, Valov, Lu, Electrochemical dynamics of nanoscale metallic inclusions in dielectrics. *Nat. Commun.*, *5*, 1, 4232, 2014.
30. Van der Ven, and Ceder, Lithium diffusion mechanisms in layered intercalation compounds. *J. Power Sources*, *97*, 529–531, 2001.
31. Zhou, Peng, Yin, Zhou, Liu, Ling, Lei, Zhou, Tang, Modulating memristive performance of hexagonal WO3 nanowire by water-oxidized hydrogen ion implantation. *Sci. Rep.*, *6*, 1, 32712, 2016.
32. Slesazeck, Mähne, Wylezich, Wachowiak, Radhakrishnan, Ascoli, Tetzlaff, Mikolajick, Physical model of threshold switching in NbO 2 based memristors. *RSC Adv.*, *5*, 124, 102318–102322, 2015.
33. Zhu, Fu, Meng, Wu, Gong, Ding, Gustafsson, Trinh, Jin, Zhu, Lead halide perovskite nanowire lasers with low lasing thresholds and high quality factors. *Nat. Mater.*, *14*, 6, 636–642, 2015.
34. Green, Ho-Baillie, Snaith, The emergence of perovskite solar cells. *Nat. Photonics*, *8*, 7, 506–514, 2014.
35. Fang, Shi, Zhang, Wu, An, Wei, Lu, Liu, Impact of Ta/Ti electrodes on linearities of TaO x-based resistive random-access memories for neuromorphic computing. *Sci. China Phys., Mech. Astron.*, *63*, 1–6, 2020.
36. Jo, Chang, Ebong, Bhadviya, Mazumder, Lu, Nanoscale memristor device as synapse in neuromorphic systems. *Nano Lett.*, *10*, 4, 1297–1301, 2010.
37. Zucker, and Regehr, Short-term synaptic plasticity. *Annu. Rev. Physiol.*, *64*, 1, 355–405, 2002.
38. Berridge, Lipp, Bootman, The versatility and universality of calcium signalling. *Nat. Rev. Mol. Cell Biol.*, *1*, 1, 11–21, 2000.
39. Long, Lian, Cagli, Cartoixa, Rurali, Miranda, Jiménez, Perniola, Liu, Suñé, Quantum-size effects in hafnium-oxide resistive switching. *Appl. Phys. Lett.*, *102*, 18, 2013.

40. Mehonic, Vrajitoarea, Cueff, Hudziak, Howe, Labbe, Rizk, Pepper, Kenyon, Quantum conductance in silicon oxide resistive memory devices. *Sci. Rep.*, *3*, 1, 2708, 2013.
41. Zhu, Su, Liu, Hu, Pan, Lu, Zhang, Li, Observation of conductance quantization in oxide-based resistive switching memory. *Adv. Mater.*, *24*, 29, 3941–3946, 2012.
42. Sellmyer, D. and Skomski, R. (Eds.), *Advanced magnetic nanostructures.* Springer US, Boston, MA, 2006.
43. Tan, Huang, Avci, Büttner, Mann, Hu, Mazzoli, Wilkins, Tuller, Beach, Magneto-ionic control of magnetism using a solid-state proton pump. *Nat. Mater.*, *18*, 1, 35–41, 2019.
44. Yang, Zhan, Zhu, Liu, Yang, Hu, Shang, Pan, Chen, Li, Tunneling magnetoresistance induced by controllable formation of Co filaments in resistive switching Co/ZnO/Fe structures. *Europhys. Lett.*, *108*, 5, 58004, 2014.
45. Wang, Joshi, Savel'ev, Jiang, Midya, Lin, Hu, Ge, Strachan, Li, Memristors with diffusive dynamics as synaptic emulators for neuromorphic computing. *Nat. Mater.*, *16*, 1, 101–108, 2017.
46. Midya, Wang, Zhang, Savel'ev, Li, Rao, Jang, Joshi, Jiang, Lin, Anatomy of Ag/Hafnia-based selectors with 1010 nonlinearity. *Adv. Mater.*, *29*, 12, 1604457, 2017.
47. Shahsavari, Memristor technology and applications: An overview, 2013.

10

Commercialization and Industrial Aspects of Nanoionics: Lab to Market

M. Rizwan[1]*, H. Hameed[1], A. Ayub[2], G. Nabi[2] and M. Tanveer[2]

[1]*School of Physical Sciences, University of the Punjab, Lahore, Pakistan*
[2]*Department of Physics, University of Gujrat, Gujrat, Pakistan*

Abstract
Embark on a journey through the realm of nanoionics, where the fusion of ionic conduction and nanotechnology opens doors to groundbreaking innovations. From micron-sized sensors and batteries to the fabrication of nanometer-scale devices, microionics directs a path of scientific marvels. Explore the pivotal role of nanoionics in energy storage, electronics, and sensors, revolutionizing industries. Uncover the challenges in commercialization, emphasizing safety and regulatory frameworks. Nanoionics emerges as a transformative force, influencing the future of batteries, memory devices, sensors, and bioelectronics.

***Keywords*:** Energy storage, nanoionics, microionics, next-generation electronics, sensor technology, commercialization challenges

10.1 Introduction

Solids having fast ionic conduction are the basis of micron-sized sensors, batteries, and devices which are fabricated by the technical and scientific direction given by microionics. Solid state ionics have separated from this new direction, independent of the definite physical parameter; i.e., without transformation to the novel phenomenon inferred. The main target behind this idea is to minimize the extent of devices further in the form of tablets [1]. Furthermore, new possibilities are expected by using ionic conductors in ICs for the reason that microionics are fairly comparable

**Corresponding author*: rizwan.sps@pu.edu.pk

Inamuddin, Tariq Altalhi, Mohammad Luqman and Jorddy Neves Cruz (eds.) Nanoionics: Fundamentals and Applications, (263–276)

to semiconductor devices. This advancement in size should result in nanometer-sized devices that derive the possibility of new phenomena. In reality, in all structures having ionic conductors, there are nanometer regimes of précised substantial characteristics [2]. These regions are of space charges, of Debye Radius having size-lambda, which are found in the free ionic conductor spaces and electrodes and various materials value λ varied in waveguiding. At 300K conductive, ionic conductors have a value $\lambda \sim 60nm$, whereas for high conducting solid electrolytes such as $RbAg_4I_{5-}$ family (having $\sim 10^{20}\ cm^{-3}$ concentration of Ag^+ mobile ions) its value of $\lambda \sim 0.5nm$. Potential outcomes to impact on these particular regimes in a manageable way might be displayed in tiny-sized devices. With particular physical properties, the [1] width of the region may exceed the value of λ considerably when in solid electrolyte electronic and ionic defects are introduced simultaneously the width of λ exceed at the metallic cathode dissolution, for the reason to keep the native electric neutrality same.

Here the word nanoionics can be presented based upon the rigid visible parameters as the proportion of the girth of the region having explicit effects on the specific girth of devices of nanometer size and smallness of lateral size. Right now, to fabricate nanometer scaled devices, EB lithography is one of the important approaches. Feature development is accomplished with the utilization of delicate to electron illumination exposed uncovered by a filtered centered electron light emission 20-200 keV energy. Lately, the highest sensitivity electron beam of $RbAg_4I_5$ and $CsAg_4Br_{3-x}I_{2+x}$ has been published; the thickness of solid electrolyte is $\sim 150nm$ which is deposited on the thin substrate of carbon [3]. It was observed that by focusing the electron beam, only feature on the film can be formed directly which makes the other technological operations, etching process, and resists unnecessary. When the beam current density increases up to the threshold one $\sim 10^2\ A/cm^2$ fast feature formation has been observed. In comparison to the inorganic resists, the ratio of the exposure process is about ten times larger which can be utilized in totally dry technology [4].

10.1.1 Importance in Emerging Technologies

In the realm of emerging technologies, nanoionics has played a crucial role in the field connecting ionics and nanotechnology. Nanoionics offers a unique perception of material science as it explores ion transport behavior or ion transportation on a nanoscale. This escalating field is not only revealing material's fundamental properties but also clearing the path for revolutionary advancements in various regions. The potential of nanoionics is very important in redefining the landscape of developing technology, which offers solutions for persistent challenges and unrevealing new possibilities [5].

10.1.1.1 Advancements in Energy Storage Applications

In the advancement of energy storage applications, nanoionics has shown significant development. In portable electric automobiles and electronics traditional Lithium-ion batteries are used extensively which is the result of nanoionics advancements. At the nanoscale; tailoring the ion transport properties the performance, life span and performance of batteries can be enhanced [6]. Research is being conducted to find nanostructured materials and interfaces to improve the lithium-ion movements which have enhanced the energy density and faster charging-discharging rates. This improvement not only extends the life of the battery but increases the battery efficiency as a whole. This has significant ramifications for feasible energy solutions, tending to the developing interest for productive and harmless ecosystem power storage [7].

10.1.1.2 Next-Generation Electronics

Nanoionics led to an exemplary shift in the evolution of electronics. At the nanoscale, the manipulation of ions leads to the development of new electronic components which makes more energy-efficient and faster devices. Memristors or memory resistors, which are the invention of nanoionics research, imitate the synaptic way of behaving of the human brain and can possibly alter neuromorphic computing. This connection between nanoionics and electronics is not only escalating the abilities of traditional computing but also leading to the new frontlines in AI and data processing [8].

10.1.1.3 Transformation in Sensor Technologies

In the advancement of sensor technology nanoionics is very instrumental. Nanomaterial's exclusive properties of ion transportation make it possible to create highly sensitive sensors, that have fast response time and enhanced selectivity. This has broad applications in medical services, environmental observation, and various industries. Nanoionics sensors can reshape clinical diagnostics, giving ongoing and exceptionally precise data for infection location and observation [9].

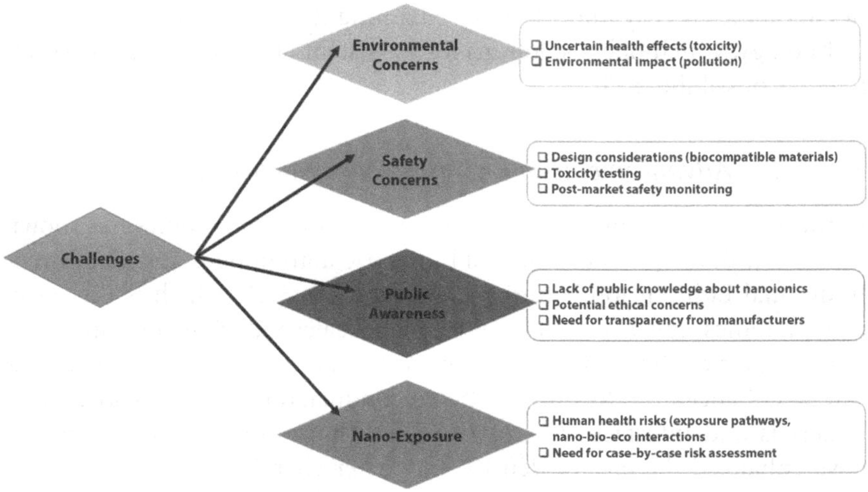

Figure 10.1 Commercialization challenges in nanoionics.

10.2 Commercialization Challenges

Nanoionics (transportation of ions at the nanoscale) has revolutionized numerous industries from electronics to energy storage applications. However, the transformation from theoretical revolutions to practical applications displays an exceptional set of challenges for commercialization which needs to be understood from a market perspective. These challenges are explained in the preceding section, and summarized in Figure 10.1.

10.2.1 Navigating Health and Environmental Concerns

Nanotechnology is a rapidly growing field due to its small size and is becoming the way of life in cosmetics, drug delivery systems, food and packaging, biosensors and therapeutics, and many more. Since its smaller size, order fighting and anti-bacterial properties make it to be utilized in various commercial products as detergents, wound dressing, or antimicrobial coating [10].

Consumer goods are being transported to the market at a rate of 3-4 per week according to reports. Thousands of tons of ceria, alumina, and silica as a mixture of ultrafine abrasive particles plus nanoionics are being used in slurries for accurate polishing of silicon wafers each year according to the National Nanotechnology initiative [11].

Hence, as the applications are expanding the population, exposure to nanomaterials is increasing [11]. Despite the evident advantage of the power of the small size of the particle, there are open inquiries concerning how the nanoparticles utilized for everyday life might influence the environment. One of the pivotal issues that must be promoted in the not-so-distant future, before the massive creation of nanomaterial, is their toxicity to humans and their effect on the environment [12]. The property of nanomaterial which could have adverse biological effects having the potential to initiate toxicity is under debate. It needs to be observed that when the size of nanoparticles is degraded biologically in the cellular environment, then what cellular response will be? For example; degraded nanoparticles biologically may be stored inside the cells which leads to intercellular alternation such as gene alteration or interruption of organelle reliability [13]. Some important questions are: Are nanoparticles more toxic in comparison to their non-nano counterparts? Will nanoparticles be transported into the environment having more toxicity? Nanotoxicology research is important to understand and reveal the influence of nanoparticles on the environment before they are allowed to be utilized in daily life so that their unwanted properties can be dodged [13b].

10.2.2 Ensuring Safety in Nanoionic Applications

In various industries, nanoionics holds immense applications, from energy storage to healthcare. However, the integration of nanomaterials at the molecular and atomic scale promotes genuine safety concerns. The main concern about safety in the manufacturing of nanoionics lies in the thoughtful design. As in the previous section, it is discussed that the toxicity of nanoparticles must be inherently low and should have a minimal environmental effect. For medical applications implementation of biocompatible materials and energy storage applications employing eco-friendly nanomaterials are crucial steps toward attaining safety objectives. Also, during the phase of material development, toxicity studies are essential to recognize and ease potential risks in a timely manner in the process of innovation [14].

Evolution protocols and rigorous tests must be conducted before nanoionic applications hit the market. A comprehensive valuation of the physical, chemical, and biological interactions of nanomaterials is included. To evaluate the performance and safety of nanoionic devices standard testing methodologies need to be established. To reflect the improvement in the nanotoxicology and safety valuation methods this procedure should regularly be updated. Risk assessment conduction to ensure safety in nanoionics is the key component [15]. In the process of risk assessment identification of potential hazards and implementation of effective modification strategies are important factors. This involves assessing the impact of nanomaterials on the ecosystem, human health, and occupational safety. Cooperative efforts among regulatory bodies, researchers, and industry stakeholders are important for the development and implementation of a strong risk assessment framework. To ensure the safety of applications of nanoionics strict regulatory framework and standards are dominant. International and Government authorities should work in collaboration to establish clear rules for the production, usage, and disposal of nanoionic material and devices [16].

To adjust to developing an understanding of nanomaterial safety it should be mandatory to implement these regulations and constant monitoring and updates. It is important to monitor the effects of nanoionics that may not be visible immediately to reveal potential environmental and health concerns. For nanoionic devices establishment of post-market investigating programs, specifically in medical applications for the assurance of ongoing safety evaluations. This practical approach makes the identification of any unexpected issue and protects human health and environmental security it permits timely interventions [17].

10.2.3 The Role of Public Awareness and Acceptance

The addition of silver NPs into the packaged materials for the reason of antimicrobial activity was reported several times and in much food packaging these materials had been extensively used. These products were being used by the general public and they were unaware of the addition of NPs which could be a moral and judicial issue [18]. Now proper labeling is mandatory so that people will be aware of what they consume. Manufacturers must ensure the labeling for transparent information for the public. It is important that the general public is aware of and accepts the addition of NPs but product manufacturers ignore this fact [19]. Product manufacturers keep this information secret and new product development underground and don't share this information with the public (trade secrets and competition might be the possible reasons). This conflict is the fact that the public has the right to know what and why product manufacturers are marketing. In Singapore, in the case study, it is already proved the general public is unaware of nanotechnology and the increase in the severe effect of nanotechnology makes the perception of the public negative. A survey conducted on the island of Ireland concluded that even agri-food organizations are unaware of nanotechnology [20]. In food and other industries, public opinion is twofold, assent and dissent (or altruism and skepticism) with nanotechnology. Public attitude is highly dependent upon précised applications. In 2005 in food nanotechnology received 49% support and in 2008 it reduced to 32% (among all applications of nanotechnology this is the lowest value) while in 2008 food packaging applications that screens state were braced by 73% of people. Particularly in food, the highest concern regarding nanotechnology is 28% [21].

10.2.4 Mitigating Risks Associated with Nanoproduct Exposure

Over the past decade, public attention regarding nanotechnology-based products has increased. Nanomaterial is being added to the products intentionally or unintentionally. As the spread of nano-products expands environmental and human health concerns arise accordingly owed to the nanomaterial's unique physiochemical properties. Environmental health concerns are directly caused by nanomaterial usage interactions such as nano-pesticides, nano-fertilizers, and nano-herbicides, and are less probable in powerless nanosensors [22]. In the environment fate and behavior of NPs are dependent upon the physiochemical properties of nanomaterials per se. Also, the fate and behavior of NPs are limited by the complexity of the environmental conditions. Nano-bio-eco-interactions make it difficult to monitor and trace the nano-materials dispersal. In a connected eco-system, for the understanding of nano-bio-eco interactions among nano-materials, biotic and abiotic interactions holistic approach has been recommended. For the certain valuation of environmental nano-toxicity case-to-case studies are needed [23].

In a human digestive system, the exposure route of aluminum NPs can be quoted for the assessment of nano-product-associated risks through the illustration of complexity and probable experimental workflow. In artificial saliva, the samples remain unchanged at the initial stage of digestion. Al NPs start to dissolve partially and release Al ions in the fluid of the stomach. During the digestion process particle accumulation might arise anytime most probably in the stomach fluid. In the intestinal liquid, agglomerates incline to deagglomerate into main particles [24]. Also, NP structures form because of free ions. Similarly, this happens to silver NPs. Silver colloids happen to be in any medium as complex mixtures with many diverse species captivated on the surface. For drug-resistant bacteria antimicrobial agents and culprits for toxicity release of Ag^+ is triggered.

Gold NPs can also discharge Au (I and III) ions in ambient conditions and donate to downstream exchanges [25].

10.3 Nano-Ionic Memory: Implications for the Economy

The energy demand is large enough globally and has been increasing at an alarming rate. According to the survey, by 2030 this demand will be increased by 40%. 20% of the global energy supply has approximately been used by the US, annually at the cost of 1trillion$ which is 5% more than the GDP of the US [26]. The energy demand is increasing globally, as practically all energy generation is currently from non-renewable sources with limited supply, like oil, flammable gas, and coal. The US is dependent upon foreign energy resources heavily, importing more than 60% of its oil. Also, US produces the greenhouse gases, the world's largest amount according to its energy use, which makes the US the largest contributor to climate change and global warming. Hence for security, economic, national, and environmental benefits US needs to improve its energy generation and energy usage efficiency [27].

One explicit reason for expanded energy interest in the US is the expanded utilization of chips and memory in family gadgets like PCs, computerized video recorders, and high-level cell phones like PDAs. PCs use approximately 5% energy of the total household electricity usage, and progressively increasing, as over 60 million PCs are sold in the US. Additionally, DVR's similar to PCs but they need disk drives, and microprocessors are in 25% of households in the US approximately. Globally, the popularity of smartphones with advanced memory and microprocessors is growing, the sale of smartphones has overtaken the sale of desktop PCs [28].

Chip and related memory are generally liable for huge energy demand in the US. This energy demand will keep on increasing not just on account of the expanded number of devices utilizing progressed semiconductor innovation, but also because the hidden semiconductor innovation is turning out to be less energy proficient as the circuits shrink to nano-size. The transistor's leakage current is responsible for the growing share of power used by smart processors and impartial memory chips. Over the next decade, it is expected to 10 times increase the leakage power per SRAM cell, as shown by the ITRS roadmap [28]. Cache memories based upon SRAM are becoming undeniably an enormous part of microprocessors for enterprise-class networking and commuting. For, the SRAM cache is 70% of the StrongARM12 processor and 32% of the 45nm 8-core Intel Xeon processor13 which has a collective 8MB-L3 cache heaving 731 million transistors, which makes the memories and their allied leakage in such chips one of the main power dissipation sources. For SRAM cache is commonly responsible for over 10% consumed energy and sometimes, it reaches as high as 40%. Different plans incorporate the utilization of on-chip microcontrollers exclusively for power the executives and numerous voltages to permit the utilization of low-spillage shut-down, reserve, and sluggish modes [29].

10.4 Future Prospects

Based on the broad definition of nanoionics, this phenomenon can also be explained as an indirect effect produced by the local movement of transportable ions at the interface of heterogeneous; whereupon, heterogeneous connection with higher carrier density materials; a layer of space charge forms in ion or mixed conducing materials, such as semiconductors and metals due of the work function difference of two materials [30]. This is similar to the Schottky barrier which is usually invoked for the explanation of heterogeneous band structure intersection at the interface of metal/semiconductor for the understanding of numerous electronic functions of boundary (obstacle), for instance, rectification. In relaxation mobile ions play a significant part in the case of mixed/ionic conductors at the interface for the development of the space charge layer, which in the liquid electrolyte system might be similar to the electrochemical double layer [31]. This is because the controlled meditation of both ionic and electronic imperfections can donate to the charge transmission reaction kinetics at the boundary region but also the space charge layer; enhancement of ionic and electronic conductivity. It is additionally predictable for nanoionics peculiarities at the heterogeneous connection point that, due to the diminished girth of the layer, the expanded surface charge girth might regulate the surface adsorption conduct of encompassing gases [32].

These remarkable properties, as described, have significant implications for various applications, as illustrated in Figure 10.2. In the fabrication of high-enactment batteries, nanoionics could contribute to improved energy density, longer life cycle, and faster charging time, and can work in various applications such as portable electronics and electric vehicles. Additionally, in solid-state batteries which have the potential to replace custom lithium-ion batteries nanoionics play a crucial role. Solid-state batteries have improved energy, density, safety, and performance. In the development of improved memory devices such as ferroelectric memory,

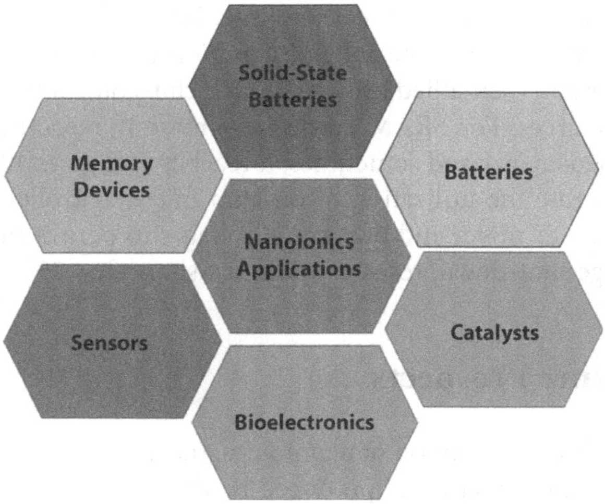

Figure 10.2 Future prospective applications in nanoionics.

low power and non-volatile storage, resistive switching memory, and offers high speed [33]. In highly sensitive sensors nanoionics may be applied to detect various environmental factors, biomolecules, and gasses, which can be applied to the healthcare environmental monitoring and different areas. In bioelectronics devices, nanoionic materials can be integrated which enables better interface among biological systems and electronic devices, which is advancement in the human-machine interfaces and medical devices. Nanoionics has an expanding future in several other fields, as computing, nano-electromechanical system catalysts, and many more [34].

10.5 Conclusion

Nanoionics is an advanced field of science in which fast ion transport has made it to be utilized in information technology, advanced electronics, and the medical industry with high efficiency. It has been commercialized for the last two decades; people are using nanotechnology-based products without identifying its hazardous effects on human health in the environment. The toxicity of the field is as strong as its advantage; research is being conducted to reduce the toxicity of the nanoparticles so the technology makes the world advance at ease. As it can be said nanotechnology could be the main technology created with aversion to environmental, moral, and social issues. Today if the aspects of technology are examined responsibly and fearlessly, it can be anticipated that our future will be augmented with benefits. There are enormous advantages of nanotechnology with many thrilling and potential products in the market. However, there are few challenges that need to be addressed, and fundamental queries need to be answered regarding low toxicity and high performance for the applications and development of nanotechnology. In our understanding, there are major gaps regarding the environmental, health, social, and safety impacts of nanotechnology. Hence, for the long-term achievement of nanotechnology, these gaps need to be bridged. For the manufacturing, dispensing, and applications along with disposal of nanomaterial legislations and regulations are paramount. Questionable issues relating to nanotechnology have proactively started public interest in the field. Laying out open trust, and creating and keeping up with the validity of nanoscience will require a sound and sane methodology for the benefit of the logical venture, cautious preparation and vital coordination, and the uniting of the concerned multidisciplinary groups with a systems administration mentality. We propose that nanotechnology should offer plenty of chances, by giving a novel and maintainable option in different areas.

References

1. Despotuli, A.L. and Nikolaichik, V.I., A step towards nanoionics. *Solid State Ionics*, 60, 275, 1993.
2. Pham Thi, M., Velasco, G., Colomban, P., Novak, A., Vibrational study of hydrogenated uranyl phosphates hydrates (HUP). *Solid State Ionics*, 9–10, 1055, 1983.
3. Despotuli, A.L. and Andreeva, A.V., Nanoionics: New materials and supercapacitors. *Nanotechnol. Russ.*, 5, 506, 2010.
4. Maier, J., Defect chemistry and ionic conductivity in thin films. *Solid State Ionics*, 23, 59, 1987.
5. Bastani, B. and D.F., Intellectual Property Rights in Nanotechnology. *Inf. Technol. J.*, 4, 69, 2004.
6. Pomerantseva, E., Bonaccorso, F., Feng, X., Cui, Y., Gogotsi, Y., Energy storage: The future enabled by nanomaterials. *Science*, 366, 6468, 1–12, 2019.
7. Serrano, E., Rus, G., García-Martínez, J., Nanotechnology for sustainable energy. *Renew. Sustain. Energy Rev.*, 13, 2373, 2009.
8. Payal, and Pandey, P., Role of Nanotechnology in Electronics: A Review of Recent Developments and Patents. *Recent Pat. Nanotechnol.*, 16, 45, 2021.
9. Silver, J.K., Preface. Postpolio Syndr., ix, 2004.
10. Asharani, P.V., Lian Wu, Y., Gong, Z., Valiyaveettil, S., Toxicity of silver nanoparticles in zebrafish models. *Nanotechnology*, 19, 255102, 2008.

11. Hauck, T.S., Ghazani, A.A., Chan, W.C.W., Assessing the effect of surface chemistry on gold nanorod uptake, toxicity, and gene expression in mammalian cells. *Small*, 4, 153, 2008.
12. Doshi, R., Braida, W., Christodoulatos, C., Wazne, M., O'Connor, G., Nanoaluminum: Transport through sand columns and environmental effects on plants and soil communities. *Environ. Res.*, 106, 296, 2008.
13. (a) Hussain, S.M., Javorina, A.K., Schrand, A.M., Duhart, H.M.H.M., Ali, S.F., Schlager, J.J., The interaction of manganese nanoparticles with PC-12 cells induces dopamine depletion. *Toxicol. Sci.*, 92, 456, 2006.
(b) Lewinski, N., Colvin, V., Drezek, R. Cytotoxicity of nanoparticles. *Small*, 4, 1, 26–49, 2008.
14. NSM, J., SBRS, A., Ahmad, N., Sodium-ion nanoionic hybrid solid electrolyte: Extended study on enhanced electrical and electrochemical properties. *Solid State Ionics*, 377, 115882, 2022.
15. Lal, B., Qasim, A., Mohammad Shariff, A., Application of Ionic Liquids in Wax, Scale and Asphaltene Deposition Control, in: *SpringerBriefs Pet. Geosci. Eng.*, p. 49, 2021.
16. Das, G., Choi, J.H., Nguyen, P.K.T., Kim, D.J., Yoon, Y.S., Anion Exchange Membranes for Fuel Cell Application: A Review. *Polymers (Basel)*, 14, 1197, 2022.
17. Zhang, X.D., Yue, F.S., Liang, J.Y., Shi, J.L., Li, H., Guo, Y.G., Structure Design of Cathode Electrodes for Solid-State Batteries: Challenges and Progress. *Small Struct.*, 1, 3, 2000042, 2020.
18. Chun, A.L., Will the public swallow nanofood? *Nat. Nanotechnol.*, 4, 790, 2009.
19. George, S., Kaptan, G., Lee, J., Frewer, L., Awareness on adverse effects of nanotechnology increases negative perception among the public: survey study from Singapore. *J. Nanopart. Res.*, 1–11, 16, 2014.
20. Cormick, C., Why do we need to know what the public thinks about nanotechnology? *Nanoethics*, 3, 167, 2009.
21. Yadav, G.L., Singh, S., Yadav, M., Yadav, G.K., Yadav, B., Effect of bioregulators and bio-fertilizers on growth of garlic (Allium sativum L.). *Int. J. Agric. Food Sci.*, 5, 18, 2023.
22. Hwang, H.M., Ray, P.C., Yu, H., He, X., Toxicology of designer/engineered metallic nanoparticles, in: *RSC Green Chem.*, p. 190, 2013.
23. He, X., Aker, W., Huang, M.-J., Watts, J., Hwang, H.-M., Metal Oxide Nanomaterials in Nanomedicine: Applications in Photodynamic Therapy and Potential Toxicity. *Curr. Top. Med. Chem.*, 15, 1887, 2015.
24. He, X., Aker, W.G., Hwang, H.M., An *in vivo* study on the photo-enhanced toxicities of S-doped TiO_2 nanoparticles to zebrafish embryos (Danio rerio) in terms of malformation, mortality, rheotaxis dysfunction, and DNA damage. *Nanotoxicology*, 8, 185, 2014.

25. He, X., Aker, W.G., Fu, P.P., Hwang, H.M., Toxicity of engineered metal oxide nanomaterials mediated by nano-bio-eco-interactions: A review and perspective. *Environ. Sci. Nano*, 2, 564, 2015.
26. Butzbaugh, J., Hosbach, R., Meier, A., Miscellaneous electric loads: Characterization and energy savings potential. *Energy Build.*, 241, 110892, 2021.
27. *2021 IEEE Int. Conf. Nanoelectron. Nanophotonics, Nanomater. Nanobioscience & Nanotechnol. 5Nano 2022*, p. i, 2021.
28. Noori, H., Goudarzi, M., Inoue, K., Murakami, K., Temperature-aware configurable cache to reduce energy in embedded systems. *IEICE Trans. Electron.*, E91-C, 418, 2008.
29. Kumar, R. and Hinton, G., A family of 45nm IA processors. *Dig. Tech. Pap. - IEEE Int. Solid-State Circuits Conf.*, p. 58, 2009.
30. Waters, B.M. and Troupe, G.C., Natural variation in iron use efficiency and mineral remobilization in cucumber (Cucumis sativus). *Plant Soil*, 352, 185, 2012.
31. Du, C.W., Wang, Y.H., Xu, F.S., Yang, Y.H., Wang, H.Y., Study on the physiological mechanism of boron utilization efficiency in rape cultivars. *J. Plant Nutr.*, 25, 231, 2002.
32. Guidong, L., Cuncang, J., Yunhua, W., Distribution of boron and its forms in young "Newhall" navel orange (Citrus sinensis Osb.) plants grafted on two rootstocks in response to deficient and excessive boron. *Soil Sci. Plant Nutr.*, 57, 93, 2011.
33. Zhang, J., Liu, W., Dai, J., Xiao, K., Nanoionics from Biological to Artificial Systems: An Alternative Beyond Nanoelectronics. *Adv. Sci.*, 9, 23, 2200534, 2022.
34. Wan, T., Zhang, L., Du, H., Lin, X., Qu, B., Xu, H., Li, S., Chu, D., Recent Developments in Oxide-Based Ionic Conductors: Bulk Materials, Nanoionics, and Their Memory Applications. *Crit. Rev. Solid State Mater. Sci.*, 43, 47, 2018.

11

Ion Migration and Defects in Nanostructures: Implications for Device Performance and Reliability

M. Rizwan[1]*, M. Aqeel[2], I.-M. Arshad[2], A. Ayub[2] and G. Nabi[2]

[1]*School of Physical Sciences, University of the Punjab, Lahore, Pakistan*
[2]*Department of Physics, University of Gujrat, Gujrat, Pakistan*

Abstract
Nanoionics is a fascinating field that explores the behavior of ions in nanoscale materials. It focuses on the migration of ions within these structures and the defects that can occur. Ion migration refers to the movement of ions, such as atoms or molecules with an electric charge, within the nanomaterial. Defects are imperfections or irregularities in the structure of these materials. Understanding ion migration and nanostructure defects is crucial for developing advanced technologies like energy storage, electronics, and sensors. It allows scientists to design and optimize materials with enhanced properties and functionalities. This chapter discourses the chemistry of defects in bulk and interfacial areas. Scenarios where the defect chemistry depends on the interface spacing are given special attention. These proper side effects occur in the nanoregion and include structural and electrical modifications. Several instances of nanocrystalline ion conductors are examined, including hetero layers of CaF_2/BaF_2, CeO_2 β-AgI: Al_2, O_3, and so on.

Keywords**:** Ion migration, nanoionics, energy storage, limitations, nanoionics, defects, nanocrystalline systems, thermodynamics

11.1 Introduction

Information and communication equipment uses a lot of electronic devices; enhancing the functionality of these devices which are mostly dependent

**Corresponding author*: rizwan.sps@pu.edu.pk

Inamuddin, Tariq Altalhi, Mohammad Luqman and Jorddy Neves Cruz (eds.) Nanoionics: Fundamentals and Applications, (277–308) © 2025 Scrivener Publishing LLC

on semiconductor performance that requires managing the movement of electrons and holes within semiconductors. It has been observed that conventional semiconductor devices' fine processing sizes have been getting smaller every year. There is a growing concern that traditional fine processing technology is approaching its maximum capabilities, and there is an impending risk of reaching the physical operational limits of semiconductor devices. By applying nanotechnology, it may be feasible to make advances within the materials used in devices and the principles governing their operation, which could help overcome these technological and physical constraints. Principle of gadget operation employing nanotechnology: The devices based on nanoionics and built upon the principles of solid-state ionics and nanotechnology, operate through ion/vacancy local migration and electrochemical reactions, diverging from conventional electron and hole migration. It holds significant promise as an advanced type of nanodevice. High performance and innovative functionality things that traditional semiconductor devices cannot match can be anticipated from nanoionic devices. Due to the previously mentioned factors, nanoionic devices have recently attracted more interest as cutting-edge Nanodevices designed for information and communiqué applications function on a novel premise based on the findings of our research.

The term "ion migration" describes the flow of ions through an interface or within a material as a result of chemical gradients, mechanical stress, applied electric fields, and temperature changes. Many disciplines, including solid-state physics, materials science, electrochemistry, and semiconductor technology, are affected by this phenomenon. In solids, liquids, or gasses, ions can move across conductive routes because they are electrically charged atoms or molecules. Depending on the situation, this migration can result in modifications to the material's functionality, structure, or qualities. It can also be desired. The period migration refers mainly to the motion of ions because of an externally-implemented electrostatic discipline. The resultant temperature upward push reasons a boom in tan σ, and while there's a massive number of ionic additives inside the insulation layer of the substrate, ionic migration takes place inside the insulation layer.

In nanoionics, ion migration refers to the movement of ions in nanoscale materials. This movement can be achieved by introducing an electric field to the material. The electric field helps propel the ions, causing them to move from one place to another. This ion migration is important because it allows the material and its properties to be manipulated and controlled, which is useful in a variety of applications such as energy storage and electronic devices. By empathy and exploiting ion transport in nanoscale materials, scientists and engineers can develop innovative technologies with improved efficiency and functionality.

The common thermal strength at temperatures inside the water's liquid range (given with the aid of using RT) is satisfactorily massive to dominate the motion of ions even with the presence of an implemented electric-powered discipline. This way, the ions collectively with the water molecules surrounding them, are engaged in a wild dance as they're buffeted approximately with the aid of using thermal motions (which encompass Brownian movement) [1].

If we now follow an outside electric-powered discipline to the answer, the chaotic movement of every ion is supplemented with the aid of using an occasional bounce inside the course dictated with the aid of using the interplay among the ionic fee and the discipline. But that is a pretty tiny effect.

The conductance of electrolytic answer outcomes from the motion of the ions it incorporates, as they migrate towards the suitable electrodes. But the photograph we tend to have in our minds of those ions shifting in an orderly, direct march toward an electrode is wildly mistaken. The thermally-triggered random motions of molecules are called diffusion. Diffusion is the term describing the erratic motion of molecules, as they move from areas of high concentration to those with lower concentrations.

Migration refers to the motion of charged particles prompted by a local electric field. The extent to which migration contributes is influenced by factors such as charge, concentration, diffusion coefficient, and the size of the electric field gradient experienced by the ion [2].

11.2 Ion Migration in Nanoionics

Ion migration describes the movement of ions at the nanoscale within materials, usually in the form of nanoscale surfaces or structures. Nanotechnology, energy storage, electronics, and other fields all heavily rely on nanoionics, which studies the behavior of ions in nanoscale systems, such as thin films, interfaces, nanomaterials, and nanoparticles. Nanoscale local ion migration has a profound impact on the characteristics and functionality of materials and electronics. Important details about local ion migration in nanoionics are as follows:

Because of confinement effects, surface contacts, defects, or quantum mechanical phenomena, ions may flow in nanostructures differently than they do in bulk materials. The material's overall conductivity, diffusion rates, and electrochemical characteristics may be impacted by this movement.

The development of electrochemical devices, including supercapacitors, hydrogen cells, and batteries, depends heavily on nanoionics. Gaining more knowledge about and managing ion migration at the nanoscale is crucial to enhancing these devices' effectiveness, capacity, and lifespan.

Additionally, nanoionics has shown to be relevant in the domains of actuators and sensors, particularly concerning chemical sensors and resistive switches [3].

Some nanomaterials have high ionic conductivity, which makes them useful for use as membranes in different electrochemical devices or as solid-state electrolytes. The performance of these materials is determined by their local ion migration characteristics. Resistive switching and non-volatile memory devices in nanoelectronics are made possible by the exploitation of ion migration phenomena. To build these technologies, it is essential to comprehend and regulate ion mobility at the nanoscale.

Ion migration behavior, nanostructures can be greatly influenced by defects, grain boundaries, and interfaces. To create materials with desirable ion transport capabilities, control over these traits is crucial. In nanoionics, the motion of ions on the nanoscale is studied. This entails expertise in how ions migrate inside tiny substances like nanoparticles or skinny films. This migration can affect the material's properties, together with its capacity to behavior strength or save strength. By investigating those processes, scientists can broaden new technology with improved overall performance. Expert control is essential for operating nanoionic devices, managing solid-state electrochemical processes, and regulating local ion migration. This involves grasping and managing the reactions occurring at the interface where solid-state electrolytes, blended conductors, and other materials like semiconductors come into contact. Thus, it's imperative to create how to regulate solid-state ion migration of the hetero-interfaces on the electrochemical processes at the nanoscale. The best way to maintain control is to make use of a "nano-space," which is a nanostructure. It is feasible to use nanostructures, like films on a nanoscale. We've created a range of nanoionic two- and three-terminal gadgets made using straightforwardly layered nanoscale sheets techniques for vapor deposition, like graphene or metallic oxide sheets of oxide/ionic conductors, where the movement of Redox reactions involving oxygen/hydrogen ions can occur at the hetero-interfaces with the use of bias voltage.

There are four ways to achieve multi-functionality such as by managing ion transfers and electrochemical reactions at hetero-interfaces, one can alter local responses: Figure 11.1 illustrates four applications for *in-situ* hetero-interface control using solid-state ionics and nanotechnology. Diverse nanoionic devices have been developed, some capable of adjusting

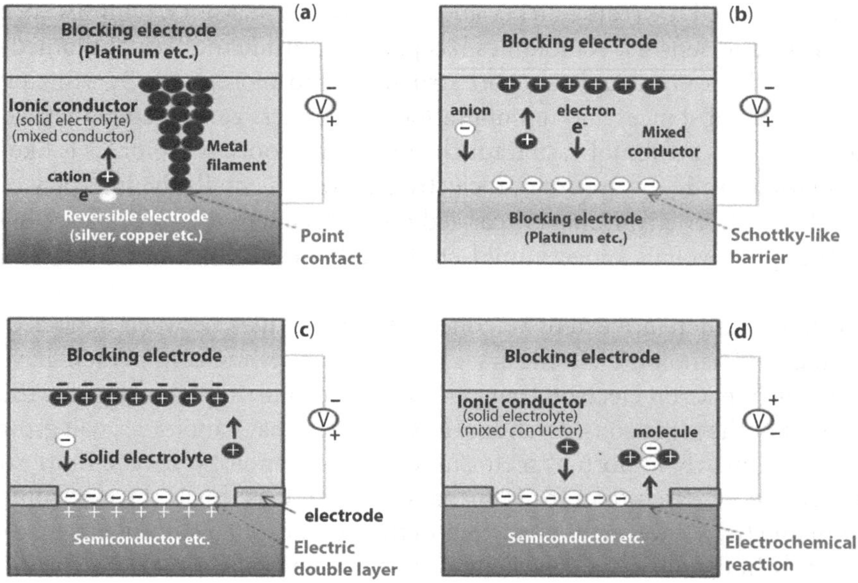

Figure 11.1 Illustration of four approaches for managing local ion transfers and electrochemical reactions at hetero-interfaces to acquire multifunctionality: (a) point contact; (b) Schottky-like barrier; (c) solid-state EDL (electric double layer); and (d) electrochemical reaction [9].

transition temperatures, exemplified by an atomic switch [4], a brain-like device [5], a superconducting device [6], a decision-making device [7], an all-solid EDL transistor [8], versatile on-need device [9]. High performance and new capabilities made feasible by these devices were not achievable with semiconductor devices.

We use nanoionics as it permits us to recognize and control the motion of ions on the nanoscale. This information is critical for growing superior technology like high-overall performance batteries, supercapacitors, and digital devices. Nanoionics allows us to enhance the strength garage, decorate conductivity, and create extra green digital components.

Nanoionics has numerous programs throughout extraordinary fields. Some examples include: nanoionics performs a critical function in growing high-overall performance batteries and super capacitors with advanced strength garage capabilities. This can result in longer-lasting and extra green strength garage answers for transportable electronics, electric-powered vehicles, and renewable strength systems.

Nanoionics allows the improvement of superior digital gadgets with more advantageous conductivity and progressed overall performance.

It may be used to create quicker and extra green transistors, reminiscence gadgets, and sensors. Nanoionics has packages inside the area of biomedicine, inclusive of drug transport structures and biosensors. By utilizing nanoscale substances and ion migration, researchers can broaden focused drug transport techniques and touchy diagnostic tools. Nanoionics is likewise used inside the area of optoelectronics, which entails the interplay of mild and electricity. It may be carried out to broaden high-overall performance photovoltaic gadgets, mild-emitting diodes (LEDs), and displays.

Beyond the apparent distinctions (classical vs. quantum-mechanical treatment), there are extensive conceptual similarities between nanoelectronics and nanoionics. In the first example, surfaces and confinement have a major impact on electron transport; in the second, ionic transport is the subject of the same considerations. It is expected that nanoionics will grow to be a similarly important sector in terms of technology, maybe more so for energy research than for information research as this is the electronic domain [10]. However, be conscious of the significant advancements made recently in the domains of switches, actuators, and sensors [11].

11.2.1 Types of Migration

All ionic answers include at the least varieties of ions (a cation and an anion); however, they can also additionally include others as well. In the 1870s, physicist Friedrich Kohlrausch observed that salts sharing a common ion exhibit consistent differences in their limiting equivalent conductivities. These differences reflect the variations in conductivity of the ions that are not shared between the two salts.

11.2.1.1 Cation Migration

The movement of positively charged ions, or cations, within a material or medium is referred to as cation migration. This migration may take place in several situations, such as cations can move from one electrode to another, carrying an electric charge, in electrolytic solutions, or during electrochemical processes such as battery operation.

In ion exchange resins, cation migration is also a crucial mechanism. The fixed negatively charged sites in these resins draw and swap out cations in a solution. Ionic conductors, semiconductors, ceramics, and other solid-state materials can all experience cation migration. It influences material properties, for example, when flaws or ions migrate within a crystal lattice. Cation migration is important in soil science and environmental studies because it affects soil fertility and nutrient availability to plants by moving cations in the soil due to water flow and other factors. Cation migration is the motion of definitely charged ions inside a material. It's vital as it influences the material's conductivity and different properties. For example, in solid-country batteries, cation migration allows the glide of ions among the electrodes, taking into account strength garage and release. Researchers observe cation migration to broaden substances with more advantageous ion conductivity, which could result in extra green strength garage gadgets.

Cation migration has numerous crucial roles in diverse fields. Cation migration is essential for the operation of batteries and gasoline cells. It lets in for the motion of ions among the electrodes, permitting the garage and launch of strength. Cation migration is applied to the improvement of solid-country digital gadgets, consisting of memristors and resistive random-get entry to memory (RRAM). These gadgets depend upon the managed motion of the cation to keep and manage data. Cation migration can play a position in catalytic processes, in which ions flow inside a catalyst cloth to facilitate chemical reactions. This can beautify the performance and selectivity of the catalytic process. Cation migration may be harnessed in sensors and actuators. For example, in certain types of gas sensors, cation migration is used to detect and measure the concentration of specific gases. These are only a few examples of the use of cation migration. It is a flexible phenomenon that reveals programs in diverse fields, contributing to improvements in strength garage, electronics, catalysis, and sensing.

Numerous elements can affect cation migration in materials. The association of atoms in a fabric's crystal shape can affect the pathways and simplicity of cation migration. Different crystal systems may have various levels of cation mobility. Cation migration is regularly temperature-dependent. Higher temperatures can grow the mobility of cations, making an allowance for quicker migration inside the fabric. The presence of an electric-powered discipline can affect cation migration. A carried-out electric-powered discipline can force the motion of cations in a particular direction, affecting their migration behavior.

The awareness of cations inside the fabric can affect their migration. Higher concentrations can result in improved interactions and hindered migration, even as decreased concentrations can sell quicker migration. The presence of defects or impurities inside the fabric can affect cation migration. These defects can create extra pathways or obstacles for cation motion. These elements, among others, can affect the fee and volume of cation migration in materials. Understanding and controlling those elements are critical for optimizing fabric residences and designing purposeful devices. If you follow an electrical current, the wonderful ions migrate to the cathode even as the terrible ions will migrate to the anode. Positive ions are known as cations and are all metals. Because in their valence they misplaced electrons and are capable of selecting out up electrons.

All metals are positive ions, also referred to as cations. They can absorb electrons and lose electrons due to their valence. Inverse ions are called anions. They can give up some of the extra electrons they carry. The cations regain their lost electrons and return to their elemental state if they come into contact with the cathode. The anions respond in the other direction. They relinquish their excess electrons and become elemental if they come into contact with the anode. The anions will be oxidized and the cations will be reduced at the electrode [12].

11.2.1.2 Anion Migration

The movement of negatively charged ions, or anions, within a material, medium, or solution is referred to as anion migration. Numerous factors contribute to this migration, which is important in some scientific fields. Because anions are negatively charged, they can move toward the positive electrode (anode) in electrolytic cells or during electrochemical reactions like those in batteries. The key component of anion exchange resin-based ion exchange processes is anion migration. The fixed positively charged sites in these resins draw in and swap out anions from a solution.

Ionic conductors, ceramics, semiconductors, and other solid-state materials can all experience anion migration. It affects the flow of ions or defects within a crystal lattice, which modifies the properties of the material. Anion migration is a factor in soil nutrient movement as it relates to soil science and environmental studies. Anions that migrate through the soil profile, such as nitrate (NO3-) or sulfate (SO4^2-), can have an impact on groundwater quality and plant uptake [13].

Anion migration refers to the motion of negatively charged ions inside a fabric. Similar to cation migration. In battery technology, anion migration is critical for the motion of ions among the tremendous and terrible electrodes at some stage in charging and discharging. This permits the garage and launch of electrical energy. Fuel cells additionally depend on anion migration. Anions like hydroxide (OH-) or carbonate (CO32-) ions pass *via* an electrolyte to facilitate the electrochemical reactions that generate electricity.

Anion migration is applied in membrane separation methods which include anion trade membrane (AEM) and anion trade chromatography (AEC). These strategies assist in separating and purifying ions in numerous industries, which include water remedy and pharmaceutical manufacturing. Additionally, anion migration performs a position in environmental remediation. Techniques like electro-kinetic soil remediation mobilize anions to resource with inside the elimination of contaminants from soil or groundwater. These are only a few examples of ways anion migration is carried out in one-of-a-kind fields [14].

Numerous elements could have an effect on anion migration: Higher temperatures usually boost anion mobility, taking into consideration quicker migration. The association of atoms in a fabric's crystal shape can affect the pathways and simplicity of anion migration. The presence of an electric-powered area can affect the route and pace of anion migration, just like cation migration. The attention of anions inside the fabric can affect their migration behavior. The presence of defects or impurities inside the fabric can create extra pathways or limitations for anion motion. These elements play a sizeable position in figuring out the fee and volume of anion migration in materials.

Anions move toward the anode because the half-cell gradually becomes more positive, balancing the anion charge. Similarly, at the cathode, cations migrate to neutralize the increasing negative charge.

In the field of electrode material design science, essential concepts revolve around ion migration and mobility, closely tied to compound chemistry and crystal structure. To facilitate rapid cation mobility in materials, especially in compounds involving Li-intercalation and solid-state electrolytes, various design principles based on structure and electronics have been proposed. The fundamental ideas of ion diffusion as well as the critical elements influencing multivalent ion diffusion have been covered in section [15].

11.3 Effect of Local Ion Migration on Device Performance

Finding ways to alter and regulate the concentrations of point defects is crucial because they are the pertinent ionic carriers in charge of mass and charge transport. The correct ground structure must be selected first, since it dictates the window of opportunity for altering the defect chemistry. The equilibrium or metastable state is defined by some control factors given this ground structure (temperature, component potential conditioning factors, trial, and doping content). Beyond that, defect chemistry is significantly and controllably affected by the introduction of interfaces [16].

This contribution mostly focuses on the impact of interface spacing, which covers the nano state [17] the realm of mesoscopic effects where physics and chemistry collide intriguingly. The performance of many different electronic devices, particularly those built using solid-state components like semiconductors, batteries, memristors, and more, can be greatly impacted by local ion migration. Among the consequences of ion migration are:

11.3.1 Modified Electrical Properties

Conductivity, resistance, and capacitance are just a few of the electrical characteristics that can be altered by ion migration within semiconductor devices. Unintentional current flow pathways created by uncontrolled migration can result in short circuits or changed behavior in devices.

11.3.2 Material Degradation

Unintended ion migration in integrated circuits and transistors can lead to material degradation, which reduces the lifetime and dependability of the device. Migration-induced defects may change the physical structure of the device, making it less functional.

11.3.3 Memory Devices and Resistive Switching

Controlled ion migration is utilized in some memory devices, such as memristors, and resistive random-access memory (RRAM) or memristors are used to store and modify data. On the other hand, unchecked migration may result in non-uniform resistance states, which could impair device performance and data reliability.

11.3.4 Battery Performance

Over time, ion migration within battery electrodes or electrolytes may cause a decline in performance. This may result in diminished energy storage capacity, capacity fading, and eventually shorter battery life.

11.3.5 Corrosion and Chemical Reactions

Ion migration may result in unintentional chemical reactions that cause materials to corrode. Electronic device performance and structural integrity may be impacted by this.

In the engineering and design of electronic devices, managing and comprehending ion migration is essential. Material engineering, device architecture design, adding barriers or interfaces to control ion movement, and streamlining manufacturing procedures to reduce migration-related problems are some of the strategies to lessen its negative effects.

Materials at the nano-scale could exhibit unconventional chemical transport, comparable to chemical diffusion seen in processes like chemical filters. Additionally, they may display unique storage characteristics along with conductivity effects. It is anticipated that the altered defect chemistry will have equal or greater significance concerning catalytic, photochemical, and reaction-kinetic characteristics [17a–c, 18].

11.3.6 Effect of Neighbourhood Ion Migration on Tool Overall Performance

When ions migrate inside a tool, it may motivate numerous issues. For example, it may result in tool degradation, wherein the substances inside the tool go to pot over time, resulting in decreased overall performance and a shorter lifespan. Additionally, ion migration can unintentionally create pathways for current flow, leading to increased leakage current and reduced device efficiency. This can motivate instability inside the tool's electric properties, resulting in fluctuations in overall performance. In a few cases, ion migration may even motivate brief circuits, disrupting the everyday operation of the tool.

To cope with those effects, tool designers and engineers appoint numerous techniques. They optimize fabric properties, put in force barrier layers, and manage working situations to reduce undesirable ion migration, making sure the choicest tool has overall performance.

Local ion migration could have a sizable effect on the tool's overall performance. When ions migrate inside a tool, it may result in numerous effects. Uncontrolled ion migration can motivate the degradation of tool substances, leading to decreased overall performance and shorter lifespan. Ion migration can create accidental paths for present-day flow, resulting in expanded leakage present day and reduced tool efficiency. Ion migration can motivate modifications inside the tool's electric properties, leading to instability and fluctuations in overall performance [19].

If ions migrate to accidental locations, it may bring about brief circuits, disrupting the everyday operation of the tool. To mitigate those effects, tool designers and engineers appoint numerous techniques together with

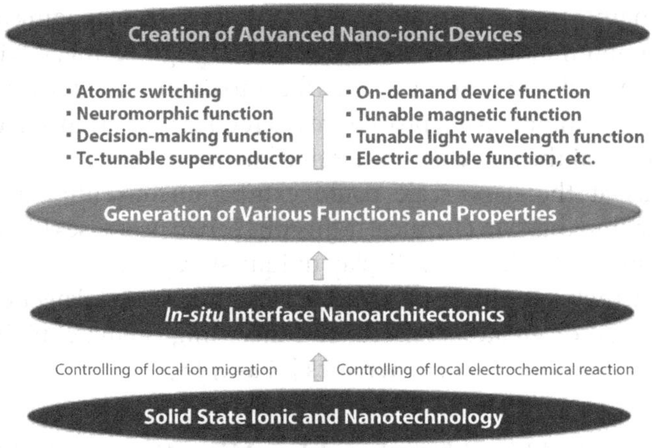

Figure 11.2 Illustration of a conceptual diagram for the creation of advanced nanoionic devices. These devices are based on the integration of nanotechnology and solid-state ionics, enabling a diverse array of functionalities [9].

optimizing fabric properties, enforcing barrier layers, and controlling the working situations to reduce undesirable ion migration.

With the use of local ion migration at hetero-interfaces and interfacial nano-architectonics, Figure 11.2 illustrates the process of creating a nanoionic device with a range of functionalities.

Solid-state Fuel cells and batteries are examples of electrochemical technologies that are essential for low-power energy production and storage that is economically feasible [10a, 20]. Since ion transport *via* solid electrolytes, which are often polycrystalline, is essential to their function, ion blockage at grain boundaries has a significant impact on performance. In the case of nanoionics, ion transport across grain boundaries is much more significant since surfaces have a greater impact than the bulk as device size decreases. But it's unclear where the obstacles to ionic transport at grain boundaries come from [20c, 21].

11.4 Limitations of Ion Migration

Ion migration can cause several restrictions and difficulties in a variety of applications, particularly in materials and electronic devices. Several of the main restrictions consist of: Over time, uncontrolled ion migration may cause devices to deteriorate, which may have an impact on the dependability and longevity of electronic components. It may result in device failure, changed electrical characteristics, and structural damage. Variability in ion migration can cause inconsistent device performance and compromise the integrity of stored data in devices (like memristors and RRAM) that depend on controlled ion migration for functionality.

Unintended ion migration in semiconductor devices can cause leakage paths, short circuits, or changes in electrical properties, which can result in a malfunction or improper operation of the device.

Ion migration in some solid-state materials can result in phase shifts and structural instabilities that impact the performance and characteristics of the material. Excessive ion migration in batteries and energy storage devices can cause capacity fading, lower energy density, and shorter battery life.

Ion migration can cause unintentional chemical reactions that impair material performance in some applications and cause corrosion and material degradation. It can be difficult to control and mitigate ion migration during the fabrication of devices. This requires precise material engineering, specialized manufacturing processes, and expensive precautions to avoid or reduce migration-related problems [22].

Here are a few key factors approximately the constraints of ion migration:

- Device Degradation
- Instability and Fluctuations
- Increased Leakage Current
- Potential for Short Circuits

These barriers can affect the tool's overall performance and lifespan. However, engineers plan to mitigate them through fabric choice and running conditions. Excessive ion migration can result in the deterioration of the tool's overall performance and capability over time. Uncontrolled ion migration can cause fluctuations in tool behavior, resulting in instability and unreliable operation. Unwanted ion migration can result in improved leakage current that can negatively affect tool performance and strength consumption. If ions migrate to unintentional areas, they can create pathways for brief circuits, main to tool failure or damage. The study of ion migration and defects in nanostructures is essential for advancing technologies in various fields. Understanding these drawbacks spotlights the significance of coping with and controlling ion migration in tool layout and operation [23].

11.5 Defects in Nanoionics

Exploring the intricate physics of the transportation of ions in substances at the nanoscale, the field of nanoionics is emerging at the intersection of ionics and nanoscience. As we investigate the realm of nanomaterials for applications such as fuel cells, batteries, and sensors, understanding the significance of defects becomes crucial. It also depends on the enhanced techniques for manufacture and characterization. Most notably, a large number of the findings have significant implications for the storage of the energy sector, which is seeing a massive thrust of its own for motives that do not require to be addressed. Furthermore, nanoionics has shown to be relevant in the domains of actuators and sensor technologies, particularly concerning chemical sensing and resistivity switching [3]. Some factors interrupt the crystalline structures and that's all due to these defects. In the following sections, we will go through the understanding of how these defects alter the behavior of nanoionic materials.

11.5.1 Point Defect Chemistry

First, let's analyze the importance of defect chemistry for research in the field of solid state [24]. Point defects are not only the important carriers of charges in the transfer of charge and mass, but they also serve as chemical excited states. Knowing not only the ideal structure but also, and often to an even greater extent, the point defect chemistry is necessary to comprehend the chemistry of substances. Additionally, this suggests that in terms of responsiveness, the key areas are the point deficiencies. For ionized crystals, even a significant ionotropic acid-base idea needs to be grounded in point-defect.

11.5.2 Size Defects

In ionic systems, side effects have a remarkable impact on both the storage of matter and ion transit. This makes up the discipline of nanoionics. Size can have a variety of effects, but the surface-to-surface symmetry breakdown and carrier rearrangement that results from it are the main ones. The contribution extends beyond the realm of microscopic sizes, surpassing the mesoscopic case, where all evidence of unaffected mass defect structures disappeared. In conclusion, the performance of nanomaterials in ion transport applications depends disapprovingly on size defects in nanoionics. Comprehending and adjusting these faults related to size is crucial in customizing the characteristics of nanoionic materials to fulfill certain technological demands, including better energy retention in batteries or more advanced sensing in nanoscale devices.

11.5.3 Bulk Defects and Interfacial Thermodynamics

Using "chemical language," we formulate defect properties to address the thermodynamics of defects. We apply the following condition on '*r*' which denotes the defective chemical reactions.

$$\sum v_{jr}\tilde{\mu}_j = 0\ldots \tag{11.1}$$

The stoichiometric value for the anomalous reaction associated with particle 'j', which is denoted by the letter 'r', is indicated by the variable. v_{jr} In eq. (11.1). For products, the sign '*m*' is positive; for educts, it is negative. In terms of the electrochemical potential of defect '*j*', the term '$\tilde{\mu}_j$' has the following definition:

$$\tilde{\mu}_j = \left(\mu_j^{\,o} + zF\phi + \mu_j^{\,ex,o}\right) + \mu_j^{\,ex,c} + RTlnc_j\ldots \tag{11.2}$$

The typical chemical potential, which takes into account structure-specific factors and includes 'elastic effects' which is defined by different parts [24]. The electrical potential is denoted by the word $zF\phi$ while the ideal configurational contribution is represented by $RTlnc_j$. The correction word, μ^{ex}, is separated into an excess term describing energetic interactions and a configurational excess component. Within 'physical language', the first three components in Equation (11.2) together denote what is referred to as the 'energy level' [22], which is represented by the letter 'e' in the level diagrams illustrated below:

Here, we inspect the energy stability of an explicit kind (e.g., $Mr\,e^-$) in MX. Figure 11.3 display the electron stimulation from the valence to the conduction region, which is intentionally shown reversed. It also depicts its ionic equivalent, that is the transference of a M^+ From a normal spot to an interstitial position (inside a silver halide). There are gaps and openings created by the procedure in both cases. Eq. (11.1) describes a thermochemical technique that is more effective in treating more complex disease kinds. The neutral substance's chemical capability affects the association between the two 'Fermi levels', which in Figure 11.3 reflect the ions' and electrons' electrochemical potential vendors. This is competently held by

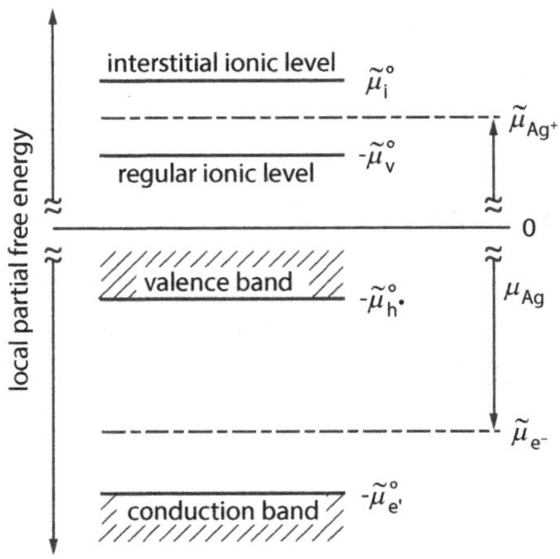

Figure 11.3 The level diagram of a normal silver halide shows electron production and Frenkel disorder [24].

familiarizing the reaction of M(g) or X_2(g) into MX in the reaction structure. The cathe lculation of electroneutrality creates the ultimate relation. The following expression for the defect concentration is determined using the Brouwer approximation:

$$the\ c_j = \alpha_j \left(\prod_p P_p^{N_{pj}}\right)\left(\prod_q C_q^{M_{qj}}\right)\left(\prod_r K_r^{\gamma_{rj}}(T)\right)\ldots \quad (11.3)$$

The dependence of c_j on temperature, component potential, and doping content is demonstrated by eq. (11.3). In general, the '*in situ* parameters' are designated in the initial brace, while the '*ex situ* parameters' are described in the next bracket [25]. While MO is coated by $M_2 * O$ at elevated temperatures, especially when C = [M*], $P = PO_2$, M* can experience enhanced mobility. As a result, the oxide displays ternary behavior, with PM* being added to the P-list and the second element being eliminated from the C-list. On the other hand, for oxides found in water, PO_2 and PH_2O are both important independent *in situ* factors at elevated temperatures. Lower temperatures, however, cause the water content to stay constant, which means that PH_2O is now not mentioned in the P-list and $[H_2O]$ is a constant in the C-list [25, 26]. Apart from the presence of equilibrium, flawed relations are quite significant. The treatment of active connections involves the use of activity coefficients more generally or associates, particularly in the case of short-range interactions. This results in shortening of the energetic level ranges or intragap placements. The latter is seen in the premelting effect, which leads to the superionic phase transition. A degree of degeneracy characterizes the final condition. As [27] shows, the process described can be concisely expressed as a cubic root law with regard to concentration, where the disorder enthalpy and dielectric constant define the prefactor.

$$\mu_j^{ex,o} = -Jc^{\frac{1}{3}} \ldots \tag{11.4}$$

(See Figure 11.4). Lower concentrations are better suited for Debye-Hückel adjustments [28].

In the ensuing discussion, we omit alteration factors. We conceptualize the interfacial scenario as an unexpected mechanical transition from the majority's point to the specific central interfacial coating, represented by a step function for simplicity. The electrochemical potentials μ_o in both the majority and interfacial center are characterized with help of terminologies similar to Eq. (11.1). Such inevitably results in a charging phenomenon, delineated by the transmission of charges from the majority to the center, both of which are in a balance with the gas segment. Despite such

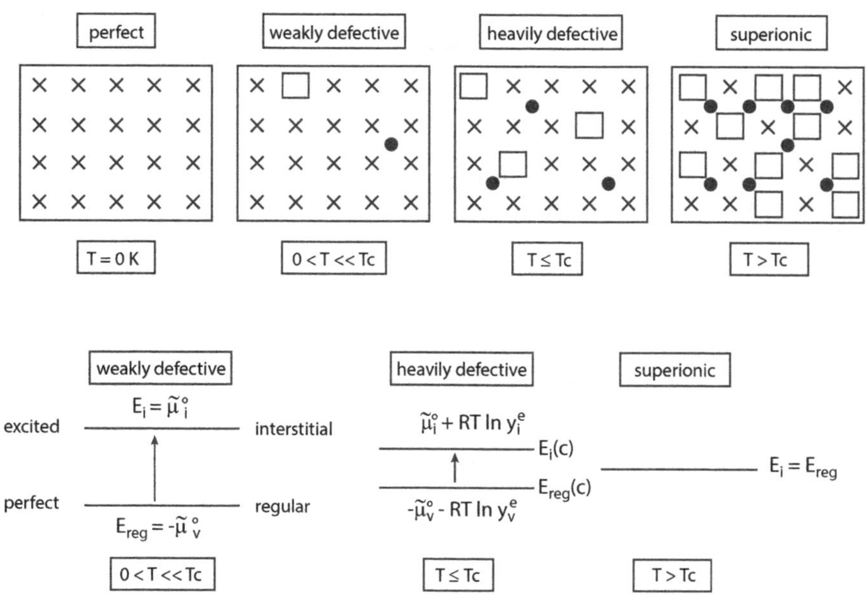

Figure 11.4 A Frenkel denatured object's thermal destiny: from T<0 to the superionic phase [29].

approximation involved in modeling the center as independent interphase, the central space charge model provides as depicted in Figure 11.5 a robust outline for addressing defect interaction. ([30] is emphasized for comprehensive details). It is crucial to highlight that the impact degree ϑ [23] can capture the effects. This degree functions as a degree of the space charge potential by characterizing the surface to bulk concentration ratio. To be more precise, there is no effect when $\vartheta_j = 0$. The highest enrichment effect is indicated by $\vartheta_j = +1$, while the maximum depletion effect is indicated by $\vartheta_j = -1$. At a distance x from the interface, the change in defect concentration is represented by the following expression:

$$\left[\frac{c_j(x)}{c_j(\infty)}\right]^{1/z_j} = e^{-F\Delta\phi/RT} \ldots \tag{11.5}$$

It is intrinsically evident that in materials where interfacial dominance is critical, the accumulation/depletion process strongly affects ion conductivity. The impact of heterogeneous doping, which is best exemplified by the "rule of heterogeneous doping," is one such instance.

$$z_j\delta c_j \Big/ \delta\sum j < 0 \ldots \tag{11.6}$$

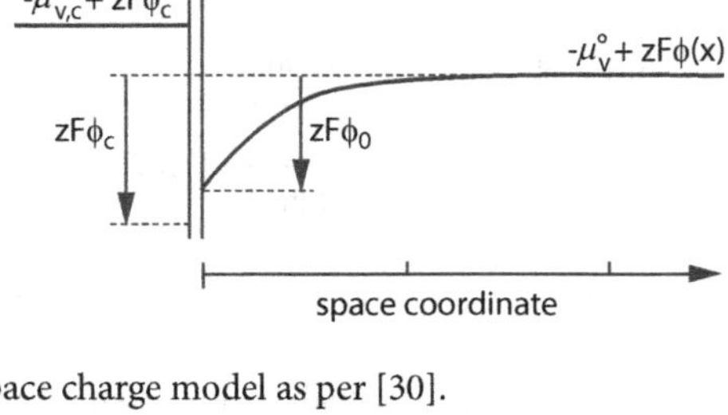

Figure 11.5 Core-space charge model as per [30].

The quantities of all favorably noticeable abnormalities decrease and those of negatively detectable defects rise within the space charge areas of the core (A) upon positive charging. Similar thing is done for the negative charging. This is similar to homogenous principle of doping.

$$z_j.\delta c_j \Big/ z_c \delta C < 0 \ldots \tag{11.7}$$

As further elucidated in [31], the reference to both the majority and softened contaminants (c) are not coincidental, highlighting its importance in materials research. If a comprehensive model is available, interfacial chemistry can be used to calculate the space charge potential, represented by the symbol ϑ [23]. It is usually regarded as a parameter. We shall assume significant impacts in the discussion that follows ($|\vartheta| \cong 1$). The concentration profile is then obtained by combining Eq. (11.5) with Poisson's equation gives

$$c_1 \approx \frac{c_{10}}{\left(1+\sqrt{C_0 / C\infty}\, x/(2\lambda)\right)^2} \ldots \tag{11.8}$$

At a distance of 2k, it roughly converges to the majority value, where the following relation determines λ:

$$\lambda = \sqrt{\varepsilon RT \Big/ 2 z_1{}^2 F^2 c_{1\infty}} \ldots \tag{11.9}$$

For more information, check Ref. [10]. Even if one of the two dominant flaws becomes stationary, eq. (viii) is still true if the mobile majority defect is enhanced. If the mobile one is depleted, the charge concentration has a nonequilibrium value that is almost constant. In a Mott–Schottky situation, the depleted anti carrier (2) contour is (half) Gaussian c_j due to a lack of screening, and the postponement (λ^*) into the majority is unfathomable, where the Gaussian is given by

$$c_j = e^{-\left|\frac{z_2}{z_c}\right|\left(\frac{x-\lambda^*}{2\lambda}\right)^2} \ldots \tag{11.10}$$

Also,

$$\lambda^* = \lambda\sqrt{\frac{4z_cF}{RT}ln\frac{c_{20}}{c_{2\infty}}} \ldots \tag{11.11}$$

Note that the bulk charge density, z_cFC, is incorporated into λ. For the Mott–Schottky and Gouy–Chapman scenarios, Table 11.1 provides data on the conduction of space charges' efficient depth and its efficient intensity. Look into [23, 32, 33] for further details on conductance effects related to enriched and depleted carriers.

In polycrystalline materials, the combination of many routes holds particular significance. Space charge layers that are parallel to the current direction can also bypass the bulk channel, but they are broken up by layers that are perpendicular to it. [34] contains a detailed discussion of the superposition inside a brick-layer model and its detectability in impedance spectroscopy. Note that the properties are very anisoptropic and dependent on the direction of the current due to the composite structure of the boundary and the profile character of the space charge conditions. This might be a very basic example. The impedance spectrum has two arcs: the first addresses bulk plus space charge (due to parallel routes) and the second shows the influence of the blocking core (perpendicular boundaries) if the space charge is a strong accumulation layer and the core is strongly insulating. However, take note that space charge profiles are themselves anisotropic. The superposition of many paths in a bricklayer model is expressed as (φ, volume fraction).

$$\sigma_m = \frac{\sigma_\infty \sigma_{gb}^{\perp} + {}^{2}\!/_{3}\, \varphi_{gb} \sigma_{gb}^{\parallel} \sigma_{gb}^{\perp}}{\sigma_{gb}^{\perp} + {}^{1}\!/_{3}\, \varphi_{gb} \sigma_\infty} \tag{11.12}$$

Table 11.1 Optimal electrostatic effect rates in the Schottky-Mott and Gouy-Chapman cases for $|\vartheta| \to 1$.

	Actual thickness	**Actual concentration**
Conductance in Gouy-Chapman case	2λ	$\sqrt{c_o c_\infty}$
Resistance in the Schottky-Mott case	λ^*	$2c_o ln\left(\frac{c_\infty}{c_o}\right)$

where space charge (sc) effects and core effects $\left(\text{i.e. } \varphi_{gb}\sigma_{gb}^{\perp -1} = \varphi_{core}\sigma_{core}^{\perp -1} \varphi_{sc}\sigma_{sc}^{\perp -1}; \varphi_{gb}\sigma_{gb}^{\parallel} = \varphi_{core}\sigma_{core}^{\parallel} + \varphi_{sc}\sigma_{sc}^{\parallel}\right)$ can be separated out to explain $\sigma_{gb}^{\perp}$ and $\sigma_{gb}^{\parallel}$. The intricate values must be used in place of the r values for a.c. measurements, and eq. (11.12) provides the spectrum of impedance right away [23, 25, 34]. The volume fraction rises in cubic grains in an inverse relationship with grain size.

Now let's look at an oxide depletion layer that is normal. Figure 11.6 displays the defect chemical reactions of an acceptor enriched oxide close to a connection with a positive surplus charge. While $[e']$ increases and $[h]$ decreases, the vacancy concentration decreases. Keep in mind that, in comparison to doubly charged vacancies, the effect on individual charged transporters is less pronounced (see Eq. (11.5)). More information about the insinuations for diverse transmission can be found in Ref. [2]. Additionally take note of the Mott-Schottky case's more profound penetration. The displayed Figure 11.6 applies to $SrTiO_3$,CeO_2, and probably also ZrO_2 in which Fe and Y are doped respectively [19–24]. The likelihood of n-type

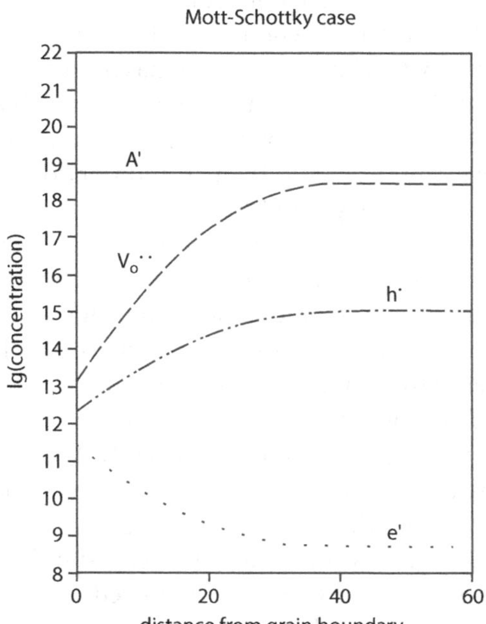

Figure 11.6 Mott-Schottky pattern close to the grain boundary, applicable to a range of oxides doped with acceptors. At the borders, the conduction would turn out to be of type n for small PO_2 significantly elevated space charge potential or, more commonly, compounds having a relatively higher (e') [24].

conduction at p-type $SrTiO_3$ grain limitations is discussed in Ref. [25]. Furthermore, the appropriate core transport (refer to Figure 11.6) may also be significant because of the altered structure and defect chemistry therein.

11.5.4 Chemical Flaws in Nanocrystalline Structures and Systems

Let's look at a few typical examples to highlight the existence of "true" and "trivial" size effects, which alter both the local attributes and the overall effect depending only on the proportion of interfaces that increase. First, let's talk about CeO_2 [35–38]. Apart from ZrO_2 [39–42] and TiO_2 [42] it is one of the model materials for nanocrystalline oxides.

Because of the weaker bonding state at surfaces (or in grain boundaries), formation energies of defects are expected to be dramatically altered and typically decreased. We have calculated the resulting cost of VO and eV in CeO_2 has been computed in [35]. Both the electron and vacancy concentrations would rise in accordance with the growing intrinsic instability if the surface layer were pure and electroneutral [43]. The surface layer's dual logarithmic O_2 fractional pressure dependency would be around -1/6, and it would be electrically conductive. In truth, in a number of tests, though not all of them, this kind of behavior has been observed in [36–38]. Naturally, minimal charging effects are required for such a treatment. On the other hand, CeO_2 exhibits notable diminution results for $V_o^{\cdot\cdot}$ connected to a positive central charge. Then, in addition to (h) being depleted, eV also has to rise (look at Figure 11.6).

Therefore, a strong n-type conduction gain is anticipated for CeO_2, whose border behavior is assumed to be similar. Subsequently, perpendicular space charge layers should either completely prevent or severely restrict ionic transfer from bulk to bulk, whereas parallel space charge zones should offer extremely conductive channels for electrons. Tschöpe has demonstrated the significance of space charge effects for nanocrystalline CeO_2 formations [44] and has recently been confirmed [45]. The interfacial area to which ions segregate, however, may be larger for extremely tiny grain sizes than the potential segregants; this could lead to an L-dependence (L=grain size) of bulk characteristics [46].

Regarding nanocrystalline CaF_2, the improved conductivity of nanocrystalline materials appears to be entirely explained by space charge effects [47]. Because the increased interface-to-bulk percentage is the only factor causing the impacts outlined thus far, they can be considered "trivial" size effects. If the thickness is less than 4λ, a true size impact takes place.

As illustrated in Figure 11.7, a thin film's parallel conductance $\left(Y^{\parallel}\right)$ should rise with thickness. With conductivity as the gradient and the involvement of space charge as the intercept, the relationship between $Y^{\parallel}$ and $L \gg 4\lambda$ is linear. The more complex behavior, if $L \lesssim 4\lambda$, is indicative of a shift in the c-profiles in the space charge zones [48]. Figure 11.8 a displays the mean recorded excessive conductance $\left(\Delta Y^{\parallel}\right)$, which is the surplus conductance $\Delta\sigma_m^{\parallel}$ (area normalized) distributed by the material's thickness. Because noncontributing bulk is removed, a drop in film thickness causes a linear increase in $\Delta\sigma_m^{\parallel}$. Almost no electroneutral bulk remains at $L \simeq 4\lambda$. A further rise is now caused by the modification of the c-profiles between the interconnected space charge layers, which is described by the nanosized factor [48].

$$g = \frac{4\lambda}{L}\left[\frac{c_{10} - c_1^{*}}{c_{10}}\right]^{\frac{1}{2}} \ldots \tag{11.13}$$

where the concentration in the center of the sample is indicated by c_1^{*}, which is higher than $c_{1\infty}$ and reaches at c_{10} for $L \to 0$. Elliptical integrals are used to show how dependent c_1^{*} is on L (look [48, 50]).

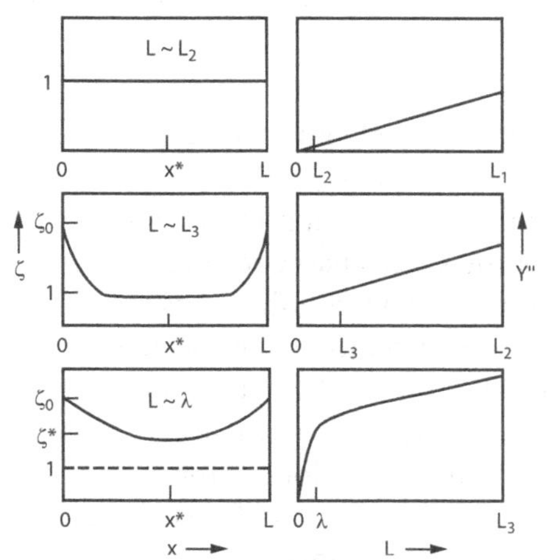

Figure 11.7 Parallel conductance and conductive characteristics for different thin-film thicknesses. The length scales are characterized by arbitrary thicknesses L_1, L_2, and $L_2(L_1 << L_2 << L_3)$. f: the division of concentration by the bulk concentration [49].

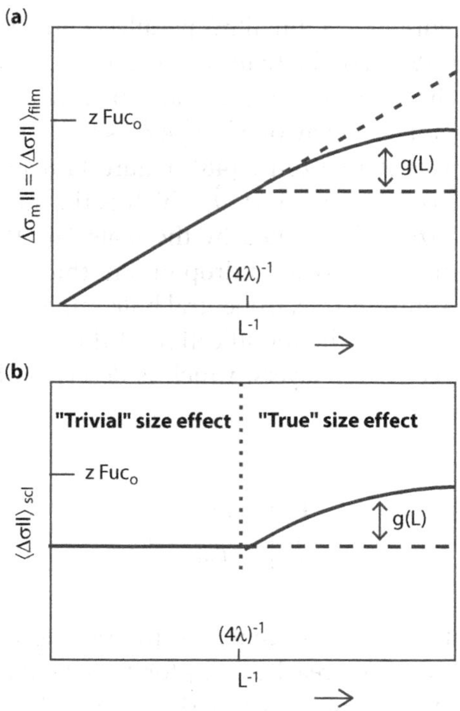

Figure 11.8 (a) The area-related space charge transmittance per film thickness represents the calculated excessive space charge conduction as an indicator of inverse separation. (b) The area-related space charge conductance as an indicator of inverse separation, expressed in the form of mean specific excess space charge conductivity per border thickness [24].

The extra increase resulting from space charge overlap when compared to $\Delta\sigma_m^{\|}(L = 4\lambda)$ is expressed in eq. (11.13). Even more clearly seen is this change from an insignificant to an appropriate size impact in Figure 11.8(b), where the breadth of the space charge regions determines the surplus conductance rather than the overall film thickness. This means that the average surplus space charge's conductance is represented by $\Delta\sigma^{\|}_{scl}$, which rises in accordance with eq. (11.13) if $L \leq 4\lambda$ but stays consistent as long as $L > 4\lambda$. When c* gets close to c_0, the limit is reached. The sample as a whole may then exhibit a 'superionic' condition in the case of a significant border effect; whether or not this is understood asynchronously or inconsistently depends on a variety of factors, including interactions, the reputation of gradient effects, and the dependency of dielectric constants. In this context, more comprehensive research is being conducted.

A very intriguing experiment was reported by Sata *et al.* [51], demonstrating that artificial hetero structured materials with abnormal conductance characteristics may be created by ionic heterostructures $CaF_2 / BaF_2 / \ldots$, where the spacing ranged from $\sim 1\,nm$ to $\sim 1\mu m$ (Figure 11.8). $\Delta\sigma_m^{\|}$ increases as expected at first when the heterostructures' spacing is reduced. $\Delta\sigma_m^{\|}$ initially increases as expected when the heterostructures' spacing is reduced. Rather than rising for significantly thinner films, it does not saturate at 4λ. It is worth looking into why the slope is significantly steeper than one would anticipate for a constant boundary concentration. In any case, what we are referring to is a material where an artificial conductor has been created and the constituent layers have lost their individuality (Figure 11.9). At minuscule thicknesses, the conductivity decreases once again. Currently, it is unclear if percolation issues are to blame for this. Such research clearly shows that artificial ion conductors have a lot of promise for use in activities at lesser or medium temperatures.

The above-discussed phenomena of heterogeneous doping may be more significant in terms of applications [35, 42]. A very useful tool for adjusting transport properties is provided by the creation of suitable nanocomposites. In contrast to the behavior of heterolayers the size of nanometers, the benefit of simpler preparation and increased morphological freedom is countered by more percolation issues. The recent explanation of *AgI*: Al_2O_3 composites' extraordinarily high conductivity, which also highlights the significance of boundary phase transitions, is an extreme example: At the edges that corresponds to $\gamma / \beta / \gamma\ldots$ "heterolayers," mounting faults are formed in an essentially atomic size order [52]. Significant adjustments

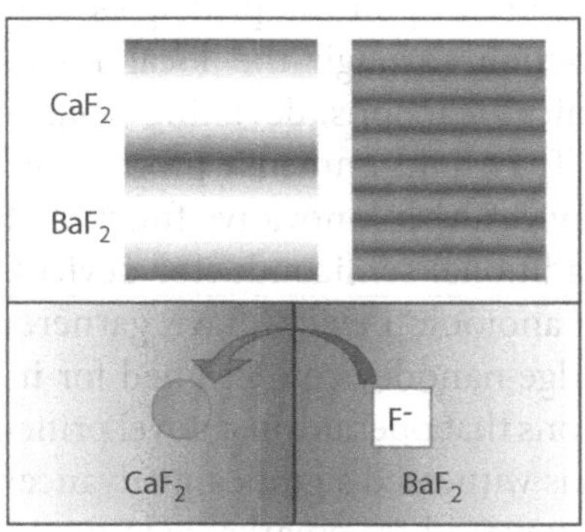

Figure 11.9 Ionic heterostructures composed of CaF_2 and BaF_2 are used to induce mesoscopic ion conduction [24].

of the defect density are anticipated to be at very tiny dimensions due to structural variations, which also affect the standard chemical potential (Look at eq. (11.2)).

11.6 Key Advances in Nanoionics and Improvements in Device Quality

Nanoionics is an interesting field that deals with ionic behavior on a nanoscale. It revolves around the migration of ions within these structures and consequent defects. The migration of ions refers to the charged atoms or molecules that can move within the nanomaterial. The comprehension of ion migration and defects within nanomaterials is crucial for development of advanced technologies centered on energy storage, sensors, and more.

Information and communication equipment relies heavily on electronic devices. Enhancing the functionality of these devices, which depend significantly on semiconductor performance, necessitates managing the movement of electrons and holes within semiconductors. It has been observed that conventional semiconductor devices exhibit a continuous decrease in processing size year after year. There is growing concern that traditional fine processing technology is nearing its peak capabilities, and there's a looming risk of reaching the physical operational limits of semiconductor devices. Nanotechnology offers a potential solution to overcome these technological and physical limitations by enabling advancements in the materials used in devices and the principles governing their operation.

Nanoionic devices, built upon the principles of solid-state ionics and nanotechnology, function through the local migration of ions/vacancies and electrochemical reactions, deviating from conventional electron and hole migration. They hold immense promise as a new class of high-performance nanodevices with innovative functionalities that are beyond the capabilities of traditional semiconductor devices. Due to these aforementioned factors, nanoionic devices have garnered significant interest recently as cutting-edge nanodevices designed for information and communication applications that operate on a novel principle. Nanoionics is an emerging field and has witnessed significant advancement in recent times. These advancements have led to remarkable improvements in device performance and quality. These advancements allow precise engineering of materials at the atomic level, which is instrumental in the field of nanoelectronics. However, these electronic devices have certain limitations;

therefore, scientists are exploring bioinspired nanoionics, which utilizes ions as charge carriers. This is a promising field offering potential flexible, low-power consumption devices. Using ions as charge carriers, the devices could mimic biological systems. Nanoionic systems can be categorized as biological nanoionics, artificial nanoionics, and hybrid nanoionics [25].

11.6.1 Biological Nanoionics

Living organisms depend on controlled ion transport across cell membranes for functions such as muscle contraction, and sending signals. Ion-channels provide them with a pathway for this process. The resulting ion current of this movement plays a great role in many bio functions. Biological nanoionic-inspired ion transport system offers selectivity, efficiency, and control. This inspires applications in designing energy-efficient computing that is based on ion transport instead of electrons. Conventional computing that is based on Moore's law is approaching its limits due to miniaturization limits and increased power usage. Neuromorphic computing is inspired by the human brain and is an energy-efficient alternative. Memristors are devices with two terminals, which have electric stimuli-dependent resistance that allows them to mimic biological synapses. Neuromorphic computing systems based on memristors as artificial synapses can be realized and will work as the human brain [26].

Photo-ionics is another field revolutionized by nanoionics that offers interesting solutions for reconfigurable optical devices. Traditionally, mechanical and thermal methods are used to alter optical device properties, which can be slow and energy-intensive. Nanoionic uses ion movement to manipulate light. Through careful design of electrical and magnetic fields, researchers can manipulate the path of light. This will allow reconfigurable optical devices which will have an input dependent function [27].

11.7 Conclusion

Materials at the nanoscale could offer unusual chemical transportation and peculiar storage qualities as well as the consequences of conduction. It is expected that the changed defect chemistry will be just as important, if not more so, for the stimulating, photochemical, and reaction-kinetic properties [17a, 17b, 18]. However, it is important to remember that nanostructuring might cause issues with stability and repeatability (Part I). In any case, nanostructured material's defect chemistry is still mainly unexplored, even though it presents new, useful degrees of freedom for materials research.

Sophisticated nanoelectronic gadgets, employing ion migration, have garnered interest as a focal point in developing next-generation information and communication technologies. Notably, non-ionic devices utilizing ionic and vacancy migration exhibit intriguing practical features and functions that were previously unattainable with traditional semiconductors. We posit that the nanoionic devices, rooted in solid-state ionics and nanotechnology, emerges as a promising candidate for the upcoming era of advanced devices. The final result of ion migration in nano ions can vary depending on the specific application and materials used. However, some possible results include the formation of desired structures, changes in material properties, and the creation of functional devices. For example, in energy storage devices such as batteries, ion migration allows ions to move between electrodes, allowing energy to be stored and released. In electronic devices, ion migration can be used to control the conductivity or optical properties of materials, thus improving performance. In general, ion migration plays a crucial role in achieving the desired functions and properties of nanoscale materials and devices.

References

1. Bielenberg, and Brenner, A hydrodynamic/Brownian motion model of thermal diffusion in liquids. Physica A, 356, 2–4, 279–293, 2005.
2. Samson, Marchand, Snyder, Calculation of ionic diffusion coefficients on the basis of migration test results. *Mater. Struct.*, *36*, 156–165, 2003.
3. (a) Rothschild, and Komem, The effect of grain size on the sensitivity of nanocrystalline metal-oxide gas sensors. *J. Appl. Phys.*, *95*, 11, 6374–6380, 2004; (b) Waser, and Aono, Nanoionics-based resistive switching memories. *Nat. Mater.*, *6*, 11, 833–840, 2007.
4. Terabe, Hasegawa, Nakayama, Aono, Quantized conductance atomic switch. *Nature*, *433*, 7021, 47–50, 2005.
5. Ohno, Hasegawa, Tsuruoka, Terabe, Gimzewski, Aono, Short-term plasticity and long-term potentiation mimicked in single inorganic synapses. *Nat. Mater.*, *10*, 8, 591–595, 2011.
6. Tsuchiya, Moriyama, Terabe, Aono, Modulation of superconducting critical temperature in niobium film by using all-solid-state electric-double-layer transistor. *Appl. Phys. Lett.*, *107*, 013104, 2015.
7. Tsuchiya, Tsuruoka, Kim, Terabe, Aono, Ionic decision-maker created as novel, solid-state devices. *Sci. Adv.*, *4*, 9, eaau2057, 2018.
8. (a) Tsuchiya, Terabe, Aono, All-solid-state electric-double-layer transistor based on oxide ion migration in Gd-doped CeO_2 on $SrTiO_3$ single crystal. *Appl. Phys. Lett.*, *103*, 073110, 2013; (b) Tsuchiya, Terabe, Aono, *In situ* and non-volatile bandgap tuning of multilayer graphene oxide in an all-solid-state electric double-layer transistor. *Adv. Mater.*, *26*, 7, 1087–1091, 2014.

9. Terabe, Tsuchiya, Tsuruoka, Kim, Aono, Current progress of solid state ionics on information and communication device technology, in: *2017 17th International Workshop on Junction Technology (IWJT)*, IEEE, pp. 38–39, 2017.
10. (a) Maier, Nanoionics: ion transport and electrochemical storage in confined systems. *Nat. Mater.*, *4*, 11, 805–815, 2005; (b) Maier, Size effects on mass transport and storage in lithium batteries. *J. Power Sources*, *174*, 2, 569–574, 2007; (c) Kern, and Maier, Nanoionics and nanoelectronics. *Adv. Mater.*, Wiley Online Library, 21, 2569–2569, 2009; (d) Maier, Thermodynamics of nanosystems with a special view to charge carriers. *Adv. Mater.*, *21*, 25-26, 2571–2585, 2009.
11. Maier, Pushing nanoionics to the limits: charge carrier chemistry in extremely small systems. *Chem. Mater.*, *26*, 1, 348–360, 2014.
12. Pak, and Hwang, Charging rate dependence of ion migration and stagnation in ionic-liquid-filled carbon nanopores. *J. Phys. Chem. C*, *120*, 43, 24560–24567, 2016.
13. Vieira, Dos Santos, Botelho, Vieira, Alves, Hoinaski, Micke, The novel Outer Plug Standardization (OPS) calibration strategy applied in a new method for the high-throughput determination of Cl–, NO_3– and SO_4– 2 in rainwater samples. *Talanta*, *235*, 122820, 2021.
14. Yuan, and Huang, Ion migration in organometal trihalide perovskite and its impact on photovoltaic efficiency and stability. *Acc. Chem. Res.*, *49*, 2, 286–293, 2016.
15. (a) Muldoon, Bucur, Oliver, Sugimoto, Matsui, Kim, Allred, Zajicek, Kotani, Electrolyte roadblocks to a magnesium rechargeable battery. *Energy Environ. Sci.*, *5*, 3, 5941–5950, 2012; (b) Wang, Sun, Shadike, Hu, Ji, Gao, Yang, Xu, Wang, How water accelerates bivalent ion diffusion at the electrolyte/electrode interface. *Angew. Chem.*, *130*, 37, 12154–12157, 2018; (c) Huang, Guo, Ma, Bin, Wang, Xia, Recent progress of rechargeable batteries using mild aqueous electrolytes. *Small Methods*, *3*, 1, 1800272, 2019.
16. Maier, Ionic conduction in space charge regions. *Prog. Solid State Chem.*, *23*, 3, 171–263, 1995.
17. (a) Schoonman, Nanostructured materials in solid state ionics. *Solid State Ionics*, *135*, 1–4, 5–19, 2000; (b) Tuller, Ionic conduction in nanocrystalline materials. *Solid State Ionics*, *131*, 1–2, 143–157, 2000; (c) Chiang, Introduction and overview: physical properties of nanostructured materials. *J. Electroceram.*, *1*, 3, 205–209, 1997; (d) Maier, Point defect thermodynamics: Macro-vs. nanocrystals. *Electrochemistry*, *68*, 6, 395–402, 2000.
18. (a) Jamnik, and Maier, Charge transport and chemical diffusion involving boundaries. *Solid State Ionics*, *94*, 1–4, 189–198, 1997; (b) Jamnik, and Maier, Generalised equivalent circuits for mass and charge transport: chemical capacitance and its implications. *PCCP*, *3*, 9, 1668–1678, 2001.
19. Nandal, and Nair, Predictive modeling of ion migration induced degradation in perovskite solar cells. *ACS Nano*, *11*, 11, 11505–11512, 2017.

20. (a) Steele, and Heinzel, Materials for fuel-cell technologies. *Nature, 414*, 6861, 345–352, 2001; (b) Arico, Bruce, Scrosati, Tarascon, Van Schalkwijk, Nanostructured materials for advanced energy conversion and storage devices. *Nat. Mater.*, *4*, 5, 366–377, 2005; (c) Litzelman, Hertz, Jung, Tuller, Opportunities and challenges in materials development for thin film solid oxide fuel cells. *Fuel Cells*, *8*, 5, 294–302, 2008.
21. (a) Pergolesi, Fabbri, D'Epifanio, Di Bartolomeo, Tebano, Sanna, Licoccia, Balestrino, Traversa, High proton conduction in grain-boundary-free yttrium-doped barium zirconate films grown by pulsed laser deposition. *Nat. Mater.*, *9*, 10, 846–852, 2010; (b) Balke, Jesse, Morozovska, Eliseev, Chung, Kim, Adamczyk, Garcia, Dudney, Kalinin, Nanoscale mapping of ion diffusion in a lithium-ion battery cathode. *Nat. Nanotechnol.*, *5*, 10, 749–754, 2010; (c) Hayashi, Noi, Sakuda, Tatsumisago, Superionic glass-ceramic electrolytes for room-temperature rechargeable sodium batteries. *Nat. Commun.*, *3*, 1, 856, 2012.
22. Hill, and Simpson, Capabilities and limitations of ion mobility spectrometry for field screening applications. *Field Anal. Chem. Technol.*, *1*, 3, 119–134, 1997.
23. Bi, Song, Li, Wu, Yan, Mitigating ion migration in perovskite solar cells. *Trends Chem.*, *3*, 7, 575–588, 2021.
24. Kröger, *The Chemistry of Imperfect Crystals*, North-Holland Publ. Comp., Amsterdam, 1964.
25. Zhang, Liu, Dai, Xiao, Nanoionics from biological to artificial systems: An alternative beyond nanoelectronics. *Adv. Sci.*, *9*, 23, 2200534, 2022.
26. Wang, Wang, Nagai, Xie, Yi, Huang, Nanoionics-enabled memristive devices: strategies and materials for neuromorphic applications. *Adv. Electron. Mater.*, *3*, 7, 1600510, 2017.
27. Elfarash, and Gholipour, Reconfigurable nanoionic and photoionic material and device platforms. *Adv. Phys.: X*, *9*, 1, 2338285, 2024.
28. Maier, Defect chemistry and ion transport in nanostructured materials: Part II. Aspects of nanoionics. *Solid State Ionics, 157*, 1–4, 327–334, 2003.
29. Maier, and Münch, Thermal destiny of an ionic crystal. *Z. Anorg. Allg. Chem.*, *626*, 1, 264–269, 2000.
30. Jamnik, Maier, Pejovnik, Interfaces in solid ionic conductors: Equilibrium and small signal picture. *Solid State Ionics, 75*, 51–58, 1995.
31. Maier, J., Composite electrolytes. *Mater. Chem. Phys.*, 17, 5, 485–498, 1987.
32. Schoonman, J., Nanostructured materials in solid state ionics. *Solid State Ion.*, 135, 1–4, 5–19, 2000.
33. Sze, S., *Semiconductor Devices Physics and Technology*. p. 482, Wiley & Sons, New York, 1985.
34. Maier, J., On the conductivity of polycrystalline materials. *Berichte der Bunsengesellschaft für physikalische Chemie*, 90, 1, 26–33, 1986.

35. Sayle, T., Parker, S., Catlow, C., The role of oxygen vacancies on ceria surfaces in the oxidation of carbon monoxide. *Surf. Sci.*, 316, 3, 329–336, 1994.
36. Lavik, E., *et al.*, Nonstoichiometry and electrical conductivity of nanocrystalline CeO. *J. Electroceram.*, 1, 1, 7–14, 1997.
37. Hwang, J.-H. and Mason, T.O., Defect chemistry and transport properties of nanocrystalline cerium oxide. *Zeitschrift für Physikalische Chemie*, 207, 1–2, 21–38, 1998.
38. Fleig, J., Microelectrodes in solid state ionics. *Solid State Ion.*, 161, 3–4, 279–289, 2003.
39. Mondal, P. and Hahn, H., Investigation of the complex conductivity of nanocrystalline Y2O3-stabilized zirconia. *Berichte der Bunsengesellschaft für physikalische Chemie*, 101, 11, 1765–1768, 1997.
40. Jiang, S., *et al.*, Electrical properties of ultrafine-grained yttria-stabilized zirconia ceramics. *J. Mater. Res.*, 12, 9, 2374–2380, 1997.
41. Brossmann, U., *et al.*, 18O Diffusion in nano crystalline ZrO2. *Nanostruct. Mater.*, 12, 5–8, 871–874, 1999.
42. Knauth, P., *et al.*, Emf measurements on nanocrystalline copper-doped ceria. *J. Solid State Chem.*, 140, 2, 295–299, 1998.
43. Tuller, H.L., Ionic conduction in nanocrystalline materials. *Solid State Ion.*, 131, 1–2, 143–157, 2000.
44. Tschöpe, A., Grain size-dependent electrical conductivity of polycrystalline cerium oxide II: Space charge model. *Solid State Ion.*, 139, 3–4, 267–280, 2001.
45. Kim, S. and Maier, J., On the conductivity mechanism of nanocrystalline ceria. *J. Electrochem. Soc.*,149, 10, J73, 2002.
46. Chiang, Y.-M., Introduction and overview: physical properties of nanostructured materials. *J. Electroceram.*, 1, 3, 205–209, 1997.
47. Puin, W., *et al.*, Local and overall ionic conductivity in nanocrystalline CaF2. *Solid State Ion.*, 131, 1–2, 159–164, 2000.
48. Maier, J., Defect chemistry and ionic conductivity in thin films. *Solid State Ionics*, 23, 1–2, 59–67, 1987.
49. Maier, J., Defect chemistry: composition, transport, and reactions in the solid state; part I: thermodynamics. *Angewandte Chemie International Edition in English*, 32, 3, 313–335, 1993.
50. Maier, J., Thin Film Conductance. Evaluation and Correlation with Defect Formation Parameters. *Physica Status Solidi(a)*, 112, 1, 115–122, 1989.
51. Sata, N., *et al.*, Mesoscopic fast ion conduction in nanometre-scale planar heterostructures. *Nature*,408, 6815, 946–949, 2000.
52. Lee, J.S., Adams, S., Maier, J., Transport and Phase Transition Characteristics in AgI: Al2O3 Composite Electrolytes Evidence for a Highly Conducting 7-Layer AgI Polytype. *J. Electrochem. Soc.*, 147, 6, 2407, 2000.

12

Nanofluidics and Ion Transport at the Nanoscale: Manipulation and Sensing Applications

Nadia Akram[1]*, Ameer Hamza[1], Muhammad Ibrahim[2], Muhammad Usman[1], Akbar Ali[1] and Khalid Mahmood Zia[1]

[1]*Department of Chemistry, Government College University Faisalabad, Faisalabad, Pakistan*
[2]*Department of Applied Chemistry, Government College University Faisalabad, Faisalabad, Pakistan*

Abstract

In this chapter we explore the detailed interaction between nanofluidics and ion transport at nanoscale, studying their significant impact on manipulation and sensing applications. Among the distinctive outlook of the nanoscale extent, fluidic conduct, ion interactions, and electrostatic interface go through significant modifications due to confinement effects, offering a productive foundation for innovative technologies. Nanofluidic frameworks, well-known for their unique precision and selectivity in ion transport, enable innovative applications, such as highly sensitive sensing and detection of distinct molecules. The integration of nanofluidics and ion transport marks the inception of a biosensing revolution, as indicated by nanochannel-based systems that demonstrate remarkable sensitivity in biomolecule detection, as well as potential advances in medical diagnostics and biotechnology. Nanofluidics has practical applications outside laboratory settings, including the creation of small sensors for point-of-care diagnostics, environmental monitoring, and energy storage. Despite advancements, issues in scalability and repeatability remain, spurring an investigation into innovative materials and multidisciplinary collaborations. This chapter captures the cutting edge of scientific and technological innovation, arguing for ongoing research and joint efforts to push the boundaries of nanoscale manipulation and sensing capabilities.

**Corresponding author*: nadiaakram@gcuf.edu.pk

Inamuddin, Tariq Altalhi, Mohammad Luqman and Jorddy Neves Cruz (eds.) Nanoionics: Fundamentals and Applications, (309–344) © 2025 Scrivener Publishing LLC

Keywords: Nanofluidics, transport phenomena, electrokinetic, nano channels, nanofluidic devices

12.1 Introduction

The nanofluids are the colloidal particle suspensions in base fluids with typical particle sizes varying from 10 to 100 nm. The commonly observed nanoparticles in the environment are metal/metal oxides/carbon-based nanoparticles. The problem of keeping very small particles suspended in a fluid was first examined well over a century ago by Maxwell [1].

In recent years there has been tremendous growth of nanofluidics as a specialized field within the field of microfluidics. This thriving progress has been driven in large part by the invention of nanoelectromechanical systems (NEMS) technologies, allowing the creation of nanostructures with at least one dimension in the nanoscale. NEMS technologies are, for instance, constructed by top-down manufacturing of nanostructures with 10-1000 nm in size.

In addition to its reduced size, individual nanospaces also exhibit dissimilar physicochemical attributes from that of the macro-scale, such as molecular diffusion and electrical and fluid properties, including viscosity, condensation, and proton mobility, all of which stem from the high surface-to-volume ratio of nanospaces [2, 3].

The term nanofluidics refers to the study and use of fluids in nanostructures [4]. Nevertheless, the physical behavior of fluids in nanoscale structures is different from that of bulk or micrometer-scale materials. Therefore, using chip-based nanofluidic devices and nanofluidic structures such as nanopipettes, nanopores, nanogaps, nanotubes, it is possible to reveal new physical phenomena that dominate the field, e.g. Nonlinear transport phenomena, like concentration polarization, correction of ion currents, Modified liquid characteristics, including reduced dielectric constant, increased viscosity, and enhanced proton mobility [5]. The confined region of one nanometer in size, where Van der Waals and electrostatic forces over long ranges, as well as steric interactions at short ranges and hydration begin to significantly affect the situation, is the subject of nanofluidics [6]. In the last two decades, there has been a notable increase in research interest in nanofluidics, which finds applications in diverse fields like engineering, chemistry, biology, and physics [7]. Asymmetrically formed ion channels embedded in lipid bilayers in living organisms control physical, geometric and chemical aspects at the nanoscale or molecular level to effectively control transmembrane ion flux [8].

In nature, nanofluidic transport is widespread, and living things interact with their environment through complex membranes. They are primarily engaged in neurotransmission, osmotic control, and sensory detection (including audition, touch and thermoception) [9].

Ion transport is an important phenomenon with diverse applications in energy storage, materials science, biology, and electrochemistry. The concentration gradient, electric field, and type of material through which the ions travel can all affect how the ions behave during transport [10].

The field of ion transport in nanofluidics is relatively new but has benefited from recent advances in the synthesis of both synthetic and inorganic materials, which provide control over pore length and diameter in the nanoscale range [11]. Various approaches have been used to fabricate a variety of nanopores, nanochannels, and nanopipettes within different materials and with different morphologies inspired by different biological ion channels [12], and work was done to investigate the fundamental functions of ion transport, such as ion pump properties, ion optimization, and ion selectivity [13]. In addition, ion transport in nanofluidics has been investigated for various technological applications, including energy conversion and storage, ultrafiltration, sensing, and purification [14].

12.2 Fundamentals of Nanofluidics

12.2.1 Nanoscale Fluid Behavior

Experimental investigation of nanoscale fluid behavior is of particular importance as it advances not only theoretical knowledge but also engineering frontiers in the fields of clean energy [15], lab-on-a-chip [16], chemical analysis, [17] oil and gas industries and laboratory applications [18].

In addition to experimental investigations, computer simulations have proven to be an effective tool to be used for the analysis of fluid flow at nanoscale, which is still difficult to do at the atomic scale. More precisely, the best approach for nanoflow simulations is atomistic modeling using molecular dynamics (MD), which presents information at the atomic level. Current investigations show the existence of a critical dimension at the atomic scale, below which all fluid properties (temperature, slip length, diffusion coefficient, maximum velocity, density, shear viscosity, and thermal conductivity) change. is done and deviates from their larger values. It has been shown that the critical size is further influenced by the geometric roughness and wettability of the nanochannel wall [19].

12.2.1.1 Surface Effects

In the realm of nanofluidics, the surface properties of nanochannel walls are of great interest. Surface functionalization and modification are important for the development of new devices as well as for the description and investigation of specific nanofluidic phenomena. Surface properties in nanofluidic devices affect the "surface charge governance" and "geometry governance" transport processes, as well as the capabilities and performance of various nanofluidic applications due to the extremely high density of nanochannels. surface-to-volume ratio [20].

The heat transfer process can be improved at the micro/nanoscale stage by nanostructures on the heat transfer surfaces. Simulations of molecular dynamics were utilized to examine the evaporation behavior of liquid films at nanoscale on periodically sinusoidal shaped surfaces, to understand the expansion mechanism introduced by the nanostructures. By plotting the curve describing the amplitude (A) and period (T) of the sinusoidal shape, which possess shapes like.

$$j(x)=h_0+A\sin(2\pi x/T)$$

At macroscopic and microscopic scales, heat transfer surfaces can be effectively optimized for pool boiling heat transfer processes using conventional heat transfer fluids [21], Cubic nanostructures with different parameters influencing the evaporation rate were found to reduce the interfacial thermal resistance by calculation. When cubic nanostructures of different heights were examined for their effect on boiling [22], it was found that increasing nanostructure height based on density and liquid film behavior facilitates rapid boiling based on number phase change [23].

12.2.1.2 Viscosity Changes

It is essential to consider the impact of liquid viscosity in nanoconfinement from a practical point of view. Random kinetics are prevalent in a wide range of nanofluidic application fields. Therefore, during the controlled infiltration or extrusion of fluids into nanoporous materials, high-pressure (P, V, T) calorimetry facilitates the direct measurement of volume, temperature, pressure and thermal effects and it has been shown that adjusting the fluid viscosity can control the fluid discharge pressure and the resulting thermal consequences [24].

There exists a connection between shear rate and shear stress that governs the rheological dynamics. The force of shear per unit area is known as shear stress. The change in stress over time is known as shear rate. Because viscosity can be described as the relation between strain rate and shear stress, it is recognized as a fundamental metric for evaluating the rheological dynamics of fluids.

Nanofluid behavior can be classified as non-Newtonian or Newtonian based on various factors, including volume fraction, shear rate, and size of nanoparticles. These are the outcomes:

- Nanofluid viscosity rises when the volume percentage of nanoparticles grows. Newtonian behavior is noted at low volume fractions, whereas non-Newtonian conduct is predominantly observed at enhanced volume fractions in nanofluids.
- When shear rates are low, behaviors show Newtonian characteristics; when shear rates are high, they acquire non-Newtonian properties.
- While most nanofluids using carbon nanotubes are non-Newtonian, they do not exhibit nearly Newtonian behavior.
- Compared to low viscosity nanofluids, high viscosity nanofluids are more Newtonian [25].

Nanoparticle transport phenomena in confined flows in high-viscosity systems engineering applications are a major source of concern because they require more power to pump the nanofluid through the system, which increases operating costs and nanofluids. The overall effectiveness in the design of nanofluids, determines the ideal combination of viscosity and thermal conductivity; it's essential to assess the simultaneous enhancement of both these properties. There is still scope for further research on this topic. The dependence of heat transfer coefficients on thermal conductivity and viscosity for energy systems is well known in research on thermal kinetics of nanofluids [26].

12.2.1.3 Transport Phenomena

One of the fundamental problems in nanofluidics is the phenomenon of transport in the nano space [27].

Physics governs many aspects of nature and technology, playing a role in human physiology, vegetation, energy models, water collection, treatment systems, chemical processes, material structure, medicine, and the movement of fluids and ions at the nanoscale. Improved performance techniques using ion diodes, transport of liquids in a chain-like manner, transport of confined gases, ion transport driven by surface charge, and iron gates with physical or chemical characteristics [28] are also included.

At least one feature of an object smaller than 100 nm allows fluid to move around and within it, allowing phenomena that would be impossible at larger length scales. Because of this characteristic, which comes from Brownian motion and the nanoscale action of nanofluids, nanofluids have favorable mass transfer and heat transfer properties, making them useful in nano/micro heat exchangers. Nanofluids are important, especially in heat exchangers, which are recognized as traditional areas of energy transfer. Khanlari, a prominent researcher, has made many contributions in this area [29].

It has been demonstrated that nanofluids are innovative working fluids with superior thermophysical properties that enhance the heat dissipation capacity of micro/nanochannels. Here, we use molecular dynamics modeling to provide atomistic insights influencing the flow dynamics considering the properties of nanofluids' heat transfer in nanochannels. Suspension as well as aggregation of nanoparticles are observed during the flow process of the nanochannel. In nanofluids, the wall-fluid interface is where the temperature rises, and the temperature growth can be accelerated. Fixed nanoparticles near the wall cause the fluid velocity to change. Nanofluids can facilitate heat transfer more efficiently than base fluids, and those with higher concentrations of nanoparticles have higher heat transfer capacity [30].

12.2.1.4 Electrokinetic Phenomena

Many applications, including biological sensing, bioelectronics, and separation of molecules, are driven by the ions migrating electrokinetically between ion-permeable solid mediums and electrolyte solutions. To combine electrical and ionic systems for these applications, a special class of materials and tools is required. The devices are designed to operate in an aqueous environment and based on micro/nanofluidic pathways or ion-permeable materials. They can convert charged species by applying electrical fields or the conversion of electrical impulses from biological processes [31].

The fascinating potential of induced charge electrokinetic (ICEK) makes it a popular topic of discussion in micro- and nanofluidics. Researchers have made significant progress in basic research and applied research in the last ten years. In induced-charge electroosmosis (ICEO), they observed (i) a flow reversal which they attributed to the ions' phase delay effect; (ii) a stochastic ICEO, which they associated with the bulk solution's concentration polarization; (iii) they associated the non-quadratic relationship in ICO regarding non-Newtonian fluids with the influence of viscosity power-law. (iv) a cycle of Janus doublets using induced charge electrophoretic with other phenomena. In addition, several ICEK-based micro- and nanofluidic devices, including nanopore ion diodes, particle concentrators, traps, and micropumps, have been developed [32].

Due to its many applications, including the creation of lab-on-a-chip (LOC) devices, drug administration, and macromolecule separation, electrocatalytic ion transport and fluid flow have attracted much attention [33]. When in contact with an electrolyte solution, the majority of surfaces used in studies acquire a net surface charge as a result of several electrochemical processes, such as ion adsorption or association or distribution of functional groups on the surface. The exposed charged surface of the electrolyte solution attracts the counter, evaporating the coin. In fact, a non-zero charge layer called an electric double layer (EDL) can develop near the charged surface. Electroosmotic flow (EOF) is induced by additional counterions experiencing a net non-zero electromotive force under an applied electric field's impact in extended region of the EDL [34].

12.2.2 Nanofluidic Devices

Because of their incredibly small size, that minute size makes them useful for biosensing, biochemical separation, seawater purification, nanofluidic transistors, proteins, and lab-on-a-chip (LOC) for prediction. Controlling electronic transport in a nanochannel produced by an advanced nanofabrication technique may lead to such applications as shown in Figure 12.1 [35].

In biological research, nanofluidic devices are used as analytical platforms for samples with limited volume. By precisely aligning and monitoring biological targets smaller than a single cell, nanofluidic devices utilize the lower dimensions of nanostructures as opposed to microstructures. Therefore, large macromolecules such as DNA and proteins as well as biological particles such as viruses and extracellular vesicles can be analyzed using nanofluidic devices. It is also possible to identify and characterize a single biomolecule or particle by accurately configuring and designing a nanofluidic system [3].

Because of their different transport characteristics and their analytical uses, ions in nanofluidic systems and their identification are of great interest. There are certain elements of microscale transport that transfer straight towards the nanoscale. However, diffusion, ion current repulsion, surface charge, double-layer overlap, along with entropic forces can have a large effect on such nanofluidic systems, although these phenomena are minimal or nonexistent in large microchannels [36].

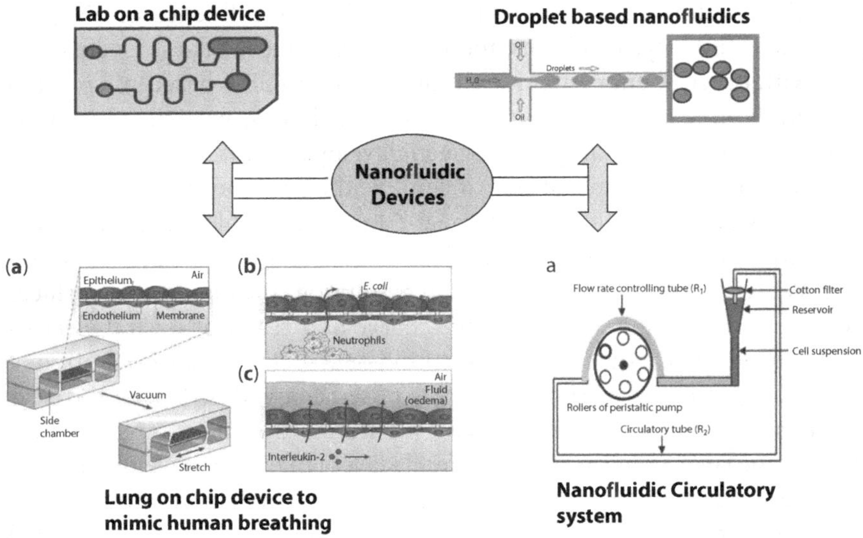

Figure 12.1 Nanofluidic devices.

12.2.2.1 Nanochannels and Nanopores

Nanopores and nanochannels are two major types of nanofluidic structures, and the names can be used to refer to systems that are comparable to each other.

12.2.2.1.1 Nanochannels

An integral part of a nanofluidic chip, the nanochannel plays an important role in determining device performance. It serves as the basis for the fabrication, detection, and application of nanofluidic chips [37].

The process of creating planar, small patterns is known as nano channeling. Structures limited to the few dimensions at the nanoscale have been fabricated using a number of traditional micro- and nanofabrication techniques, including e-beam lithography [38], nanoscale-depth microchannels photolithography [39]. The channel width is likewise limited to nanoscale dimensions by focused e-beam and ion beam techniques, while the channel depth is limited to nanometer dimensions by the first two procedures. In addition, any arbitrary two-dimensional shape can be produced by ion- and e-beam writing. For example, planar format scanning electron microscope (SEM) images depicting nanofluidic funnels featuring 5°, 10°, and 20° taper angles along with a straight channel easily form nanochannels with short tips and wide bases. These fins are fabricated using e-beam lithography, and were molded from masters using SU-8 negative-tone resist, which was used to write on high stiffness poly(dimethylsiloxane) (h-PDMS). The various ion transport pathways are discussed below. More recently, self-sealing and cell-confined nuclear layer deposition techniques produced nanochannels less than 10 nm in width and lateral dimensions [40, 41]. A different manufacturing method creates nanoscale gaps directly between microchannels in PDMS, using an electric field to disrupt junctions between neighboring channels [42]. However, this method presents more challenges in terms of dimensional control. Creating nanofluidic channels on the plane of a substrate has two main advantages: (a) it can be integrated into more complex devices; and (b) it can be used to theoretically visualize material transport, which contributes to the understanding of transport mechanisms [36].

12.2.2.1.2 Nanopores

Nanoporous channels with diameter d and length L fall into the nanoscale range [43]. A nanopore is a gap or channel in an independent membrane at the nanoscale level and is a component of nanoparticles. For biological transmembrane proteins, solid-state nanopores like silicon, SiO_2, and Si_3N_4 serve as biosensors [44].

Nanopores typically possess a conical or cylindrical form and are rounded. The pore are perpendicular to the plane of the substrate. Often, nanochannels when formed possess the identical depth and breadth in the substrate's plane as the lateral dimensions, and do not have spherical symmetry along the axis of the channel. The term "nanopore" refers to a variety of different types of pores, such as track-made synthetic nanopores, electron beam pierced through pores, femtosecond laser drilled pores [45], ion beam implanted pores [46], engineered protein pores [47], and nanopipettes. Compared to nanochannels, nanopores can be made from a wider range of materials and have smaller side-to-side measurements, typically around 1-2 nm. The main disadvantage is that integrating these holes into more intricate systems with additional nanofluidic components is more challenging. A conical porous membrane with an interior diameter of 600 nm and an outermost diameter of 50 nm was traced and imaged using a scanning electron microscope (SEM). An e-beam, or focused ion beam can be used to align the holes to achieve more precise control over their diameter. A silicon nitride membrane that is 30 nm thick with a 15-nm nanopore formed inside was imaged using a transmission electron microscope (TEM). With this manufacturing method, the pores in the membrane can be as small as 3 nm in diameter [48]. To create single-pore integrated devices at the channel junction, traceable membranes have been integrated onto microfluidic platforms [49]. Devices featuring numerous nanopores at the channel intersections are employed to trap nanoparticles and cells using electrophoresis and dielectrophoresis [50]. The number of holes in the cross-section of the device is determined by the hole density and the channel width, either of which can be individually adjusted according to the desired experiment. Interlocking multilayer structures of membranes are used in microfluidic separation systems to control sample distribution [51].

12.2.2.2 Lab-on-a-Chip Devices

A Lab-on-a-chip (LOC) is a compact device that integrates multiple instruments for high-resolution separation and detection on a single chip [52]. Its advantages are cost-effectiveness, diagnostic capabilities, and sensitivity for analyzing. DNA sequencing and biochemical detection are also provided by this assay. Small footprint lab-on-a-chip devices have indeed been made possible by microfluidic manufacturing techniques including lithography, xerography, and laser machining [53].

12.2.2.3 Fluid Transport in Nanoscale Conduits

Large slopes for unusually high permeate fluxes, estimation of kinetic mass transfer limits for vapor, modeling and measurement of vapor condensation coefficients, architecture for wall-fluid interaction [54], and understanding of surface charge-mediated aqueous electrolyte flow are all possible at the nanoscale. Despite the growing corpus of work in this area, a comprehensive assessment of transmission resistance in nanofluidics for liquid–vapor systems is still not fully understood [55].

12.2.2.4 Biological and Chemical Sensing

A fresh framework for nanofluidic devices has been made possible using nanotubes and nanowires. Surface interaction with biological molecules has an impact on the carriers' accumulation and removal across the nanostructure, as opposed to only a single portion of it, compared to planar semiconductor sensors. In chemical and biosensing, this significantly increases sensitivity and potentially opens up the possibility of identifying individual molecules [56].

12.2.2.5 *Fluidic Control and Manipulation*

In industrial applications, fluid manipulation is important. Flows at different length scales are common in the natural world. Many scientific fields as well as commercial applications rely on the ability to simulate flow at different length scales. When the dimension moves from the macroscale to the nanoscale, the flow behavior changes significantly [57]. Channel diameters at the nanoscale are similar to the short range of molecular interactions, making molecular interactions non-negative and specialized flow phenomena for ultrafast water transport [58].

12.2.2.6 *Energy Harvesting and Storage*

A major problem facing humanity and our contemporary culture is the lack of affordable, abundant and sustainable renewable energy sources. These sources include solar energy and solar energy, sometimes called "blue energy," and in theory, both are easy to collect. However, because energy exchange and storage processes in life rely on ion transport and energy storage molecules to function in concert and efficiently, nature has gifted us with a unique "ionic power." Which can provide perspective. Organisms use a variety of ion transport mechanisms to accumulate and store energy, such as ion passive transport and ion active transport. Electric gel is a system developed to improve electricity generation through ionic gradients [14].

12.2.2.7 *Two-Dimensional Materials in Nanofluidics*

Nanofluidic ion transport in bilayer membranes has two distinct advantages: (1) the construction process is simple and highly scalable; [57]. (2) since carbon materials are frictionless and have closely spaced nanochannels, the fabricated graphene-based membrane provides high-flux chemical transport; [59]. (3) In bulk solution, the characterization and chemical synthesis of 2D nano-building blocks will be performed significantly more efficiently; [60]. (4) 2D nanochannels' height may be precisely regulated, from less than one nanometer to more than 10 nanometers, structure and fundamental transport properties in two-dimensional layered materials of nanofluidics, as well as energy conversion, storage, water purification, and molecular screening, biomimetic ion transport manipulation, and advanced battery separators [61].

12.2.3 Nanofluidic Devices Performance and Reliability

Experiments, measurement methods, and computer modeling are all used together to visualize the performance and reliability of nanofluidic systems. Here is a complete plan:

12.2.3.1 Conductance Measurements

Check how stable and long-lasting the nanofluidic devices are by measuring their conductance overtime.

Check the difference in conductance (ΔG) between devices that have had their surfaces changed in different ways, like hydrophobic HMDS treatment, to see how it affects stability.

12.2.3.2 Electrochemical Measurements

Test the performance and stability of devices by cyclic voltammetry and other electrochemical approaches.

12.2.3.3 Scanning Electron Microscopy (SEM)

SEM imaging and shape and structural confirmation for a nanofluidic device because SEM can see images at the nanometer scale.

12.2.3.4 Atomic Force Microscopy (AFM)

To check the fate and also to have a better understanding of the device, the behavior of a device on the surface level (DAS) using AFM has been reported, revealing the texture and hardness of the device to assess the working and reliability.

12.2.3.5 Electron Microscopy

A typical method, for example, is to employ scanning or transmission electron microscopy to investigate internal structure and defects of nanofluidic devices.

12.2.3.6 High-Resolution Imaging

Surface and Interface Investigation - (high-resolution imaging techniques like STM or AFM for study of surface and interface properties of devices).

12.2.3.7 *Electrochemical Impedance Spectroscopy (EIS)*

Study the electrical properties of devices by using EIS to analyze the performance and reliability.

12.2.3.8 *Capillary Filling*

Research on capillary re-locate rates to assist in leave out units with insufficient potential and against a complete convolution of solvent and pressures.

12.2.3.9 *Tools for Microfabrication*

Fabrication of nanofluidic devices with control over surface properties and geometry with a focused ion beam (FIB).

12.2.3.10 *Systematic Analysis*

To find and fix possible problems compromising the devices' performance and dependability; do a methodical investigation of their manufacture, characterization, and testing.

These techniques enable researchers to thoroughly assess the dependability and performance of nanofluidic devices, therefore facilitating the creation of more durable and effective systems for a range of applications.

12.3 Advances in Nano Ionics and Improvement in Device Quality

In the past few years, there have been big steps forward in nanoionics and better device quality, especially in the creation of devices that store and change energy. Some important progresses are:

12.3.1 Graphene-Based Energy Storage Devices

Utilizing graphene's special properties—high surface area, exceptional electrical conductivity, and strong mechanical strength—energy storage systems based on graphene enhance the performance of batteries and supercapacitors. With their longer lifespans, faster charge/discharge cycles, and higher energy and power densities, these improvements are appropriate for use in consumer devices, renewable energy systems, and electric cars. Even as scale production and material integration are still difficult, developments in graphene oxide, 3D graphene structures, and composite materials are progressing the area and pointing to a future of more effective and environmentally friendly energy storage technologies [62].

12.3.2 Nanostructured Metal-Based Electrodes

Nanostructured metal-based electrodes make energy storage devices much better by increasing their surface area, electrical conductivity, ion diffusion, and strength. These enhancements enable higher energy and power density, improved charge/discharge cycles, and longer cycle life, which is an advantage for lithium-ion batteries and supercapacitors. Scalability and stability issues persist, but much of the field has developed due to advances in production, protective coatings, and integration. This gives us better, longer-term energy storage solutions for applications such as electric cars and green energy systems [63].

12.3.3 2D Nanosheets and MOF Nanosheets

Two-dimensional (2D) metal–organic framework (MOF) nanosheets have emerged as a class of two-dimensional materials, which possess appealing features such as ultrathin thickness, large surface area and a high surface to volume atomic ratio [1–6]. Over the past few years, we have witnessed substantial advances in the controllable synthesis and practical exploitation of these nanosheets. In this review, researchers discuss the latest advancement in the preparation method, including the bottom-up methods like 3-layer synthesis, interfacial synthesis, and PDMS surface assisted synthesis for 2D MOF nanosheets, continued by another important technique called chemical exfoliation, Modulated synthesis, sonication synthesis, and surfactant-assisted synthesis, while top-down methods are mechanical exfoliation, sonication exfoliation, chemical exfoliation, and Li-intercalation exfoliation. Besides these, recent advancement in 2D MOF nanosheet based nanocomposites is also discussed in the paper. It also discusses key challenges regarding the research on 2D MOF nanosheets and their composites towards energy conversion and storage, gas separation, catalysis, sensing, and biomedicine [64].

12.4 Ion Transport Mechanisms

Ion transport is a common phenomenon in cellular membranes and is essential for all important life processes as shown in Figure 12.2 [65]. Solid-state nanofluidic systems were proposed as a way to simulate and

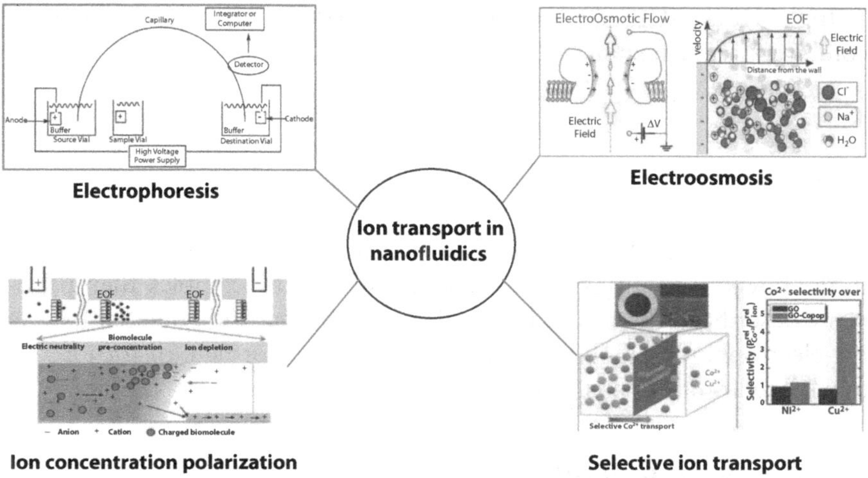

Figure 12.2 Ion transport in nanofluidics.

understand the function of ion transport systems [66]. Ion transport always exhibits significant differences in confined nanofluidic nanochannels, nanotubes and nanopores, with characteristic sizes on the order of 1–100 nm. These differences include emerging effects such as ion selectivity, ion reactivity, and ion gating properties [67]. These special properties are used in a variety of industries, including ion filtration, dialysis, and water purification. The diameters of the channels are close to the Debye screening length in salt gradient energy transitions [68], and the surface charge of the channels can significantly change the ionic behavior of the nanofluids. As a result, only target ions are able to pass through, successfully competing [69].

12.4.1 Electrophoresis

The phenomenon of movement of charged species in an aqueous medium when a constant electric field is applied is called electrophoresis. In microfluidic settings, this effect is often used to support discrete charged particles [70].

A common potential associated with the potential of a point charge is electrophoresis as

$$\zeta = a_p E = a_p q / 4\pi\varepsilon_0\varepsilon_r$$

In the Hückel-Onsager limit, the electrophoretic mobility is obtained by combining these formulas,

$$\mu_{ep} = \frac{2\varepsilon_0\varepsilon_r\zeta}{3\eta}$$

Since the molecules and coins move around it in the opposite direction to the flow of the atmosphere, the situation is more complex than previously described. The resulting polarization field has the potential to alter the electrophoretic mobility. The relaxation effect, or field distortion due to molecular mobility, is another issue that has not been considered [71].

12.4.2 Electroosmosis

When the solution moves because of an applied electric field or potential gradient, an electrokinetic phenomenon known as electroosmotic flow (EOF) occurs in electrolyte solutions. Electrohydrodynamic motion of liquids and ions in the presence of charged solids is frequently observed Within intricate biological systems like DNA and transmembrane protein pathways like α-hemolysin, as well as in naturally occurring materials like silica and clay [72]. Controlling and regulating flow rates is necessary for a variety of engineering systems, such as energy conversion and separation science, membranes for proton exchange in fuel cells, power generation, laboratory-on-a-chip technologies, and micro/nanoscale compressors [73], and synthetic pores for the filtration of water [74], is a practical reason to study electrochemical filtration (EOF). These intricate systems function at tiny length scales, where surface effects are significant. A Stokes equation can be used to describe the electroosmotic flow in a silt channel [75]

$$\frac{d}{dz}\left[\mu(z)\frac{du_{eo}}{dz}\right] + \sum_{\iota\in\pm} \rho\iota(\mathrm{z})^{\sim}\ \mathrm{zieEext\ x} = 0,$$

within boundary limitations

$$\mathrm{u}_{eo}(-\mathrm{L}/2+\delta) = \mathrm{u}_{eo}(+L/2+\delta) = \mathrm{u}_s$$

In which (x, z) represent the streaming and limiting directions, respectively, and the channel width is denoted by L. The confining walls are at ± L/2.

The flowing direction's EOF velocity field is denoted by ueo(z), the ion concentration field is denoted by ρi (z), the electronic charge is denoted by (e), the electrical field that is applied to wall is parallel to (i), and the ion concentration, can be either a cation (+) or an anion (-). To calculate the total electroosmotic driving force, the summation is applied over all ionic species (i∈±) [76].

An applied electric field can cause the liquid to move, a phenomenon known as electroosmosis. Around the charged surfaces, counter ions form and cause coupling between the external electric field and liquid mass transfer. Electroosmosis is thus extremely important in nanofluidic technology, where surfaces play an extensive role. It has complex and subtle effects on the transport parameters of both synthetic and biological nanopores [77].

12.4.3 Ion Concentration Polarization (ICP)

The ICP concept has many applications in the fields of biosensing, disinfection, molecular enrichment, and clinical diagnostics [78].

Ion concentration polarization (ICP) is a straightforward, efficient method for power generation, desalination, biosensing, and diagnostics used in microchannels in order to focus charged analytes in microfluidic devices [79].

An electrical double layer (EDL) is formed when an electrolyte solution is exposed to a charged surface within a microchannel, and the electrolyte's counterions accumulate on the surface that is charged [80]. A tangential electric field (Et) causes charged ions in the microchannel to move, resulting in movement of the bulk solution (a process known as electroosmotic flow). When ionic transport is influenced by surface charge in the nanochannels formed between the microchannels, ion selectivity develops, enabling opposing ions with overlapped EDL to pass through. The high normal electric field (En) produced by the ion selection of the nanochannels has an impact on an ion concentration gradient at the junction of the nano- and microchannels. ICP is caused by the existence of nanochannels with chargeable surface in the nanochannel within an electric field. Adjacent microchannels experience regions of enrichment and depletion because of the ICP device. A nanoporous membrane can be incorporated into the microchannel or nanofabrication techniques can be used to produce nanochannels. ICP devices are mostly used for biological applications [81], micro-mixing [82], ionic diodes [83], and biomedical applications [84]. Many microfluidic-based techniques, such as affinity-based extraction, electrophoresis [85], isoelectric focusing, and other techniques for disease diagnosis are used to predict charged analytes. For example, DNA or protein [86].

12.4.4 Capillary Electrophoresis

A type of electrokinetic separation method is capillary electrophoresis (CE). When subjected to an electric field, it facilitates the chemical distribution of charged ions in aqueous phases based on variation in their velocity. Depending on the type of separation electrolyte and capillary used, there are a variety of operational methods (e.g., zone electrophoresis, micellar electrophoresis, isoelectric focusing, and isotachiophoresis). These methods, whether in capillary format or microsystems, are well established in industrial and clinical fields due to their excellent resolution and efficiency, speed of analysis, and low sample volume required [87].

12.4.5 Ion Selectivity

Understanding how cations and anions are separately charged is fundamental to nanofluidic devices. Ion selectivity in surface-charged nanofluidic channels is favored by counterion transfer over co-ions using an electrical double layer, or EDL. The electrostatic force on the charged wall repels ions that have the same charge as the wall (co-ions), while ions that have counterions are attracted. However, only a small fraction of the counter reaches the charged wall [14].

The selectivity of ionic species transport through polymer grafted nanochannels was covered in this subsection. Ion selectivity (S), a nondimensional variable that we developed to measure the preferences that are used in each channel,

$$S = \frac{|Ic| - |Ia|}{|Ic| + |Ia|}$$

Ij stands for the average current density of the ionic species throughout the channel j th (j = c,a) [34].

12.4.6 Coupling of Ion and Fluid Transport

Ion transport in fluid transport: When charged surfaces are present, a net force from an external electric field acts on the charged Debye layer, causing the fluid to move. Electroosmotic (EO) flow is the name given to this internationally developed flow. Ionic currents in the Debye layer [43].

This is required for applications such as drug delivery and controlled release systems [88].

12.5 Manipulation Techniques at the Nanoscale

12.5.1 External Fields and Nanofluidic Control

Some of the problems facing current nanofluidic technology include its slow reaction rates, limited stability, complex preparation procedures, and unpredictable molecular and ion transport rates. Here, we present a nature-inspired ferrofluid-based nanofluid, which enables rate-controlled, multilayer, ultrafast reaction ion and molecular transport. Changes in the ferrofluid pattern on the outer surface of the nanochannels result in a velocity-controlled, ultra-fast reactive molecular transport (0.1s), and current gating ratio controlled by an external magnetic field [89].

12.5.1.1 Electric Fields

A better understanding of the confined dynamics, structure, and transport fluids at the nanoscale under locally and externally applied fields is essential for the creation of intricate nanofluidic systems including the valves, pumps, and various other flow control parts. The flow can be regulated by varying the wettability of the surface of the nanochannels using localized electric fields. Graphene exploited as a local electrode, and an appropriate electric field can be generated by adjusting the electrical charges associated with every electrode [90]. However, atomistic modeling can be used to obtain a fine-grained understanding of fluid transport processes at the nanoscale, which are essential for such improvements [91].

12.5.1.2 Magnetic Fields

In many fields, the magnetic field is indispensable because it acts as a flow controller near electrical heat transfer equipment. Different heating devices have different results when a magnetic field is present. Some devices, such as heatsinks or ducts, have covers that either increase or decrease heat transfer, much like enclosures do. Therefore, heat transfer in magnetic field is investigated by researchers. Nanoparticle deposition results from the addition of magnetic field. The magnetic field creates a force in the nanofluid along with to the Lorentz force. Fluids can be expelled by this force [92].

12.5.2 Microfluidic Integration

Microfluidic devices can effectively integrate and remove electrochemical biosensors. The applications include those in the domains of diagnostics, food and sustainability, tiny molecules, proteins, and whole cells [93].

Multiscale micro- and nanofluidic systems constructed by incorporating nanochannels into microchannels have special properties that conventional single-scale channels (microchannels or nanochannels) are unable to match [94]. Samples are given a well-defined spatiotemporal microenvironment by microfluidics, which enables quantitative observation and analysis. Convection and vortices are two examples of unpredictable fluidic disturbances that can be avoided by integrating nanochannels into microfluidics when the procedure is in progress. Additionally, it enables precise and steady mass transfer of different entities in a freely structured manner, such as molecules and ions, for accurate and convenient measurement and analysis [95]. In particular, researchers may investigate a variety of phenomena, including ion-selective transport, molecule sequestration, counter-surface transport across nanochannels, etc., by integrating nanochannels, that cannot be achieved by conventional microchannels [96].

12.5.3 Hybrid Micro-Nanofluidic Systems

Hybrid nanofluids have been created by mixing different nanoparticles with a base fluid. The stability of nanofluids is enhanced by intermolecular forces that prevent particle aggregation between different nanoparticles. Hybrid nanofluids differ from single nanofluids in their chemical and physical properties [97].

At the micro- and nanoscale, different shapes and patterns are created using different types of materials, exhibiting different properties and capabilities. Micro- and nanofluidic technologies, which offer excellent management in the micro/nanoscale manipulation of fluids with different solute, suspension, and building blocks, are advantageous for new liquid-mediated production methods [98]. The techniques used to fabricate multiscale micro- and nanofluidic devices are as follows: nanolithography, fluid pattern-assisted polymerization, sandwiching-based coating, sealing ablation (i.e., deformations), and generation (rupture engineering, self-assembly). These techniques can be categorized based on the principles of their fabrication. Direct creation of nanochannels on the substrate is possible using nanofabrication methods like FIB. nanofabrication techniques such as FIB create nanochannels directly on the substrate [96].

12.5.4 Synergy with Traditional Microfluidics

Research in the life sciences, personalized medicine [99], and air and water quality control [100]; and other fields have greatly benefited from microfluidics. The field is based on the accumulation of scientific knowledge about the behavior, control, and manipulation of fluids in organisms [101], of fluids in microsystems [102]. Synergy of Microfluidics and Ultrasound [103], Microfluidics and 3D printing combined: a powerful combination for nanomedicine [104], microfluidics and soft robotics synergizing [102], can be done.

12.5.5 Sensing Applications

Using advanced nanomaterials and new mechanisms, micro-nanotechnology has allowed sensors to advance significantly regarding dimensions, price, responsiveness, and velocity. Nanowires, nanotubes, and other microarrays have been widely used as sensor components in nanofluidic platforms to precisely target the molecules on an object's surface or control their distribution. Because of their tiny diameter and broad aspect ratio, nanofluidic devices built on nanowires or nanotubes may conduct molecule sensing even at extremely low concentrations using a variety of detection methods. Target molecule movement, reactivity, or adsorption on the surface, as well as variations in mechanical and electrical signals, all affect sensor response [56].

12.5.6 Chemical Sensing

12.5.6.1 Ion Sensing

It has been demonstrated experimentally and conceptually that nanotubes are efficient nanofluidic channels that can be used for pH and ion sensing at the molecular level. It has been reported that a SiNWs brick-based sensor with gold nanoparticle modification for extremely sensitive and selective mercury ion detection can be used for water quality monitoring [105].

12.5.6.2 Gas Sensing

The ability to detect various harmful gases including ethanol, CO, NO_2, H_2S, and acetone is becoming more and more important for human survival. With their high sensitivity, fast response, and excellent selectivity, NTs or NWs offer great promise as sensing components for large-scale gas sensing, WO_3 nanostructures have been shown to be effective materials for ethanol detection [106]. Furthermore, various structural forms have been prepared and tested, including WO_3-x, h-WO_3 rods, h-WO_3 nanotubes and h-WO_3 nanowires [107].

12.5.7 Biosensing

Ionic device advancements that are interesting include molecule isolation, biosensing, biological-inspired devices, and bio-machine interfacing technologies, are described in fundamentals of ionic transport behavior. Ionic circuits, molecular separation methods, and a new class of biosensors are developed because of electrokinetic reactions [108]. Research on the electromagnetism of ionic para molecular media and the creation of ionic gadgets that simulate brain system functions also require an understanding of the physiology of the nervous system and the underlying mechanisms of biological computing and memory. A variety of uses where interaction between living systems and machinery is necessary will be made possible by gadgets designed to convert ionic signals from organisms into electrical solid-state electronics signals and vice versa. The therapeutic platform which produces electrical stimulation or delivers drugs to facilitate the treatment of diseases associated with impaired brain signaling is an example [31].

12.5.7.1 Detection of Molecules

Different biomolecules can be recognized as well as chemical species by attaching identification groups onto the interface of nanoparticles in a targeted manner. Antigen-antibody complexes are then formed into immunosensors, proteins, drugs, hormones, viruses, and other substances, often measured by changes in electrical properties or ionic current [56].

Recent studies have shown that nanofluidic systems based on NTs and NWs can be a good option to detect tiny biological materials. The identification of pathogens is crucial among them for both disease diagnosis and treatment. Scientists developed MWCNT-based magnetic fluorescent sensors that exhibit strong linear relationship between magnetic field activity, intensity of fluorescence, and vibrio alginolyticus concentration [109].

12.5.7.2 DNA Sequencing

Real-time sequencing of DNA and detection can be facilitated by well-designed NTs or NWs [110]. Single-stranded DNA has been shown to bind precisely and persistently to the ends or surfaces of NTs or NWs. Zn-NPs filled MWCNTs detector/sensor using a probe that is immobilized [111]. The association of a small quantity of nucleic acids to the probe is turned into an electrical impulse that can be directly measured during the hybridization process. When complementary DNA is charged with electrons, the peak current is greatly reduced and three times more selective than non-complementary DNA. This is an example of the improved response properties of a straightforward, widely applicable DNA sequence detection technique. In a different example, detection of DNA at sub-nanomolar quantities using ZnO NWs and Au NPs *via* a magnetron-spectrum gas aggregation approach [112].

12.5.8 Single-Molecule Sensing

Two-dimensional (2D) membrane nanopores hold great promise for information storage, osmotic power generation, and single-molecule sensing. Recent developments in 2D nanopores, particularly on single-layer MoS2, have focused on the scalable growth and fabrication of nanopore devices. [113].

Two-dimensional (2D) materials consist of nanopores, a potential class of solid-state sensors and a flexible tool for designing biological channels and pores in cells [114]. Graphene [115], MoS_2 [116], WS_2 [117], hBN [118], and more recently MXenes [119], are the most investigated 2D materials for nanopores. A typical 2D nanopore device consists of two reservoirs separated by a supporting aperture and a single nanopore in an atomically thin membrane that stands alone. In the presence of an electrical field being applied, electrically charged biopolymers such as DNA, RNA, or proteins are forced through the nanopore and generate unique ionic current signals that move the molecules. There are signals for studies of single molecule biology, ion transport and selectivity have made 2D nanopore devices an invaluable tool [113].

12.5.8.1 Point-of-Care Diagnostics

Effective mass transfer and short diffusion distance work together to result in a feasible time in protein sensing in a nanofluidic environment. Here, we describe a disposable capsule-based point-of-care *in vitro* diagnostic platform that is fully automated and uses nanofluidic sensors to perform fluorescence immunoassays as shown in Figure 12.3. Trimodal tests of ferritin, pancreatic stone protein (PSP), and immunoglobulin E (IgE) were used to establish system performance. High forced flow and nano-confined space capture efficiency, providing a continuous maximum concentration

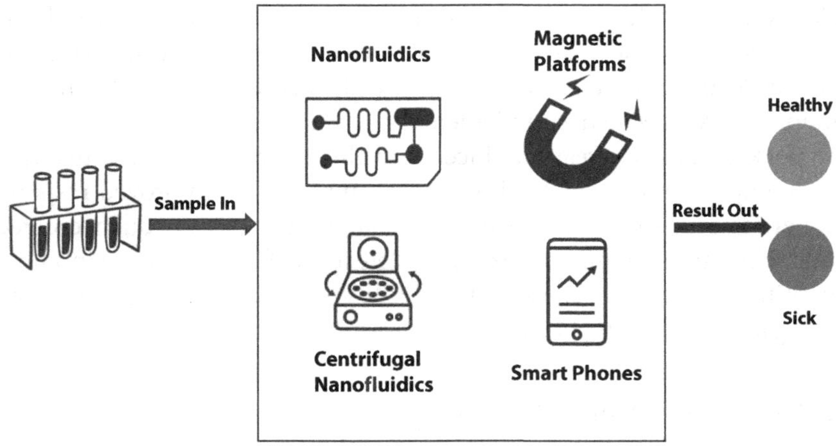

Figure 12.3 Point-of-care diagnostics.

gradient across the sensor. It is noteworthy that analytical zeptomole concentrations can be detected in just one drop of blood [120–128].

12.5.8.2 Nano Biosensors: Point-of-Care Device

Point-of-care nano biosensors are highly sensitive and can be integrated into many detection platforms, such as fluidics, and biochips, making them very promising for application in clinical diagnostics. As a result, their approach is now appealing for application in therapeutic environments (Liao, Zhang *et al.*, 2019). In addition to speeding up diagnostics, this category of nano biosensors emerged in either internal or physiological healthcare (Sharifi, Hasan *et al.*, 2020).

12.6 Emerging Trends and Challenges

12.6.1 Recent Advances in Nanofluidic Research

One of the latest cooling techniques to increase photovoltaic (PV) panel efficiency is the use of nanofluids to cool the panels. In this study, aluminum nanoparticles were used to evaluate the effects on the cooling capabilities and conversion efficiency of photovoltaic panels [129].

One of the study methods to enhance the efficiency of fluid-cooled thermally managed electric batteries is investigating efficient working media for heat transfer. The cooling capabilities of various input flow rates and nanofluid concentrations at various discharge rates and times are evaluated in an experimental system of nanofluid-based power battery liquid cooling [130].

The present work produces new results due to the combined action of an externally generated magnetic field and nonlinear thermal radiation. This work also aims to visually demonstrate the flow properties of single nanofluids in (gold), hybrid (copper-gold), and trihybrid (copper-Ag-gold) tubes with composite stenosis. According to this study, increasing the radiation parameter has an inverse effect on the flow velocity and magnetic field expansion, as the magnetic force has a significant effect on these parameters. Today, the fields of solar, biomedical, agricultural, food, and process systems engineering all make extensive use of nanotechnology. It is science and engineering [131].

Ion-based brain-like intelligence can be influenced by the way the human brain accomplishes cognitive processes such as the generation, transmission, and storage of nerve impulses by controlling the ionic conductance of ion channels in neural cells. Many researchers are interested in bioinspired nanofluidic iontronics, a developing multidisciplinary field focused on controlling the ionic conductivity of nanofluidic systems to achieve brain-like functions [132].

Significant progress in the field of mono- or hybrid nanofluid research requires greater consistency and better accuracy of results, where an extensive variety of distinct surfactants, base fluids, and nanoparticles have been identified and are continually being discovered. These innovative thermal fluids are utilized in a variety of contexts through careful documentation and adjustment of the parameters utilized in both mono- or combined nanofluid compositions for nanofluid creation. Wear out, fuel characteristics, photothermal change, and engine, maintenance, combustion, and exhaust processes, thermohydraulic and performance of the created both mono and hybrid nanofluids in terms of heat transfer can be reproducible [133].

Wettability and interfacial tension (IFT) are two important variables that affect oil recovery. A developing technique for enhanced oil recovery (EOR) is combining chemical agents including polymers, alkalis, and surfactants with reduced/low salinity water (LSW). Nanoparticle use for EOR has recently increased in popularity. This research on a lower salinity surfactant-silica nanofluid (LSS-SNF) examines the effects of different oils including n-heptane, n-decane, toluene, benzene, as well as oil of type A, oil of type B, crude oil, oil-water quartz systems, surface tension and turbidity variation [134].

Since many authors, engineering organizations, and scientific publications have published articles focusing on transfer of heat, the topic on transfer of heat and the applications of nanofluids has been the subject of much research. Ionic and biobased fluids, combined/hybrid, and environmentally derived nanoscale particles and biological convection, electrohydrodynamics, and sequential flow should be important areas of future nanofluid research, including medical, pharmaceutical, and veterinary applications of nanofluids. These should be emphasized [135].

12.6.2 Current Challenges and Limitations

It is currently unknown how human contact with nanomaterials may affect health in the long term. This is due to the wide variation in chemical composition and size of individual nanoparticles. Because of their large surface area, nanoparticles are more dangerous to people and the environment than bulk materials. Additionally, they discovered that binding or diffusion interactions allowed the particles to enter cells and access cellular DNA [136].

Although the use of nanofluids in heat pipes is widespread, little is known about their performance under long-term operation or whether their application is restricted by problems resulting from the severe condition in the pipe, primarily regarding the stability of the fluid. A significant concern for manufacturing and the researcher is the stability before and after the operating cycle The stability of the nanofluid before and after the working cycle is very important and a great concern for researchers to move this technology up to meet the industrial heat pipes. It is crucial to further stabilize the nanofluids, and additional restrictions which are necessary for condensers have been recommended by researchers [137].

12.6.2.1 The Effects of Nanofluids on Reservoirs

Aquifer applications: the use of nanofluids in aquifers may present particular challenges. As an example, nanoparticles (NPs) will agglomerate and change reservoir porosity, consequently influencing fluid flow. They may also plug the pores of the reservoir, reduce the permeability of the reservoir, and affect the migration of oil. Thus, it becomes paramount to systematically verify the characteristics of nanofluids in real-world settings and adjust the properties of injection method and concentration appropriately. In addition, various reservoirs have their own specific reaction to nanofluid displacement because of their internal attributes. A comprehensive study is required to determine the formation type, porosity, and other factors to reduce the risk of reservoir damage prior to nanofluid displacement.

12.6.2.2 The Nanofluids' Economic Viability

One downside to all this is that nanofluids are expensive to produce. In response to the growing demand for oilfields, it is necessary to design cost-effective devices for the preparation of nanofluids with the features of large areas. This is an important area of research which will help to decrease the operation cost and hence, can make them economically more viable [138].

12.7 Conclusion

This has opened up a host of possibilities in the field of nanofluidics, and in nanoscale manipulation and sensing efforts. These fields have been coupled curd to revolutionize the world with some of the most promising technologies for applicable domains such as energy storage to environmental monitoring, for better healthcare.

The diffusion phenomena at the nanoscale as a special environment affect the fluid behavior, ion transport dynamics, and also the electrostatic interactions [3]. The ability to both understand and utilize these phenomena and combine them has led to significant leaps in technology in the domains of sensing and manipulation. Nanofluidic systems can provide extremely high selectivity and accuracy in ion transport. The fact that this same effect can be used to control and manipulate ions at the nanometer scale has enabled highly sensitive, specific, and even single molecule detection.

Combining ion transport with nanofluidics is transforming biosensing. Scientists have developed systems that take advantage of the unusual properties of nanochannels for ultra-sensitive detection of biomolecules. This enabled incredibly exciting new directions in biotechnology and medical diagnostics.

Furthermore, nanofluidics and nanoscale ion transport are not limited to laboratory environments but also develop into microfluidic sensors and practical systems with applications for energy storage, environmental monitoring, and vision diagnosis. These are just some of the areas that show what practical real-world solutions can achieve.

Although a lot of headway has been made, hurdles still persist. The main concern that needs to be highlighted are concerned with the device scalability, the repeatability as well, to be worked on the integration into the real time applications. Addressing them will necessitate multidisciplinary approaches, aided by state-of-the-art techniques in advanced manufacturing and materials innovations.

Scientific and technological progress are in the vanguard of ion transport and nanofluidic research. The future is a continued collaborative effort between scientists of different disciplines and new avenues to control, and sense different ions and liquids at the nanoscale. This chapter reflects the dynamic and ever-evolving nature of this topic, and emphasizes further research, collaboration and creativity to achieve unprecedented discoveries.

References

1. Okonkwo, E.C., *et al.*, An updated review of nanofluids in various heat transfer devices. *J. Therm. Anal. Calorim.*, 145, 2817–2872, 2021.
2. Mawatari, K., *et al.*, Extended-nanofluidics: fundamental technologies, unique liquid properties, and application in chemical and bio analysis methods and devices. *Anal. Chem.*, 4068–4077, 2014, ACS Publications.
3. Yamamoto, K., Ota, N., Tanaka, Y., Nanofluidic devices and applications for biological analyses. *Anal. Chem.*, 93, 1, 332–349, 2020.
4. Chantipmanee, N. and Hauser, P.C., Determination of tobramycin in eye drops with an open-source hardware ion mobility spectrometer. *Anal. Bioanal. Chem.*, 414, 14, 4059–4066, 2022.
5. Chantipmanee, N. and Xu, Y., Nanofluidics for chemical and biological dynamics in solution at the single molecular level. *TrAC, Trends Anal. Chem.*, 158, 116877, 2023.
6. Esfandiar, A., *et al.*, Size effect in ion transport through angstrom-scale slits. *Science*, 358, 6362, 511–513, 2017.
7. Lucas, R.A., *et al.*, Ionic amplifying circuits inspired by electronics and biology. *Biophys. J.*, 118, 3, 159a, 2020.
8. Zhang, Z., Wen, L., Jiang, L., Nanofluidics for osmotic energy conversion. *Nat. Rev. Mater.*, 6, 7, 622–639, 2021.
9. Robin, P. and Bocquet, L., Nanofluidics at the crossroads. *J. Chem. Phys.*, 158, 16, 160901, 2023.
10. Zhang, B., *et al.*, Mechanisms and properties of ion-transport in inorganic solid electrolytes. *Energy Storage Mater.*, 10, 139–159, 2018.
11. Pérez-Mitta, G., *et al.*, Molecular design of solid-state nanopores: Fundamental concepts and applications. *Adv. Mater.*, 31, 37, 1901483, 2019.
12. Xiao, K., Wen, L., Jiang, L., Biomimetic solid-state nanochannels: from fundamental research to practical applications. *Small*, 12, 21, 2810–2831, 2016.
13. Zhu, Z., *et al.*, Ion/molecule transportation in nanopores and nanochannels: From critical principles to diverse functions. *J. Am. Chem. Soc.*, 141, 22, 8658–8669, 2019.
14. Xiao, K., Jiang, L., Antonietti, M., Ion transport in nanofluidic devices for energy harvesting. *Joule*, 3, 10, 2364–2380, 2019.

15. Siria, A., Bocquet, M.-L., Bocquet, L., New avenues for the large-scale harvesting of blue energy. *Nat. Rev. Chem.*, 1, 11, 0091, 2017.
16. Recktenwald, S.M., *et al.*, Erysense, a lab-on-a-chip-based point-of-care device to evaluate red blood cell flow properties with multiple clinical applications. *Front. Physiol.*, 13, 812, 2022.
17. Levin, S., *et al.*, A nanofluidic device for parallel single nanoparticle catalysis in solution. *Nat. Commun.*, 10, 1, 4426, 2019.
18. Zhong, J., *et al.*, Exploring anomalous fluid behavior at the nanoscale: Direct visualization and quantification *via* nanofluidic devices. *Acc. Chem. Res.*, 53, 2, 347–357, 2020.
19. Sofos, F., Karakasidis, T.E., Liakopoulos, A., Fluid flow at the nanoscale: How fluid properties deviate from the bulk. *Nanosci. Nanotechnol. Lett.*, 5, 4, 457–460, 2013.
20. Xu, Y., Nanofluidics: a new arena for materials science. *Adv. Mater.*, 30, 3, 1702419, 2018.
21. Li, L., *et al.*, Transient characteristics of a parabolic trough direct-steam-generation process. *Renew. Energy*, 135, 800–810, 2019.
22. Zhang, S., *et al.*, Molecular dynamics simulation on explosive boiling of liquid argon film on copper nanochannels. *Appl. Therm. Eng.*, 113, 208–214, 2017.
23. Cao, Q. and Cui, Z., Molecular dynamics simulations of the effects of surface sinusoidal nanostructures on nanoscale liquid film phase-change. *J. Therm. Sci.*, 29, 1076–1084, 2020.
24. Grosu, Y., *et al.*, Viscosity at the nanoscale: confined liquid dynamics and thermal effects in self-recovering nanobumpers. *J. Phys. Chem. C*, 122, 26, 14248–14256, 2018.
25. Nadooshan, A.A., Eshgarf, H., Afrand, M., Evaluating the effects of different parameters on rheological behavior of nanofluids: a comprehensive review. *Powder Technol.*, 338, 342–353, 2018.
26. Radhakrishnan, R., *et al.*, Nanoparticle transport phenomena in confined flows. *Adv. Heat Transfer*, 51, 55–129, 2019.
27. Kazoe, Y. and Xu, Y., Advances in nanofluidics. *Micromachines*, 427, 20, 4, 2021, MDPI.
28. Nazari, M., *et al.*, Transport phenomena in nano/molecular confinements. *ACS Nano*, 14, 12, 16348–16391, 2020.
29. Liu, C., *et al.*, Recent advances of nanofluids in micro/nano scale energy transportation. *Renew. Sustain. Energy Rev.*, 149, 111346, 2021.
30. Sun, H. and Wang, M., Atomistic insights into heat transfer and flow behaviors of nanofluids in nanochannels. *J. Mol. Liq.*, 345, 117872, 2022.
31. Cheng, L.-J., Electrokinetic ion transport in nanofluidics and membranes with applications in bioanalysis and beyond. *Biomicrofluidics*, 12, 2, 021502, 2018.
32. Feng, H., *et al.*, Recent advancement in induced-charge electrokinetic phenomena and their micro-and nano-fluidic applications. *Adv. Colloid Interface Sci.*, 280, 102159, 2020.

33. Wang, X., *et al.*, Electroosmotic pumps and their applications in microfluidic systems. *Microfluid. Nanofluid.*, 6, 145–162, 2009.
34. Barman, B., *et al.*, Electrokinetic ion transport and fluid flow in a pH-regulated polymer-grafted nanochannel filled with power-law fluid. *Soft Matter*, 16, 29, 6862–6874, 2020.
35. Ha, D., Nanochannel-Assisted Active Control Of Mass Transport In Polydimethylsiloxane-Based Micro-/Nanofluidic Systems, 2020.
36. Zhou, K., Perry, J.M., Jacobson, S.C., Transport and sensing in nanofluidic devices. *Annu. Rev. Anal. Chem.*, 4, 321–341, 2011.
37. Chen, X. and Zhang, L., Review in manufacturing methods of nanochannels of bio-nanofluidic chips. *Sens. Actuators, B*, 254, 648–659, 2018.
38. Karnik, R., *et al.*, Effects of biological reactions and modifications on conductance of nanofluidic channels. *Nano Lett.*, 5, 9, 1638–1642, 2005.
39. Fu, J., *et al.*, A patterned anisotropic nanofluidic sieving structure for continuous-flow separation of DNA and proteins. *Nat. Nanotechnol.*, 2, 2, 121–128, 2007.
40. Perry, J.M., *et al.*, Ion transport in nanofluidic funnels. *ACS Nano*, 4, 7, 3897–3902, 2010.
41. Nam, S.-W., *et al.*, Sub-10-nm nanochannels by self-sealing and self-limiting atomic layer deposition. *Nano Lett.*, 10, 9, 3324–3329, 2010.
42. Lee, J.H., *et al.*, Poly (dimethylsiloxane)-based protein preconcentration using a nanogap generated by junction gap breakdown. *Anal. Chem.*, 79, 17, 6868–6873, 2007.
43. Kavokine, N., Netz, R.R., Bocquet, L., Fluids at the nanoscale: From continuum to subcontinuum transport. *Annu. Rev. Fluid Mech.*, 53, 377–410, 2021.
44. Muthuraman, A. and Kaur, J., Antimicrobial nanostructures for neurodegenerative infections: present and future perspectives, in: *Nanostructures for Antimicrobial Therapy*, pp. 139–167, Elsevier, 2017.
45. Uram, J.D., *et al.*, Submicrometer pore-based characterization and quantification of antibody–virus interactions. *Small*, 2, 8–9, 967–972, 2006.
46. Li, J., *et al.*, Ion-beam sculpting at nanometre length scales. *Nature*, 412, 6843, 166–169, 2001.
47. Bayley, H. and Cremer, P.S., Stochastic sensors inspired by biology. *Nature*, 413, 6852, 226–230, 2001.
48. Merchant, C., DNA translocation through graphene nanopores. *Biophys. J.*, 100, 3, 521a, 2011.
49. Zhou, K., Kovarik, M.L., Jacobson, S.C., Surface-charge induced ion depletion and sample stacking near single nanopores in microfluidic devices. *J. Am. Chem. Soc.*, 130, 27, 8614–8616, 2008.
50. Kovarik, M.L. and Jacobson, S.C., Integrated nanopore/microchannel devices for ac electrokinetic trapping of particles. *Anal. Chem.*, 80, 3, 657–664, 2008.
51. Cannon, D.M., *et al.*, Nanocapillary array interconnects for gated analyte injections and electrophoretic separations in multilayer microfluidic architectures. *Anal. Chem.*, 75, 10, 2224–2230, 2003.

52. Jama, M., *et al.*, Critical review on nanofluids: preparation, characterization, and applications. *J. Nanomater.*, 2016, 6717624, 2016.
53. Verma, N. and Pandya, A., Challenges and opportunities in micro/nanofluidic and lab-on-a-chip. *Prog. Mol. Biol. Transl. Sci.*, 186, 1, 289–302, 2022.
54. Fuest, M., *et al.*, A three-state nanofluidic field effect switch. *Nano Lett.*, 15, 4, 2365–2371, 2015.
55. Rangharajan, K.K., *et al.*, Surface dependent enhancement in water vapor permeation through nanochannels. *Analyst*, 143, 18, 4256–4266, 2018.
56. Zhu, X.-Y., *et al.*, Novel nanofluidic cells based on nanowires and nanotubes for advanced chemical and bio-sensing applications. *Nanomaterials*, 11, 1, 90, 2021.
57. Xin, Y., *et al.*, Continuous flow synthesis of porous materials. *Chin. Chem. Lett.*, 31, 6, 1448–1461, 2020.
58. Chen, L., *et al.*, Millifluidics, microfluidics, and nanofluidics: manipulating fluids at varying length scales. *Mater. Today Nano*, 16, 100136, 2021.
59. Joshi, R., *et al.*, Precise and ultrafast molecular sieving through graphene oxide membranes. *Science*, 343, 6172, 752–754, 2014.
60. Wang, X. and Shi, G., An introduction to the chemistry of graphene. *PCCP*, 17, 43, 28484–28504, 2015.
61. Gao, J., *et al.*, Nanofluidics in two-dimensional layered materials: inspirations from nature. *Chem. Soc. Rev.*, 46, 17, 5400–5424, 2017.
62. Nguyen, H., *et al.*, A novel method for preparing microfibrillated cellulose from bamboo fibers. *Adv. Nat. Sci.: Nanosci. Nanotechnol.*, 4, 015016-1, IOP Publishing, 2013.
63. Baig, N., Kammakakam, I., Falath, W., Nanomaterials: A review of synthesis methods, properties, recent progress, and challenges. *Mater. Adv.*, 2, 6, 1821–1871, 2021.
64. Zhao, M., *et al.*, Two-dimensional metal–organic framework nanosheets: synthesis and applications. *Chem. Soc. Rev.*, 47, 16, 6267–6295, 2018.
65. Hille, B., *Ion channels of excitable membranes*, vol. *507*, Sinauer, Sunderland, MA, 2001.
66. Martin, C.R. and Siwy, Z.S., Learning nature's way: Biosensing with synthetic nanopores. *Science*, 317, 5836, 331–332, 2007.
67. Sparreboom, W., van den Berg, A., Eijkel, J.C.T., Principles and applications of nanofluidic transport. *Nat. Nanotechnol.*, 4, 713, 2009.
68. Zhang, Z., *et al.*, Ultrathin and ion-selective janus membranes for high-performance osmotic energy conversion. *J. Am. Chem. Soc.*, 139, 26, 8905–8914, 2017.
69. Xiao, K., *et al.*, Nanofluidic ion transport and energy conversion through ultrathin free-standing polymeric carbon nitride membranes. *Angew. Chem. Int. Ed.*, 57, 32, 10123–10126, 2018.
70. Sonker, M., *et al.*, Separation phenomena in tailored micro-and nanofluidic environments. *Annu. Rev. Anal. Chem.*, 12, 475–500, 2019.

71. Schoch, R.B., Han, J., Renaud, P., Transport phenomena in nanofluidics. *Rev. Mod. Phys.*, 80, 3, 839, 2008.
72. Noskov, S.Y., Im, W., Roux, B., Ion permeation through the α-hemolysin channel: theoretical studies based on Brownian dynamics and Poisson-Nernst-Plank electrodiffusion theory. *Biophys. J.*, 87, 4, 2299–2309, 2004.
73. Feng, J., *et al.*, Single-layer MoS2 nanopores as nanopower generators. *Nature*, 536, 7615, 197–200, 2016.
74. Heiranian, M., Farimani, A.B., Aluru, N.R., Water desalination with a single-layer MoS2 nanopore. *Nat. Commun.*, 6, 1, 8616, 2015.
75. Qiao, R. and Aluru, N.R., Surface-charge-induced asymmetric electrokinetic transport in confined silicon nanochannels. *Appl. Phys. Lett.*, 86, 14, 143105, 2005.
76. Bhadauria, R. and Aluru, N., Multiscale modeling of electroosmotic flow: Effects of discrete ion, enhanced viscosity, and surface friction. *J. Chem. Phys.*, 146, 18, 184106, 2017.
77. Gubbiotti, A., *et al.*, Electroosmosis in nanopores: computational methods and technological applications. *Adv. Phys.: X*, 7, 1, 2036638, 2022.
78. Han, W. and Chen, X., A review: Applications of ion transport in micro-nano-fluidic systems based on ion concentration polarization. *J. Chem. Technol. Biotechnol.*, 95, 6, 1622–1631, 2020.
79. Razmjou, A., *et al.*, Design principles of ion selective nanostructured membranes for the extraction of lithium ions. *Nat. Commun.*, 10, 1, 5793, 2019.
80. Moghadam, A.J., Unsteady electrokinetic flow in a microcapillary: Effects of periodic excitation and geometry. *J. Fluids Eng.*, 141, 11, 111104, 2019.
81. Ma, B., Chi, J., Liu, H., Fabric-based ion concentration polarization for pump-free water desalination. *ACS Sustain. Chem. Eng.*, 6, 1, 99–103, 2018.
82. Jännig, O. and Nguyen, N.-T., A polymeric high-throughput pressure-driven micromixer using a nanoporous membrane. *Microfluid. Nanofluid.*, 10, 513–519, 2011.
83. Lee, H., *et al.*, A concentration-independent micro/nanofluidic active diode using an asymmetric ion concentration polarization layer. *Nanoscale*, 9, 33, 11871–11880, 2017.
84. Chao, C.-C., Chiu, P.-H., Yang, R.-J., Preconcentration of diluted biochemical samples using microchannel with integrated nanoscale Nafion membrane. *Biomed. Microdevices*, 17, 1–9, 2015.
85. Hügle, M., *et al.*, A lab-on-a-chip for free-flow electrophoretic preconcentration of viruses and gel electrophoretic DNA extraction. *Analyst*, 145, 7, 2554–2561, 2020.
86. Gholinejad, M., *et al.*, Multifactorial analysis of ion concentration polarization for microfluidic preconcentrating applications using response surface method. *Phys. Fluids*, 32, 7, 072012, 2020.
87. Beauchemin, D., *Sample introduction systems in ICPMS and ICPOES*, Newnes, 2020.

88. Singh, A., *et al.*, Advances in controlled release pesticide formulations: Prospects to safer integrated pest management and sustainable agriculture. *J. Hazard. Mater.*, 385, 121525, 2020.
89. Wang, C., *et al.*, Bioinspired ultrafast-responsive nanofluidic system for ion and molecule transport with speed control. *ACS Nano*, 14, 10, 12614–12620, 2020.
90. Jo, G., *et al.*, The application of graphene as electrodes in electrical and optical devices. *Nanotechnology*, 23, 11, 112001, 2012.
91. Celebi, A.T., Barisik, M., Beskok, A., Electric field controlled transport of water in graphene nano-channels. *J. Chem. Phys.*, 147, 16, 164311, 2017.
92. Rostami, S., *et al.*, A review on the control parameters of natural convection in different shaped cavities with and without nanofluid. *Processes*, 8, 9, 1011, 2020.
93. Schmidt-Speicher, L.M. and Länge, K., Microfluidic integration for electrochemical biosensor applications. *Curr. Opin. Electrochem.*, 29, 100755, 2021.
94. Hu, G. and Li, D., Multiscale phenomena in microfluidics and nanofluidics. *Chem. Eng. Sci.*, 62, 13, 3443–3454, 2007.
95. Han, J., *et al.*, Micro/nanofluidic-electrochemical biosensors for *in situ* tumor cell analysis. *TrAC, Trends Anal. Chem.*, 116799, 2022.
96. Lee, J., Lee, J., Kim, M., Multiscale micro-/nanofluidic devices incorporating self-assembled particle membranes for bioanalysis: A review. *TrAC, Trends Anal. Chem.*, 116940, 2023.
97. Wang, J., *et al.*, A review on nanofluid stability: preparation and application. *Renew. Sustain. Energy Rev.*, 188, 113854, 2023.
98. Bae, J., *et al.*, Micro-/Nanofluidics for Liquid-Mediated Patterning of Hybrid-Scale Material Structures. *Adv. Mater.*, 31, 20, 1804953, 2019.
99. Prantil-Baun, R., *et al.*, Physiologically based pharmacokinetic and pharmacodynamic analysis enabled by microfluidically linked organs-on-chips. *Annu. Rev. Pharmacol. Toxicol.*, 58, 37–64, 2018.
100. Jokerst, J.C., Emory, J.M., Henry, C.S., Advances in microfluidics for environmental analysis. *Analyst*, 137, 1, 24–34, 2012.
101. Hébert, M., Courtney, M., Ren, C.L., Semi-automated on-demand control of individual droplets with a sample application to a drug screening assay. *Lab Chip*, 19, 8, 1490–1501, 2019.
102. Gao, R.Z. and Ren, C.L., Synergizing microfluidics with soft robotics: A perspective on miniaturization and future directions. *Biomicrofluidics*, 15, 1, 011302, 2021.
103. Rivas, D.F. and Kuhn, S., Synergy of microfluidics and ultrasound: process intensification challenges and opportunities, in: *Sonochemistry: From Basic Principles to Innovative Applications*, pp. 225–254, 2017.
104. Sommonte, F., Denora, N., Lamprou, D.A., Combining 3D printing and microfluidic techniques: A powerful synergy for nanomedicine. *Pharmaceuticals*, 16, 1, 69, 2023.

105. Huang, Z., *et al.*, Gold nanoparticle modified silicon nanowire array based sensor for low-cost, high sensitivity and selectivity detection of mercury ions. *Mater. Res. Express*, 7, 3, 035017, 2020.
106. Song, C., *et al.*, Preparation and gas sensing properties of partially broken WO3 nanotubes. *Vacuum*, 114, 13–16, 2015.
107. Zeb, S., *et al.*, Controllable synthesis of ultrathin WO3 nanotubes and nanowires with excellent gas sensing performance. *Sens. Actuators, B*, 305, 127435, 2020.
108. Sun, Y., *et al.*, Electrokinetic transport phenomena in nanofluidics and their applications. *Electrophoresis*, 44, 23, 1756–1773, 2023.
109. Liu, Y., *et al.*, Facile synthesis of multifunctional multi-walled carbon nanotube for pathogen Vibrio alginolyticus detection in fishery and environmental samples. *Talanta*, 128, 311–318, 2014.
110. Spain, E., *et al.*, Gold nanowires and nanotubes for high sensitivity detection of pathogen DNA. *Sens. Actuators, B*, 215, 159–165, 2015.
111. Tahir, M.A., *et al.*, Investigating the potential of multiwalled carbon nanotubes based zinc nanocomposite as a recognition interface towards plant pathogen detection. *J. Virol. Methods*, 249, 130–136, 2017.
112. Danielson, E., *et al.*, Gas-phase synthesis for label-free biosensors: zinc-oxide nanowires functionalized with gold nanoparticles. *Sci. Rep.*, 9, 1, 17370, 2019.
113. Thakur, M., *et al.*, High durability and stability of 2D nanofluidic devices for long-term single-molecule sensing. *npj 2D Mater. Appl.*, 7, 1, 11, 2023.
114. Jiang, X., *et al.*, Nonlinear electrohydrodynamic ion transport in graphene nanopores. *Sci. Adv.*, 8, 2, eabj2510, 2022.
115. Qiu, H., Zhou, W., Guo, W., Nanopores in graphene and other 2D materials: a decade's journey toward sequencing. *ACS Nano*, 15, 12, 18848–18864, 2021.
116. Thiruraman, J.P., Masih Das, P., Drndic, M., Stochastic ionic transport in single atomic zero-dimensional pores. *ACS Nano*, 14, 9, 11831–11845, 2020.
117. Thiruraman, J.P., *et al.*, Gas flow through atomic-scale apertures. *Sci. Adv.*, 6, 51, eabc7927, 2020.
118. Liu, K., *et al.*, Geometrical effect in 2D nanopores. *Nano Lett.*, 17, 7, 4223–4230, 2017.
119. Mojtabavi, M., *et al.*, Ionically active MXene nanopore actuators. *Small*, 18, 11, 2105857, 2022.
120. Putallaz, L., *et al.*, Nanofluidics drives point-of-care technology for on the spot protein marker analysis with rapid actionable results. *J. Nanomed. Nanotechnol.*, 10, 536, 2019.
121. Gencoglu, A. and Minerick, A.R., Electrochemical detection techniques in micro-and nanofluidic devices. *Microfluid. Nanofluid.*, 17, 781–807, 2014.
122. Wu, C., *et al.*, Fabrication of all-transparent polymer-based and encapsulated nanofluidic devices using nano-indentation lithography. *Microsyst. Nanoeng.*, 3, 1, 1–9, 2017.

123. Wang, J., *et al.*, Fabrication of polydimethylsiloxane nanofluidic chips under AFM tip-based nanomilling process. *Nanoscale Res. Lett.*, 14, 1–14, 2019.
124. Haywood, D.G., *et al.*, Fundamental studies of nanofluidics: nanopores, nanochannels, and nanopipets. *Anal. Chem.*, 87, 1, 172–187, 2015.
125. Mueller, C., *et al.*, Nanofluidic cells with controlled pathlength and liquid flow for rapid, high-resolution *in situ* imaging with electrons. *J. Phys. Chem. Lett.*, 4, 14, 2339–2347, 2013.
126. Ramirez, P., *et al.*, Electrochemical impedance spectroscopy of membranes with nanofluidic conical pores. *J. Colloid Interface Sci.*, 655, 876–885, 2024.
127. Saha, A.A., *et al.*, Experimental and numerical investigation of capillary flow in SU8 and PDMS microchannels with integrated pillars. *Microfluid. Nanofluid.*, 7, 451–465, 2009.
128. Duan, C., Wang, W., Xie, Q., Fabrication of nanofluidic devices. *Biomicrofluidics*, 7, 2, 26501, 2013.
129. Salehi, R., *et al.*, Study on the performance of solar cells cooled with heatsink and nanofluid added with aluminum nanoparticle. *Int. J. Thermofluids*, 20, 100445, 2023.
130. Liu, S., *et al.*, Experimental study of the cooling performance of γ-Al2O3/ heat transfer fluid nanofluid for power batteries. *J. Energy Storage*, 72, 108476, 2023.
131. Dolui, S., Bhaumik, B., De, S., Combined effect of induced magnetic field and thermal radiation on ternary hybrid nanofluid flow through an inclined catheterized artery with multiple stenosis. *Chem. Phys. Lett.*, 811, 140209, 2023.
132. Hou, Y., *et al.*, Learning from the Brain: Bioinspired Nanofluidics. *J. Phys. Chem. Lett.*, 14, 11, 2891–2900, 2023.
133. Yasmin, H., *et al.*, Reproduction of nanofluid synthesis, thermal properties and experiments in engineering: A research paradigm shift. *Energies*, 16, 3, 1145, 2023.
134. N. P, D., *et al.*, Synergistic Effect of Low Salinity Surfactant Nanofluid on the Interfacial Tension of Oil–Water Systems, Wettability Alteration, and Surfactant Adsorption on the Quartz Surface. *Energy Fuels*, 37, 10, 7094–7110, 2023.
135. Giwa, S.O., *et al.*, Research trends in nanofluid and its applications: A bibliometric analysis. *J. Nanopart. Res.*, 24, 3, 63, 2022.
136. Shelare, S.D., *et al.*, Preparation, applications, challenges and future prospects of nanofluid materials with a solar systems in the last decade. *Mater. Today Proc.*, 100, 2024, 96–202, 2024.
137. Singh, U. and Gupta, N.K., Stability issues and operating limitations of nanofluid filled heat pipe: A critical review. *Mater. Today Proc.*, 2023, DOI:10.1016/j.matpr.2023.01.309.
138. Tong, Q., *et al.*, Research Progress in Nanofluid-Enhanced Oil Recovery Technology and Mechanism. *Molecules*, 28, 22, 7478, 2023.

13

Bioinspired Micro/Nanorobots

Yuchao Li*, Tong He, Jiaqi Xu, Xixi Chen and Baojun Li†

Guangdong Provincial Key Laboratory of Nanophotonic Manipulation, Institute of Nanophotonics, Jinan University, Guangzhou, China

Abstract

Inspired by natural microorganisms, micro/nanorobots can navigate efficiently at low Reynolds numbers and perform various biological tasks. Conventional micro/nanorobots often rely on chemical fuel decomposition to generate propulsion which limits their suitability for use in biological environments. In contrast, fuel-free micro/nanorobots, powered by external stimuli such as magnetic, acoustic, light, or electric fields, open new possibilities for *in vivo* microrobot applications due to their biocompatibility and sustainability. This review aims to provide an overview of recent advancements in bioinspired micro/nanorobots developed by other research groups. We will explore methods for achieving high-efficiency actuation in magnetic, acoustic, and light fields by imitating biological motion strategies, such as those employed by flagella and cilia. Additionally, we will discuss the diverse applications of these bioinspired micro/nanorobots, including their roles in sensing, diagnostics, drug/cell delivery, and tumor therapy. With continuous innovation, future bio-friendly and multifunctional micro/nanorobots are expected to realize diverse medical operations within the human body that benefits from the imitation of natural organisms.

***Keywords*:** Bioinspired actuation, stimuli responsiveness, biosensing, drug delivery

**Corresponding author*: liyuchao@jnu.edu.cn

†*Corresponding author*: baojunli@jnu.edu.cn

Inamuddin, Tariq Altalhi, Mohammad Luqman and Jorddy Neves Cruz (eds.) Nanoionics: Fundamentals and Applications, (345–368)

13.1 Introduction

Humans have long dreamed of micro/nanorobots. The anticipation for the application of micro/nanorobots in the biomedical field was vividly portrayed in the 1996 science fiction film "Fantastic Voyage," where the storyline featured five doctors shrinking to micron-scale individuals and embarking on a journey into the human body for disease treatment. However, achieving precise control over the movement of micro/nanorobots within biological environments presents a formidable challenge. The movement of micro/nano-sized objects in water is substantially different from the movement of macro-sized objects, according to research on the movement of microorganisms. The primary disparity lies in the low ratio of inertial force to viscous force, rendering the movement of micro/nanorobots in water as arduous as humans swimming in glue. Consequently, a continuous propelling force becomes imperative to overcome viscous resistance. The second distinction is the pronounced impact of Brownian motion on micro-and nano-objects, making unconstrained movement challenging. Fortunately, microorganisms and micro/nanorobots share analogous motion environments, which offers valuable insights for the development of micro/nanorobots capable of operating at low Reynolds numbers [1, 2]. The locomotion of Ciliates, for instance, hinges upon the rhythmic undulation of cilia that densely cover their surfaces [3]. Similarly, spermatozoa propel themselves through the rotation of their flagella around a central axis [4]. Notably, these intricate motion patterns can be replicated through the application of external rotating or oscillating fields [5–7]. Microorganisms not only exhibit the capability to navigate through intricate fluidic environments but also possess the ability to perceive their surroundings and execute responsive behaviors, exemplified by phenomena such as the exothermicity of sperm [8], which is what the field of micro-robotics aims for in artificial systems. Constructing intricate structures at submillimeter scales is a formidable task. Therefore, one of the challenges facing micro/nanorobots is to replicate the fundamental functions of natural biology while maintaining a minimal level of structural complexity. One promising approach involves embedding sensing and motion control directly into pre-designed materials [9, 10], typically necessitating the use of external fields, such as magnetic [11–15], acoustic [16–19], and optical fields [20–23].

Micro/nanorobots can be propelled by various energy sources, with the most common being chemical decomposition-based propulsion, which generates bubbles or concentration gradients of molecules [24–26].

However, these chemical fuels often lack biocompatibility, which significantly limits their use in biomedical applications. Fuel-free micro/nanorobots, powered by external fields such as ultrasonic, light, magnetic, and electric fields, offer the advantages of biocompatibility, enabling on-demand motion control and extended operational lifetimes. The motion of these micro/nanorobots, driven by external fields, depends on materials equipped with distinct response mechanisms. Specific morphological designs (e.g., starfish [27], spiral [28], and fish shapes [29]) or surface modifications [30, 31] (involving physical absorption or chemical grafting to attach specific proteins and polymers to micro/nanorobots) can imbue these micro-robots with a diverse range of functionalities. Micro/nanorobots hold significant potential for applications in the biomedical realm, including sensing and diagnostics [32, 33], targeted delivery [34–36], and tumor therapy [37, 38].

13.2 Inspiration from Creatures of Nature

Micro/nanorobots have the ability to operate within biologically relevant environments and perform tasks of biological significance, including drug delivery [34–36], targeted therapy [37, 38], and minimally invasive surgery [39]. Achieving motion at low Reynolds numbers necessitates innovative approaches to the design and mechanics of micro/nanorobots, and bionics offers a pathway to efficiently realize their motion and functionality. In the biotic community, individual organisms often possess limited perceptual and locomotive capabilities. Complex behaviors on a large scale are achieved through the coordinated movement of numerous simple entities, such as geese migrating south in formation, fish schooling to evade predators (Figure 13.1a), or the collective migration of cells. Biological groups exhibit distributed, self-organizing, and adaptable movement capacities. Drawing inspiration from these collective behaviors can benefit the development of collective micro/nanorobots, enabling the manipulation of a group of micro/nanorobots to expand their functionality and overcome the size limitations inherent in individual microrobot applications within the field of biomedicine.

The imitation of biological forms is a commonly employed strategy for enabling the motion and functionality of micro-robots. For instance, starfish utilize their curved and flexible arm tips for feeding, and this radial morphology lends itself to effective grasping and releasing motions (Figure 13.1b). The fabrication of stimuli-responsive hydrogels into starfish-like shapes enables the capture and release of drugs and cells in artificial

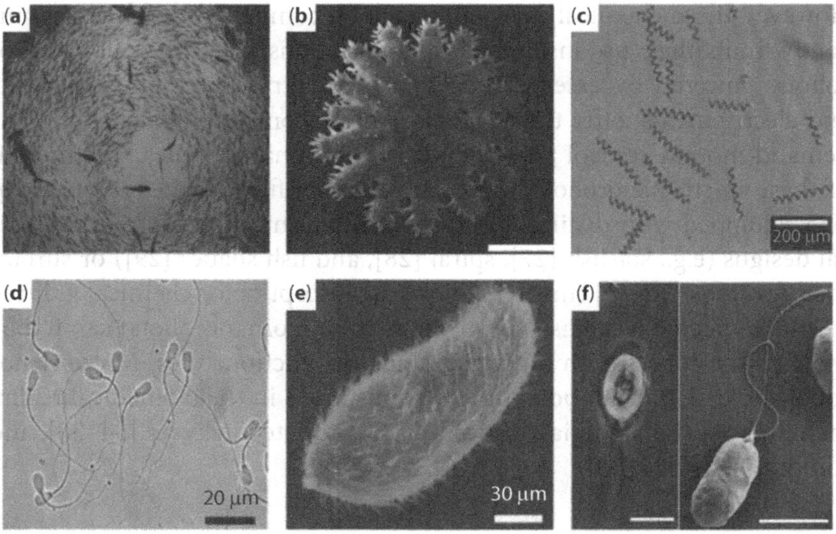

Figure 13.1 Creatures of nature. (a) Snapshot of a school of sardines [48]; (b) high-resolution image of starfish [49]; (c) microscopic view of Spirulina bacteria [47]; (d) microscopic view of mature spermatozoa [47]; (e) colored SEM micrograph of a paramecium [47]; (f) (left) phase contrast microscopy and (right) scanning electron microscopy images of a Chlamydomonas reinhardtii cell [50].

systems [9]. Emulating biological forms not only allows for the replication of intricate biological functions in artificial systems but also provides insights into their low Reynolds number motion mechanisms. The remarkable efficiency with which microorganisms navigate through biological environments, characterized by high viscosity and ionic strength, has inspired the design of numerous micro/nanorobots. Extensive research has been conducted on the flagella and cilia of microorganisms. Bacterial flagella consist of passive helical filaments propelled forward by molecular rotary motors [40]. Furthermore, beyond the spiral shape created by the rotational movement of bacterial flagella, some microorganisms themselves possess a spiral shape, as seen in spirulina bacteria (Figure 13.1c) and Spirochetes [41]. The propulsion via helical motion in artificial systems can be achieved using a rotating magnetic field. Helical flagella represent just one of the motility strategies employed by microorganisms. Sperm and eukaryotic flagella exhibit traveling wave motion in the form of beating (Figure 13.1d), which can be replicated in artificial systems through the use of a rotating magnetic field. Another common mechanism for motility involves the use of cilia for propulsion, as seen in paramecium (Figure 13.1e). Paramecium typically covers the entire surface of the organism, and the coordinated beating of their cilia generates hyper angular waves, enabling rapid movement. Artificial cilia can emulate this propulsion by employing non-reciprocal actuation to mimic the beating of paramecium cilia [42]. In addition to their remarkable locomotive abilities, microorganisms can autonomously respond to environmental cues. Bacteria can detect changes in their surroundings through surface protein receptors. When they sense a harmful concentration gradient, they exhibit movement in the opposite direction, a behavior known as chemotaxis [43]. Microorganism movement can also be influenced by environmental factors such as light intensity. For instance, Chlamydomonas reinhardtii demonstrates phototaxis (Figure 13.1f), sensing light gradients to move towards a light source. This phototactic behavior of microorganisms can also be replicated in artificial systems [44–46].

13.3 Actuation of Bioinspired Micro/Nanorobots

Precisely manipulating micro/nanorobots at low Reynolds numbers presents a considerable challenge. In the past decade, a multitude of propulsion technologies has emerged. Fuel-based micro/nanorobots rely on the decomposition of external chemical fuels, limiting their use in biological environments. In contrast, fuel-free micro/nanorobots, powered by biocompatible energy sources such as magnetic, light, and acoustic fields, offer distinct advantages, including extended lifespan, high precision in manipulation, and biocompatibility. From a bionics perspective, we explore the application of biological motion mechanisms in artificial swimmers and the realization of actuation using magnetic, acoustic, and light fields.

13.3.1 Magnetic Field-Propelled Micro/Nanorobots

13.3.1.1 Helical Magnetic Swimmers

Bacteria generate propulsive force through the helical rotation of their flagella. Typically, multiple flagella work in unison, and the localized bending creates wave-like motion, pushing from the base to the tip. In certain bacterial species, the organisms themselves have a spiral shape, such as spirulina. In the realm of artificial micro/nanorobots, many are designed in a helical configuration to achieve propulsion within low Reynolds number fluids. Rigid magnetic micro spirals can be manufactured using various techniques, including self-scrolling [5], direct laser writing (DLW) [51, 52], glancing angle deposition (GLAD) [53], template-assisted electrodeposition (TAE) [54], and biological template synthesis (BTS) [55].

DLW, short for Direct Laser Writing, is a three-dimensional lithographic printing process based on the two-photon polymerization of photoresist. Tottori *et al.*, utilized DLW to directly print artificial spirochetes and deposited a Ni/Ti bilayer onto their surfaces through physical vapor deposition (PVD), as illustrated in Figure 13.2a. This process rendered the spirochetes both magnetic and biocompatible [28]. This approach is commonly referred to as DLW-PVD and has gained widespread adoption due to its simplicity and high-throughput manufacturing capabilities. In another study, Hwang and his research team also employed DLW-PVD to create a multimodal helical microrobot featuring a conical head (Figure 13.2b). By altering the direction and frequency of the rotating magnetic field, they were able to induce rolling, rotation, and transitions between various propulsion modes [56]. GLAD technology has also been applied in the fabrication of helical micro/nanorobots. Fischer and colleagues reported the creation of a nanoscale magnetically actuated helix (Figure 13.2d) with filaments measuring around 70 nanometers in diameter, making them small enough to be ingested by cells [53]. It can penetrate the biopolymer network at a speed comparable to that of larger-sized micro-spirals in high-viscosity solutions. Following this achievement, Fischer's research team developed a microrobot capable of penetrating gastric mucin gels by mimicking the mechanism of H. pylori, as depicted in Figure 13.2e. They achieved this by immobilizing urease on the microrobot's surface, which raised the local pH of the surrounding solution, liquefying the mucus and enabling penetration of the bio-gel. This accomplishment underscores the significant potential of biomimetic approaches in penetrating bio-gels [57]. Template Electrodeposition (TEA) is another widely adopted technique for crafting spirochetes. Pané, Nelson, and their colleagues leveraged positive tone photoresist and DLW techniques to create templates, into which they deposited CoNi alloy and polypyrrole, successfully producing artificial bacterial flagella, as shown in Figure 13.2c [58]. Bio-template synthesis further enhances template-based synthesis by utilizing biological organisms as replicators. A notable example is the cobalt-based helical swimmer crafted by Schuerle *et al.*, as illustrated in Figure 13.2f. They utilized dialkynyl phospholipids as templates to form helical microstructures, subsequently coated with a multi-alloy magnetic layer [59]. Helical structures are prevalent in nature, with organisms like spirulina and the helical xylem vessels of plants serving as biological templates for the fabrication of helical swimmers [60, 61].

13.3.1.2 Flexible Magnetic Swimmers

Many unicellular eukaryotes utilize flexible flagella that propagate bending waves for locomotion [41]. Building upon this propulsion mechanism, an increasing number of soft and flexible miniature swimmers are being designed and manufactured. These swimmers feature flexible tails, joints, or entirely bendable bodies. To achieve motion, their flexible components propagate traveling waves in response to oscillating or rotating magnetic fields. Dreyfus *et al.*, pioneered the development of an early flexible artificial micromotor [11]. They used double-stranded DNA to connect magnetic colloidal particles, forming flexible artificial flagella. One end of these flagella was attached to red blood cells to break the symmetry of motion, as depicted in Figure 13.2g.

Gao proposed the fabrication of a flexible Au/Ag/Ni nanowire motor using template-assisted electrodeposition (TAE), as shown in Figure 13.2h [12]. This motor featured a gold "head" and a nickel "tail," with a central bendable silver segment. Partial dissolution of the silver segment in a hydrogen peroxide solution enabled the desired flexibility for rotational deformation under an external magnetic field. Many microorganisms possess the ability to alter their form and movement mechanisms in response to environmental changes. For instance, sperm can switch between helical and planar flagella propulsion modes in response to changes in the viscosity of the external solution [63]. Inspired by this, Khalil's team designed a microrobot capable of controllably switching between helical and planar propulsion modes (Figure 13.2i). This design allows the microrobot to maintain a high swimming speed without being significantly affected by changes in background viscosity [62]. Diller *et al.*, embedded ferromagnetic particles in an elastomer to create a magnetized flexible panel, as depicted in Figure 13.2j [13]. The magnetization distribution could be pre-designed to enable the flexible panel to actively propagate bending waves in a magnetic field. Jang *et al.*, designed a multi-link nano swimmer (Figure 13.2k), using flexible polymer bilayer hinges to connect elastic polypyrrole tails and rigid magnetic nickel chains. This design facilitated the propagation of planar waves due to the soft connecting material [14]. Li *et al.*, used three flexible silver hinges to emulate the propulsion mechanism of fish for transmitting plane waves. This design featured an 'Au' head and tail, two nickel body segments, and silver hinges connecting each segment. When subjected to an oscillating magnetic field, it generated traveling waves with velocities exceeding 30 μm/s, as seen in Figure 13.2l. The remarkable propulsion capabilities of these biomimetic nano swimmers demonstrate the potential for scaling macro-scale aquatic motions down to the low Reynolds number range.

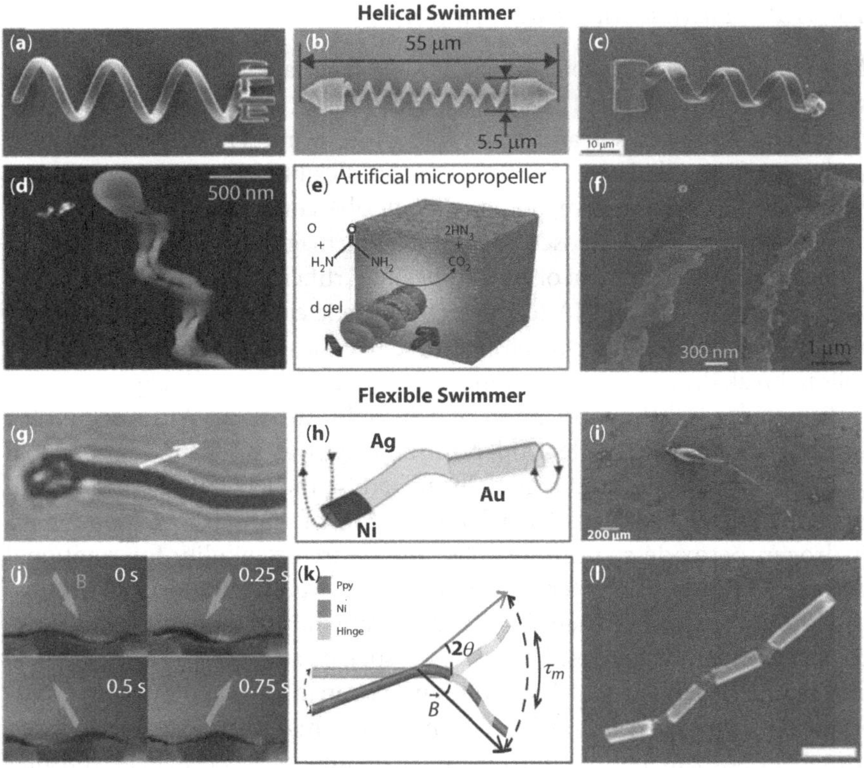

Figure 13.2 Helical magnetic swimmers. (a) the helical micromachine with a micro-holder [28]; (b) SEM image of the multimodal helical microrobot [56]; (c) SEM image of artificial bacterial flagella [58]; (d) close-up of one nano helix with both the Au-dot and the Ni-section clearly visible [53]; (e) microrobot passing through gastric mucin gel [57]; (f) helical lipid structure formed by metallization [59]; (g) DNA-linked flexible artificial flagella [11]; (h) Au/Ag/Ni nanowire motor [12]; (i) micro/nanorobots with switchable motion modes [62]; (j) a flexible panel that transmits bending waves [13]; (k) multi-link nano swimmer [14]; (l) biomimetic nano fish [15].

13.3.2 Ultrasonic Field-Propelled Micro/Nanorobots

Utilizing ultrasonic energy for micro-robot propulsion opens up avenues for the deployment of these micro/nanorobots within living organisms. Ultrasonic fields offer the benefits of minimal damage to biological tissues and the ability to penetrate deeply, making them an attractive choice for actuating and controlling micro/nanorobots. Consequently, micro/nanorobots driven by ultrasonic fields have garnered significant attention, particularly in the field of bionics.

Cephalopods, such as octopuses and cuttlefish, employ a unique propulsion mechanism where they draw in the surrounding fluid and expel it in one powerful burst to propel themselves forward. Some ultrasonically-driven micro/nanorobots have also adopted a similar strategy for their locomotion. An example is the CeFlowBot, as proposed by Misra *et al.*, (Figure 13.3a). This microrobot features an internal structure capable of capturing six vibrating bubbles, which generate directional flow through its internal channel, mimicking the movement of cephalopods and thereby enhancing propulsion. To enable directional control, a magnetic layer is coated on the CeFlowBot's surface. Through the coordinated manipulation of ultrasonic and magnetic fields, CeFlowBots can navigate in any direction, accomplish object capture, and execute site-specific delivery. Another motility strategy inspired by microorganisms is ciliary wiggling, which has been incorporated into various micro/nanorobots [42, 64]. These synthetic cilia are typically actuated by magnetic, optical, or electric fields. Ahmed *et al.*, presented an ultrasonically activated artificial ciliary ribbon (Figure 13.3b), drawing inspiration from the natural arrangement of cilia on the surface of starfish larvae. This design employs small-amplitude oscillations

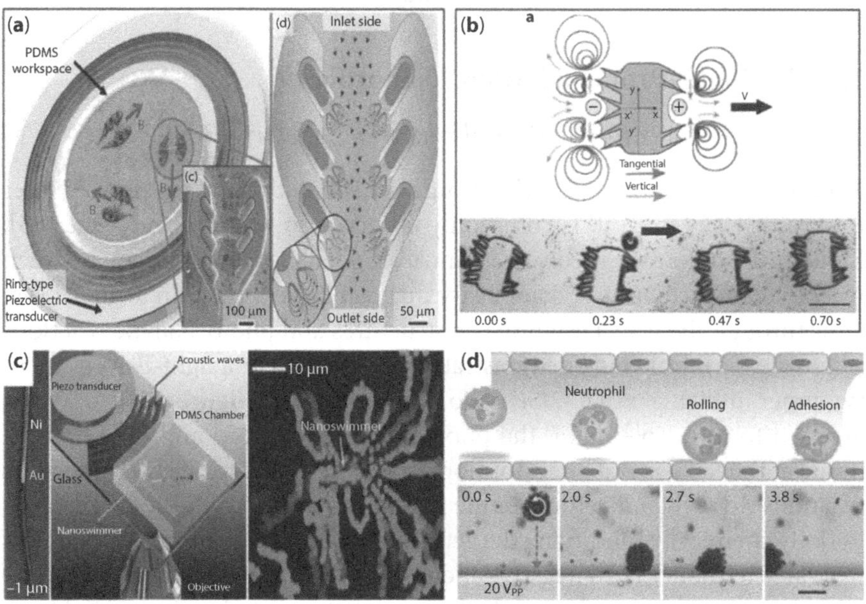

Figure 13.3 Ultrasonic field-driven bioinspired microrobot. (a) CeFlowBot microrobot driven by bubble oscillation [16]; (b) microrobot with artificial ciliary ribbons [17]; (c) design and the experimental details of the rod-shaped microrobot [18]; (d) motion of the neutrophil-inspired micro/nanorobots [19].

of the cilia to induce fluid motion. By controlling the direction of the cilia, they can manipulate the flow of the surrounding liquid, creating both a liquid source and a liquid sink, enabling propulsion and particle capture. Apart from using cilia for propulsion, microorganisms more commonly rely on their own soft flagella for movement. Figure 13.3c illustrates a rod-shaped micro swimmer composed of a bimetallic head and a flagellum-like polypyrrole tail [18]. The flagellar tail oscillates in response to ultrasonic field excitation, generating traveling waves that drive propulsion. One of the key objectives in microrobot research is to enable their application in *in vivo* drug delivery, which necessitates their ability to navigate within confined environments like blood vessels. However, much of the research to date has taken place in 2D environments, making movement along boundaries challenging. Taking inspiration from the rolling motion of neutrophils within vessels, Ahmed *et al.*, introduced superparamagnetic particles into the channel. When subjected to a rotating magnetic field, these particles aggregate into spherical shapes due to dipole-dipole interactions. Upon activation by the ultrasonic field, the particles actively migrate to the boundary and roll forward [19] (Figure 13.3d). This biomimetic motion mechanism presents a viable strategy for drug delivery within hard-to-reach narrow regions of the human body.

13.3.3 Light-Propelled Micro/Nanorobots

13.3.3.1 Photocatalytic Propulsion and Collective Behaviors

Phototaxis, a phenomenon where organisms detect light and move either toward or away from it, is commonly observed in bacteria and algae [65]. Biological phototaxis involves an intricate feedback mechanism between receptors and actuators, a complexity that can be replicated in artificial systems through the use of asymmetric structures and photocatalytic reactions (Figure 13.4a). Dai *et al.*, demonstrated the creation of a light-driven micro swimmer utilizing a TiO2/Si nano tree structure (Figure 13.4b). When exposed to ultraviolet light, the photoanode at the head and the photocathode at the tail generate negative ions and cations, respectively. These ionic reactions drive the movement of the nano trees through self-electrophoresis [44]. By chemically modifying the zeta potential of the nano tree, researchers were able to induce both negative and positive phototaxis, akin to the behavior seen in algae. In another study, Janus particles coated with a partial layer of light-absorbing carbon were exposed to a non-uniform light field. These particles exhibited movement in the direction of lower light intensity due to an oriented response generated by diffusive electrophoretic torque [66].

Collective motion is observed in various contexts, including birds, fish, and bacterial communities. Light-driven collective motion behavior in the field of micro/nanorobots was first proposed by Ibele *et al.* They observed the clustering of AgCl particles in an aqueous solution when exposed to UV light [67]. Furthermore, the use of artificial active substances has opened up novel avenues for studying collective phenomena. Singh *et al.*, employed active colloids to induce the dynamic assembly of passive SiO2 colloids [45].

Under ultraviolet (UV) illumination, self-propelled active colloids exhibit an inherent capability to attract and aggregate nearby passive SiO_2 particles, as visually demonstrated in Figure 13.4c. This phenomenon results from the diffusion effect engendered by the local chemical gradient induced by UV irradiation, and this aggregation behavior endures until the cessation of the light source. Notably, the aggregation of particles can also be achieved through the utilization of photosensitive molecules. For instance, the azobenzene molecule undergoes a reversible photoisomerization process when subjected to alternating UV and blue light, consequently transitioning from a state of hydrophobicity to one of hydrophilicity. This light-induced switching mechanism, in turn, drives the inflow and outflow movement of particles. In an intriguing study conducted by Feldmann *et al.*, they ingeniously employed simultaneous irradiation of a solution containing photosensitive molecules with both UV and blue light. This dual-light exposure caused the photosensitive molecules to aggregate at specific positions, yielding the creation of distinctive and discernible patterns, as illustrated in the Figure 13.4d [46].

13.3.3.2 Light-Driven Smart Soft Micro/Nanorobots

The emergence of intelligently responsive soft polymeric materials has opened up a new world for the development of micro-robots. These smart materials can serve as controlled actuators, responding to specific external stimuli such as changes in light or temperature. They are ideally suited to mimic the functional attributes of microorganisms. An example of such materials is poly(N-isopropylacrylamide) (PNIPAM), a widely employed hydrogel with a notably low critical solution temperature, approximately 32°C. When subjected to temperatures exceeding this threshold, PNIPAM undergoes a transition from its hydrated phase to a dehydrated phase, resulting in its contraction [69]. Importantly, this process is entirely reversible. Another class of responsive soft materials is liquid crystal elastomers (LCEs), characterized by phase transitions that can be triggered by both temperature and light. When exposed to temperatures surpassing a specific critical value or activated by light, LCEs shift from an ordered state (typically a nematic phase) to a disordered phase. Consequently, the entire material contracts along the alignment direction of the liquid crystals and expands in the vertical direction [70].

Micro-robots constructed with smart materials have the capacity to emulate biological motion strategies, enabling motion at low Reynolds numbers. In contrast to the rigid helical propellers mentioned previously, smart materials bestow micro-robots with enhanced deformability, greater flexibility, and improved biocompatibility. The principal mechanism facilitating the movement of intelligent soft micro-robots at the microscale remains the rotation of helical flagella and the beating of cilia. Möller *et al.*, proposed the application of light-responsive soft materials to craft artificial flagella [20], which undergo deformation due to the asymmetric expansion of a bilayer structure. This transformation is achieved through the resonant excitation of gold nanoparticles embedded within the hydrogel by near-infrared light, leading to a rapid temperature increase exceeding 20°C within milliseconds. Importantly, the temperature change outpaces the volume change, preventing the introduction of isotropic swelling, thus resulting in the observed asymmetric deformation as illustrated in Figure 13.5a. Following a similar methodology, Möller also detailed a light-driven two-dimensional rotor built upon a thermosensitive hydrogel [21]. In this configuration, a hydrogel ribbon is affixed to a freely rotating microsphere. The application of localized and rapid photothermal heating induces unbalanced bending deformation in the hydrogel bilayer, ultimately initiating rotation of the structure through stroboscopic irradiation (Figure 13.5b). This system emulates the rhythmic motion akin to the beating of natural cilia. In another notable endeavor, researchers harnessed the high spatiotemporal resolution of light to directly simulate the propagation of surface waves on a microorganism's body, a phenomenon attributed to ciliary beating [22]. This work leveraged structured light to propel soft micro-robots made of liquid crystal elastomers (LCEs). Specifically, only the region directly illuminated by the light undergoes deformation. The dynamic structured light engenders a traveling wave deformation, effectively propelling the robot (Figure 13.5c). Remarkably, this robot operates without pre-programming, relying solely on light control, thereby resembling a system with multiple independent actuators, capable of executing diverse gaits. Furthermore, in Reference [43], an LCE-based light-actuated artificial flytrap was demonstrated, exhibiting the ability to mimic the behavior of the natural Venus flytrap, including its closing motion (Figure 13.5d). The robot is integrated with an optical fiber to trigger LCE deformation through the reflection or scattered light generated when an object

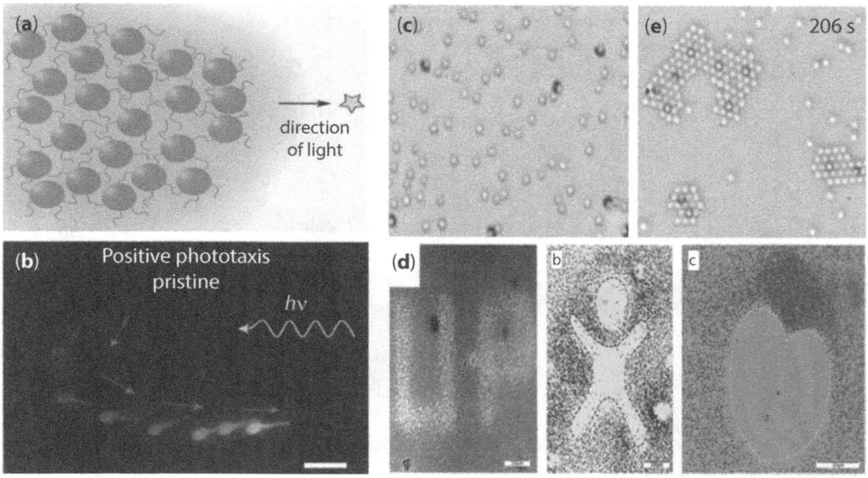

Figure 13.4 Phototaxis and collective behaviors of the microrobot. (a) Phototaxis of cyanobacteria [68]; (b) positive phototaxis motion of TiO2/Si nano-tree structure [44]; (c) active colloid-induced dynamic aggregation of passive SiO colloids under UV light irradiation [45]; (d) azobenzene molecules aggregate into unique patterns under selective irradiation of blue and UV light [46].

enters its field of view. This unique approach permits the identification of different objects based on variations in reflection and scattering intensities, enabling intelligent capture.

13.4 Biomedical Applications of Micro/Nanorobots

13.4.1 Sensing, Diagnosis and Isolation

Micro/nanorobots find applications in detecting biological macromolecules, DNA/RNA sensing, and isolating bacteria and cells in biological samples, vital for medical diagnostics. Li *et al.*, developed a fish-like motor for glucose concentration detection (Figure 13.6a), using boronic acid-modified PNIPAM hydrogel and surfactant. In a glucose solution, the hydrogel swells and releases surfactant due to glucose complexation with phenylboronic acid. This fish-shaped motor, driven by the Marangoni effect, is self-propelled and its speed is proportional to glucose concentration, offering potential applications in diabetes detection. Protein detection plays a pivotal role in medical diagnostics, and Orozco *et al.*, fabricated a tubular robot designed for selective protein transport (Figure 13.6b) [71]. This microrobot utilizes fluorescein isothiocyanate (FITC)-labeled avidin

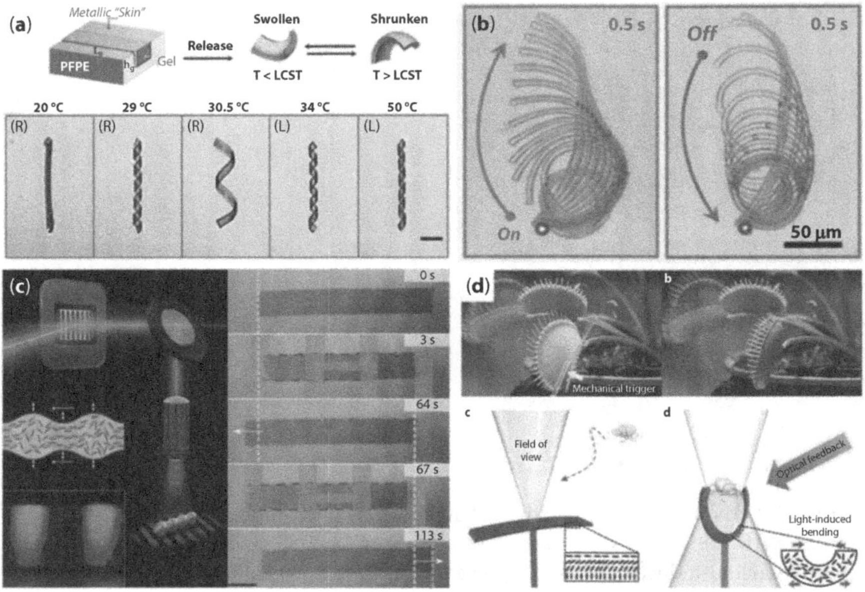

Figure 13.5 Light-driven intelligent soft micro/nanorobots. (a) PNIPAM-based artificial flagella [20]; (b) light-driven two-dimensional rotor based on thermosensitive hydrogel [21]; (c) soft micro/nanorobots of LCEs selectively driven by structured light [22]; (d) LCE-based light-driven artificial flytrap [23].

(Av-FITC) as a template, facilitating the specific binding of the microrobot's avidin-imprinted polymeric layer to fluorescently-labeled target proteins. This allows for the isolation and transport of Av-FITC in serum, making it visually detectable. The over expression of specific miRNAs is linked to cancer, underscoring the clinical significance of miRNA detection. Avila *et al.*, devised a nanomotor-based approach for miRNA sensing at the single-cell level [72]. The nanomotor is composed of graphene oxide-functionalized Au nanowires and fluorescently labeled ssDNA probes bearing attached dyes. In this configuration, graphene oxide serves as a fluorescence quencher for the dyes, as depicted in (Figure 13.6c). Driven by an acoustic field, the nanomotor can enter the interior of the cell, where the ssDNA binds to the miRNA, turning the fluorescence "on". This nanomotor-based miRNA detection method offers the advantages of speed and high sensitivity.

Precise separation of bacteria and cells holds significant importance in ensuring food safety, water quality assessment, and medical diagnosis.

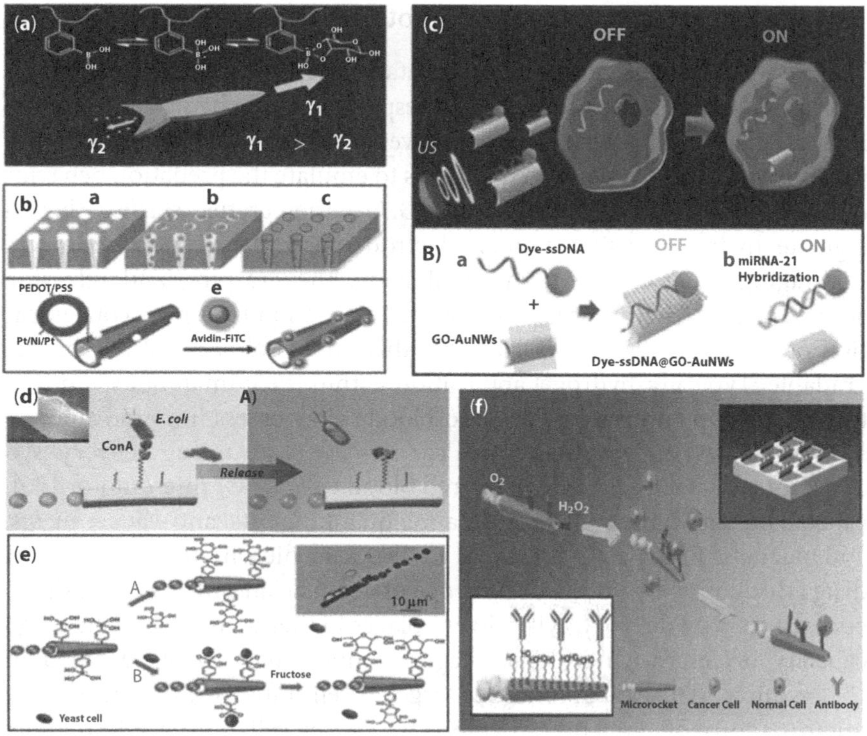

Figure 13.6 Applications of micro/nanorobots for sensing, detection and specificity separation. (a) fish-shaped motor for glucose concentration detection [44]; (b) tubular robot with selective protein transport function [45]; (c) nanomotors for RNA sensing [46]; (d) microrobot with E. coli-specific capture capability [47]; (e) tubular micro/nanorobots that be used to specifically isolate and transport yeast cells [48]; (f) micro/nanorobots with specificity for cancer cell separation and transport [49].

Modification of the microrobot's surface with Concanavalin A (ConA) lectin bioreceptor enables it to specifically capture E. coli, as ConA recognizes the surface polysaccharide of E. coli [73] (Figure 13.6d). Wang *et al.*, employed boric acid-modified micro/nanorobots for the specific separation and transport of yeast cells [74], (Figure 13.6e). Detecting and isolating cancer cells holds substantial clinical value. Balasubramanian *et al.*, utilized monoclonal antibodies on a micro rocket to bind with CEA antigens on target cancer cells, achieving highly specific cancer cell isolation and transport [75] (Figure 13.6f). This bio microfluid-based *in vitro* assay holds promise for early cancer cell diagnosis.

13.4.2 Targeted Delivery and Photodynamic Therapy

Targeted delivery of cells and drugs stands as a crucial application for micro/nanorobots. The use of stimuli-responsive smart hydrogels is instrumental in achieving efficient cargo delivery. Reference [27], devised a bio-inspired microrobot based on hydrogels to emulate the predation behavior of starfish for cargo delivery (Figure 13.7a). They crafted starfish-shaped alginate hydrogels using a microelectrode-based fabrication approach. Leveraging the ion sensitivity of alginate, the grasping and releasing actions of the structure were orchestrated by manipulating ion concentration or pH levels. It's noteworthy that this micro-robot is entirely biodegradable. Typically, hydrogel application within the human body can lead to rejection. In contrast, natural red blood cells possess immunosuppressive properties, making them ideal carriers for *in vivo* drug delivery. Wu *et al.*, introduced a multi-carrying red blood cell (RBC) micromotor [76], capable of simultaneously encapsulating quantum dots, anti-cancer drugs, and magnetic nanoparticles (Figure 13.7b). This micromotor can transport drugs through the combined influence of acoustic and magnetic fields. In contrast to drug delivery, cell delivery places higher demands on the cargo capacity of micro/nanorobots. Magnetic micro/nanorobots detailed in [77], exhibit robust capabilities for targeted cell transport. By introducing micro/nanorobots attached to colorectal cancer cells into a microfluidic chip simulating a liver tumor network, these micro/nanorobots successfully transported HCT116 cells across four liver tissues to the target tumor tissue (Figure 13.7c). Furthermore, researchers placed a microrobot loaded with human nasal mesenchymal stem cells into the abdominal cavity of nude mice, demonstrating its potential in cell delivery.

Photodynamic therapy (PDT) represents a promising approach to cancer treatment, involving the conversion of non-toxic oxygen into reactive oxygen species through photosensitizers to kill cancer cells. Micro/nanorobots play a pivotal role in facilitating PDT. Natural red blood cells serve as an ideal carrier for transporting oxygen and photosensitizers to tumor cells, thereby enhancing PDT efficacy [78]. Zhong *et al.*, fabricated a photosynthe engineer tic biohybrid nano swimmer employing magnetized spirulina (Figure 13.7d), which offers an adjunctive therapeutic effect for tumors [79]. This biological nano swimmer can photosynthesize and release oxygen within the hypoxic region of the tumor, thereby improving the efficacy of radiotherapy. Furthermore, post-radiotherapy, these swimmers release chlorophyll as a photosensitizer, which, when exposed to laser light, generates reactive oxygen species, enabling photodynamic therapy.

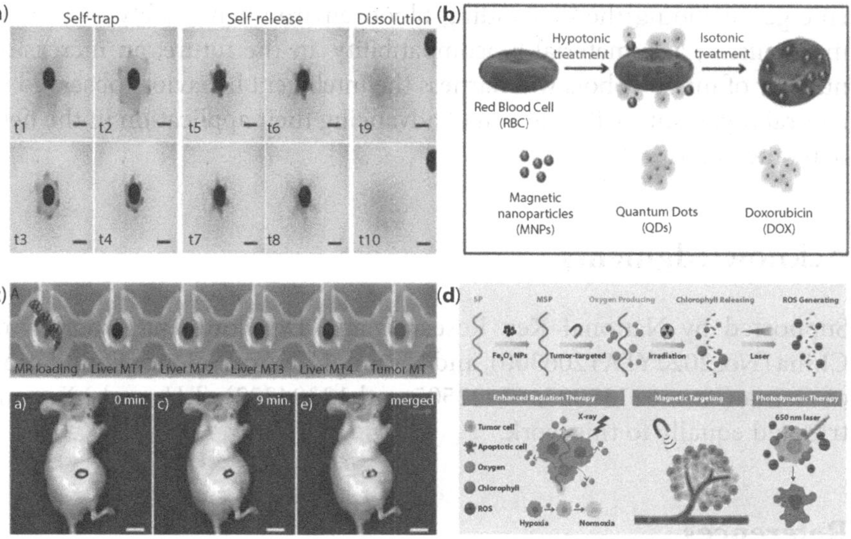

Figure 13.7 Application of micro/nanorobots for targeted delivery and photodynamic therapy. (a) a hydrogel robot capable of grasping and releasing [27]; (b) multi-loaded red blood cell micromotor [76]; (c) helical robot with targeted transport capability of colorectal cancer cells [77]; (d) biohybrid nano swimmer with adjunctive tumor therapy capabilities [79].

13.5 Conclusion

Organisms exhibit intricate and intelligent behaviors, including environmental sensing, interaction, and collective action. Bionics offers a pathway for driving and realizing the functions of micro-robots. Constructing artificial robotic systems at the microscale using conventional hardware is not a practical approach. A feasible method for achieving motion control and function realization in micro-robots relies on specialized materials, such as magnetic and light-sensitive materials. The utilization of soft, stimuli-responsive materials is central to the advancement of micro-robotics. Developing high-performance and robust micro/nanorobots necessitates a profound understanding of materials and a focus on comprehensive system design, drawing inspiration from biological behavior. In the past decade, notable progress has been made in the application of micro/nanorobots in biomedicine, particularly in areas like targeted delivery and tumor therapy. However, much of the current research remains at the proof-of-concept stage, primarily demonstrated in Newtonian fluids. When considering the application of micro-robots within the human body, numerous challenges emerge, including the demanding blood environment, safety of driving mechanisms, and material biocompatibility. In the future, an increasing number of micro-robots will harness the intelligent behaviors observed in natural organisms, with the aim of advancing their application in the field of biomedicine.

Acknowledgments

Supported by National Key Research and Development Program of China (No. 2022YFA1206300), and National Natural Science Foundation of China (Nos. 62135005, 62422505 and 12304322). T.H. and J.X. contributed equally to this work.

References

1. Purcell, E.M., Life at low Reynolds number. *Am. J. Phys.*, 45, 3, 1977.
2. Abbott, J.J., Peyer, K.E., Lagomarsino, M.C., Zhang, L., Dong, L., Kaliakatsos, I.K., Nelson, B.J., How should microrobots swim? *Int. J. Rob. Res.*, 28, 1434, 2009.
3. Hill, D.B., Swaminathan, V., Estes, A., Cribb, J., O'Brien, E.T., Davis, C.W., Superfine, R., Force generation and dynamics of individual cilia under external loading. *Biophys. J.*, 98, 57, 2010.
4. Magdanz, V., Sanchez, S., Schmidt, O.G., Development of a sperm-flagella driven micro-bio-robot. *Adv. Mater.*, 25, 6581, 2013.
5. Zhang, L., Abbott, J.J., Dong, L., Kratochvil, B.E., Bell, D., Nelson, B.J., Artificial bacterial flagella: Fabrication and magnetic control. *Appl. Phys. Lett.*, 94, 064107, 2009.
6. Alves, F.S., Cruz, J.N., de Farias Ramos, I.N., do Nascimento Brandão, D.L., Queiroz, R.N., da Silva, G.V., da Silva, G.V., Dolabela, M.F., da Costa, M.L., Khayat, A.S., de Arimatéia Rodrigues do Rego, J., do Socorro Barros Brasil, D., Evaluation of Antimicrobial Activity and Cytotoxicity Effects of Extracts of Piper nigrum L. and Piperine. *Separations*, 10, 21, 2023.
7. Khalil, I.S.M., Dijkslag, H.C., Abelmann, L., Misra, S., MagnetoSperm: A microrobot that navigates using weak magnetic fields. *Appl. Phys. Lett.*, 104, 223701, 2014.
8. Bahat, A. and Eisenbach, M., Sperm thermotaxis. *Mol. Cell. Endocrinol.*, 252, 115, 2006.
9. Jin, Q., Yang, Y., Jackson, J.A., Yoon, C., Gracias, D.H., Untethered Single Cell Grippers for Active Biopsy. *Nano Lett.*, 20, 5383, 2020.

10. Dong, M., Wang, X., Chen, X.Z., Mushtaq, F., Deng, S., Zhu, C., Torlakcik, H., Terzopoulou, A., Qin, X.H., Xiao, X., Puigmartí-Luis, J., Choi, H., Pêgo, A.P., Shen, Q.D., Nelson, B.J., Pané, S., 3D-Printed Soft Magnetoelectric Microswimmers for Delivery and Differentiation of Neuron-Like Cells. *Adv. Funct. Mater.*, 30, 1910323, 2020.
11. Camp, C.H., Lee, Y.J., Heddleston, J.M., Hartshorn, C.M., Walker, A.R.H., Rich, J.N., Lathia, J.D., Cicerone, M.T., High-speed coherent Raman fingerprint imaging of biological tissues. *Nat. Photonics*, 8, 627, 2014.
12. Gao, W., Sattayasamitsathit, S., Manesh, K.M., Weihs, D., Wang, J., Magnetically powered flexible metal nanowire motors. *J. Am. Chem. Soc.*, 132, 14403, 2010.
13. Diller, E., Zhuang, J., Zhan Lum, G., Edwards, M.R., Sitti, M., Continuously distributed magnetization profile for millimeter-scale elastomeric undulatory swimming. *Appl. Phys. Lett.*, 104, 174101, 2014.
14. Jang, B., Gutman, E., Stucki, N., Seitz, B.F., Wendel-García, P.D., Newton, T., Pokki, J., Ergeneman, O., Pané, S., Or, Y., Nelson, B.J., Undulatory Locomotion of Magnetic Multilink Nanoswimmers. *Nano Lett.*, 15, 4829, 2015.
15. Li, T., Li, J., Zhang, H., Chang, X., Song, W., Hu, Y., Shao, G., Sandraz, E., Zhang, G., Li, L., Wang, J., Magnetically Propelled Fish-Like Nanoswimmers. *Small*, 12, 6098, 2016.
16. Mohanty, S., Paul, A., Matos, P.M., Zhang, J., Sikorski, J., Misra, S., CeFlowBot: A Biomimetic Flow-Driven Microrobot that Navigates under Magneto-Acoustic Fields. *Small*, 18, 2105829, 2022.
17. Dillinger, C., Nama, N., Ahmed, D., Ultrasound-activated ciliary bands for microrobotic systems inspired by starfish. *Nat. Commun.*, 12, 6455, 2021.
18. Ahmed, D., Baasch, T., Jang, B., Pane, S., Dual, J., Nelson, B.J., Artificial Swimmers Propelled by Acoustically Activated Flagella. *Nano Lett.*, 16, 4968, 2016.
19. Ahmed, D., Baasch, T., Blondel, N., Läubli, N., Dual, J., Nelson, B.J., Neutrophil-inspired propulsion in a combined acoustic and magnetic field. *Nat. Commun.*, 8, 770, 2017.
20. Zhang, H., Mourran, A., Möller, M., Dynamic Switching of Helical Microgel Ribbons. *Nano Lett.*, 17, 2010, 2017.
21. Zhang, H., Koens, L., Lauga, E., Mourran, A., Möller, M., Microgel Rotors: A Light-Driven Microgel Rotor (Small 46/2019). *Small*, 15, 1970246, 2019.
22. Palagi, S., Mark, A.G., Reigh, S.Y., Melde, K., Qiu, T., Zeng, H., Parmeggiani, C., Martella, D., Sanchez-Castillo, A., Kapernaum, N., Giesselmann, F., Wiersma, D.S., Lauga, E., Fischer, P., Structured light enables biomimetic swimming and versatile locomotion of photoresponsive soft microrobots. *Nat. Mater.*, 15, 647, 2016.
23. Wani, O.M., Zeng, H., Priimagi, A., A light-driven artificial flytrap. *Nat. Commun.*, 8, 15546, 2017.

24. Sanchez, S., Soler, L., Katuri, J., Chemically powered micro- and nanomotors. *Angew. Chem. Int. Ed.*, 54, 1414, 2015.
25. Muzammil, S., Neves Cruz, J., Mumtaz, R., Rasul, I., Hayat, S., Khan, M.A., Khan, A.M., Ijaz, M.U., Lima, R.R., Zubair, M., Effects of Drying Temperature and Solvents on *In Vitro* Diabetic Wound Healing Potential of Moringa oleifera Leaf Extracts. *Molecules*, 28, 710, 2023.
26. Ebbens, S.J., Active colloids: Progress and challenges towards realising autonomous applications. *Curr. Opin. Colloid Interface Sci.*, 21, 14, 2016.
27. Zheng, Z., Wang, H., Dong, L., Shi, Q., Li, J., Sun, T., Huang, Q., Fukuda, T., Ionic shape-morphing microrobotic end-effectors for environmentally adaptive targeting, releasing, and sampling. *Nat. Commun.*, 12, 411, 2021.
28. Tottori, S., Zhang, L., Qiu, F., Krawczyk, K.K., Franco-Obregõn, A., Nelson, B.J., Magnetic helical micromachines: Fabrication, controlled swimming, and cargo transport. *Adv. Mater.*, 24, 811, 2012.
29. Li, M., Zhang, H., Liu, M., Dong, B., Motion-based glucose sensing based on a fish-like enzymeless motor. *J. Mater. Chem. C*, 5, 4400, 2017.
30. Restrepo-Pérez, L., Soler, L., Martínez-Cisneros, C., Sánchez, S., Schmidt, O.G., Biofunctionalized self-propelled micromotors as an alternative on-chip concentrating system. *Lab Chip*, 14, 2914, 2014.
31. Kagan, D., Campuzano, S., Balasubramanian, S., Kuralay, F., Flechsig, G.U., Wang, J., Functionalized micromachines for selective and rapid isolation of nucleic acid targets from complex samples. *Nano Lett.*, 11, 2083, 2011.
32. Park, S. and Yossifon, G., Micromotor-Based Biosensing Using Directed Transport of Functionalized Beads. *ACS Sens.*, 5, 936, 2020.
33. García, M., Orozco, J., Guix, M., Gao, W., Sattayasamitsathit, S., Escarpa, A., Merkoçi, A., Wang, J., Micromotor-based lab-on-chip immunoassays. *Nanoscale*, 5, 1325, 2013.
34. Gao, W. and Wang, J., Synthetic micro/nanomotors in drug delivery. *Nanoscale*, 6, 10486, 2014.
35. Mhanna, R., Qiu, F., Zhang, L., Ding, Y., Sugihara, K., Zenobi-Wong, M., Nelson, B.J., Artificial bacterial flagella for remote-controlled targeted single-cell drug delivery. *Small*, 10, 1953, 2014.
36. Patra, D., Sengupta, S., Duan, W., Zhang, H., Pavlick, R., Sen, A., Intelligent, self-powered, drug delivery systems. *Nanoscale*, 5, 1273, 2013.
37. Han, J., Zhen, J., Du Nguyen, V., Go, G., Choi, Y., Ko, S.Y., Park, J.O., Park, S., Hybrid-Actuating Macrophage-Based Microrobots for Active Cancer Therapy. *Sci. Rep.*, 6, 28717, 2016.
38. Wang, D., Gao, C., Wang, W., Sun, M., Guo, B., Xie, H., He, Q., Shape-Transformable, Fusible Rodlike Swimming Liquid Metal Nanomachine. *ACS Nano*, 12, 10212, 2018.
39. Kagan, D., Benchimol, M.J., Claussen, J.C., Chuluun-Erdene, E., Esener, S., Wang, J., Acoustic droplet vaporization and propulsion of perfluorocarbon-loaded microbullets for targeted tissue penetration and deformation. *Angew. Chem. Int. Ed.*, 51, 7519, 2012.

40. Jarrell, K.F. and McBride, M.J., The surprisingly diverse ways that prokaryotes move. *Nat. Rev. Microbiol.*, 6, 466, 2008.
41. Bray, D., Cell Movements In Embryos, in: *Cell Movements*, p. 333, 2020.
42. Kim, S., Lee, S., Lee, J., Nelson, B.J., Zhang, L., Choi, H., Fabrication and manipulation of ciliary microrobots with non-reciprocal magnetic actuation. *Sci. Rep.*, 6, 30713, 2016.
43. Constantin, G. and Laudanna, C., Leukocyte chemotaxis: From lysosomes to motility. *Nat. Immunol.*, 11, 463, 2010.
44. Dai, B., Wang, J., Xiong, Z., Zhan, X., Dai, W., Li, C.C., Feng, S.P., Tang, J., Programmable artificial phototactic microswimmer. *Nat. Nanotechnol.*, 11, 1087, 2016.
45. Silva, L.B., Ferreira, E.F.B., Maryam, Espejo-Román, J.M., Costa, G.V., Cruz, J.V., Kimani, N.M., Costa, J.S., Bittencourt, J.A.H.M., Cruz, J.N., Campos, J.M., Santos, C.B.R., Galantamine Based Novel Acetylcholinesterase Enzyme Inhibitors: A Molecular Modeling Design Approach. *Molecules*, 28, 1035, 2023.
46. Feldmann, D., Maduar, S.R., Santer, M., Lomadze, N., Vinogradova, O.I., Santer, S., Manipulation of small particles at solid liquid interface: Light driven diffusioosmosis. *Sci. Rep.*, 6, 36443, 2016.
47. Ebrahimi, N., Bi, C., Cappelleri, D.J., Ciuti, G., Conn, A.T., Faivre, D., Habibi, N., Hošovský, A., Iacovacci, V., Khalil, I.S.M., Magdanz, V., Misra, S., Pawashe, C., Rashidifar, R., Soto-Rodriguez, P.E.D., Fekete, Z., Jafari, A., Magnetic Actuation Methods in Bio/Soft Robotics. *Adv. Funct. Mater.*, 31, 2005137, 2021.
48. Terayama, K., Habe, H., Sakagami, M., Multiple fish tracking with an NACA airfoil model for collective behavior analysis. *IPSJ Trans. Comput. Vis. Appl.*, 8, 4, 2016.
49. Kamya, P.Z., Byrne, M., Mos, B., Dworjanyn, S.A., Enhanced performance of juvenile crown of thorns starfish in a warm-high CO2 ocean exacerbates poor growth and survival of their coral prey. *Coral Reefs*, 37, 751, 2018.
50. Xin, H., Zhao, N., Wang, Y., Zhao, X., Pan, T., Shi, Y., Li, B., Optically Controlled Living Micromotors for the Manipulation and Disruption of Biological Targets. *Nano Lett.*, 20, 7177, 2020.
51. Kim, S., Qiu, F., Kim, S., Ghanbari, A., Moon, C., Zhang, L., Nelson, B.J., Choi, H., Fabrication and characterization of magnetic microrobots for three-dimensional cell culture and targeted transportation. *Adv. Mater.*, 25, 5863, 2013.
52. Peters, C., Ergeneman, O., García, P.D.W., Müller, M., Pané, S., Nelson, B.J., Hierold, C., Superparamagnetic twist-type actuators with shape-independent magnetic properties and surface functionalization for advanced biomedical applications. *Adv. Funct. Mater.*, 24, 5269, 2014.
53. Schamel, D., Mark, A.G., Gibbs, J.G., Miksch, C., Morozov, K.I., Leshansky, A.M., Fischer, P., Nanopropellers and their actuation in complex viscoelastic media. *ACS Nano*, 8, 8794, 2014.

54. Li, J., Sattayasamitsathit, S., Dong, R., Gao, W., Tam, R., Feng, X., Ai, S., Wang, J., Template electrosynthesis of tailored-made helical nanoswimmers. *Nanoscale*, 6, 9415, 2014.
55. Yan, X., Zhou, Q., Vincent, M., Deng, Y., Yu, J., Xu, J., Xu, T., Tang, T., Bian, L., Wang, Y.X.J., Kostarelos, K., Zhang, L., Multifunctional biohybrid magnetite microrobots for imaging-guided therapy. *Sci. Rob.*, 2, eaaq1155, 2017.
56. Barbot, A., Decanini, D., Hwang, G., On-chip Microfluidic Multimodal Swimmer toward 3D Navigation. *Sci. Rep.*, 6, 19041, 2016.
57. Walker, D., Käsdorf, B.T., Jeong, H.H., Lieleg, O., Fischer, P., Biomolecules: Enzymatically active biomimetic micropropellers for the penetration of mucin gels. *Sci. Adv.*, 1, e1500501, 2015.
58. Zeeshan, M.A., Grisch, R., Pellicer, E., Sivaraman, K.M., Peyer, K.E., Sort, J., Özkale, B., Sakar, M.S., Nelson, B.J., Pané, S., Hybrid helical magnetic microrobots obtained by 3D template-assisted electrodeposition. *Small*, 10, 1284, 2014.
59. Schuerle, S., Pané, S., Pellicer, E., Sort, J., Baró, M.D., Nelson, B.J., Helical and tubular lipid microstructures that are electroless-coated with conirep for wireless magnetic manipulation. *Small*, 8, 1498, 2012.
60. Yan, X., Zhou, Q., Yu, J., Xu, T., Deng, Y., Tang, T., Feng, Q., Bian, L., Zhang, Y., Ferreira, A., Zhang, L., Magnetite Nanostructured Porous Hollow Helical Microswimmers for Targeted Delivery. *Adv. Funct. Mater.*, 25, 5333, 2015.
61. Gao, W., Feng, X., Pei, A., Kane, C.R., Tam, R., Hennessy, C., Wang, J., Bioinspired helical microswimmers based on vascular plants. *Nano Lett.*, 14, 305, 2014.
62. Khalil, I.S.M., Tabak, A.F., Seif, M.A., Klingner, A., Sitti, M., Controllable switching between planar and helical flagellar swimming of a soft robotic sperm. *PLoS One*, 13, e0206456, 2018.
63. Ishijima, S., Mechanical constraint converts planar waves into helices on tunicate and sea urchin sperm flagella. *Cell Struct. Funct.*, 37, 13, 2012.
64. Evans, B.A., Shields, A.R., Carroll, R.L., Washburn, S., Falvo, M.R., Superfine, R., Magnetically actuated nanorod arrays as biomimetic cilia. *Nano Lett.*, 7, 1428, 2007.
65. Lozano, C., Ten Hagen, B., Löwen, H., Bechinger, C., Phototaxis of synthetic microswimmers in optical landscapes. *Nat. Commun.*, 7, 12828, 2016.
66. Duan, W., Ibele, M., Liu, R., Sena, A., Motion analysis of light-powered autonomous silver chloride nanomotors. *Eur. Phys. J. E*, 35, 77, 2012.
67. Varuni, P., Menon, S.N., Menon, G.I., Phototaxis as a Collective Phenomenon in Cyanobacterial Colonies. *Sci. Rep.*, 7, 17799, 2017.
68. Yan, X., Wang, F., Zheng, B., Huang, F., Stimuli-responsive supramolecular polymeric materials. *Chem. Soc. Rev.*, 41, 6042, 2012.
69. Ohm, C., Brehmer, M., Zentel, R., Liquid crystalline elastomers as actuators and sensors. *Adv. Mater.*, 22, 3366, 2010.
70. Orozco, J., Cortés, A., Cheng, G., Sattayasamitsathit, S., Gao, W., Feng, X., Shen, Y., Wang, J., Molecularly imprinted polymer-based catalytic

micromotors for selective protein transport. *J. Am. Chem. Soc.*, 135, 5336, 2013.

71. Esteban-Fernández De Ávila, B., Martín, A., Soto, F., Lopez-Ramirez, M.A., Campuzano, S., Vásquez-Machado, G.M., Gao, W., Zhang, L., Wang, J., Single Cell Real-Time miRNAs Sensing Based on Nanomotors. *ACS Nano*, 9, 6756, 2015.
72. Campuzano, S., Orozco, J., Kagan, D., Guix, M., Gao, W., Sattayasamitsathit, S., Claussen, J.C., Merkoçi, A., Wang, J., Bacterial isolation by lectin-modified microengines. *Nano Lett.*, 12, 396, 2012.
73. Kuralay, F., Sattayasamitsathit, S., Gao, W., Uygun, A., Katzenberg, A., Wang, J., Self-propelled carbohydrate-sensitive microtransporters with built-in boronic acid recognition for isolating sugars and cells. *J. Am. Chem. Soc.*, 134, 15217, 2012.
74. Balasubramanian, S., Kagan, D., Jack Hu, C., Campuzano, S., Lobo-Castañon, M.J., Lim, N., Kang, D.Y., Zimmerman, M., Zhang, L., Wang, J., Micromachine-Enabled Capture and Isolation of Cancer Cells in Complex Media. *Angew. Chem.*, 123, 4247, 2011.
75. Wu, Z., Esteban-Fernández De Ávila, B., Martín, A., Christianson, C., Gao, W., Thamphiwatana, S.K., Escarpa, A., He, Q., Zhang, L., Wang, J., RBC micromotors carrying multiple cargos towards potential theranostic applications. *Nanoscale*, 7, 13680, 2015.
76. Jeon, S., Kim, S., Ha, S., Lee, S., Kim, E., Kim, S.Y., Park, S.H., Jeon, J.H., Kim, S.W., Moon, C., Nelson, B.J., Kim, J.Y., Yu, S.W., Choi, H., Magnetically actuated microrobots as a platform for stem cell transplantation. *Sci. Rob.*, 4, eaav4317, 2019.
77. Gao, C., Lin, Z., Wang, D., Wu, Z., Xie, H., He, Q., Red Blood Cell-Mimicking Micromotor for Active Photodynamic Cancer Therapy. *ACS Appl. Mater. Interfaces*, 11, 23392, 2019.
78. Correia, J.H., Rodrigues, J.A., Pimenta, S., Dong, T., Yang, Z., Photodynamic therapy review: Principles, photosensitizers, applications, and future directions. *Pharmaceutics*, 13, 1332, 2021.
79. Zhong, D., Li, W., Qi, Y., He, J., Zhou, M., Photosynthetic biohybrid nanoswimmers system to alleviate tumor hypoxia for FL/PA/MR imagingguided enhanced radio-photodynamic synergetic therapy. *Adv. Funct. Mater.*, 30, 1910395, 2020.

Index

Printed in the USA/Agawam, MA
August 5, 2026

908663.007